ALMOST PERIODIC FUNCTIONS

C. CORDUNEANU

University of Jassy
Jassy, Romania

with the collaboration of
N. GHEORGHIU
and
V. BARBU

University of Jassy
Jassy, Romania

Translated from the Romanian edition by

GITTA BERSTEIN
University of Massachusetts
Amherst, Massachusetts

EUGENE TOMER
Columbia University
New York, New York

CHELSEA PUBLISHING COMPANY
NEW YORK, N.Y.

SECOND ENGLISH EDITION

1989

Printed in the United States of America

PREFACE TO THE SECOND EDITION

Since the publication, in 1968, of the First English Edition of this book, at least eleven books have been published on the subject of Almost Periodic Functions and their applications. (See the addendum to our list of references under the names L. Amerio and G. Prouse; J. F. Berglund, H. D. Junghenn and P. Milnes; R. B. Burckel; A. M. Fink; V. H. Harasahal; M. A. Krasnoselskii, V. S. Burd and Yu. S. Kolesov; B. M. Levitan and V. V. Zhikov; O. Onicescu, G. Cenusa and I. Sacuiu; A. A. Pankov; T. Yoshizawa; S. Zaidman.) Several other books include substantial chapters on various topics or on applications of almost periodic functions to various other areas. (See the addendum to our list of references under the names B. P. Demidovich; C. F. Dunkel and D. E. Ramirez; H. D. Fattorini; A. Haraux; C. J. Harris and J. F. Miles; A. Halanay and D. Wexler; Y. Katznelson.) Most of these books treat only special aspects of the theory of almost periodic functions and their applications and therefore, as is to be expected, treat of special topics, with the treatment going deeper than a general work can into certain aspects of current research.

We have agreed to the proposal made by Chelsea Publishing Company to publish a new edition of our book in basically the form in which it was published in 1968 because the nature of our book is different. The purpose of our book is to provide an overall view of all the basic features of almost periodic functions, in the various meanings this term has acquired in modern research as well as the many applications of such functions. This feature of the book should, of course, make it useful to those readers who are interested in the connections that exist between the theory (pure!) of almost periodic functions and such areas of application as the theory of ordinary differential equations and partial differential equations. Several other areas of application are also indicated, together with an abundance of references.

The Bibliographical Notes at the end of almost every chapter have been added to, in this Second Edition. Also, some new material has been included in I.6, V.4, and VI.5. Completely new is II.3, which is devoted to a relatively new class of almost periodic functions: random functions almost periodic in probability. It is certain that these functions will find many applications in the theory of stochastic functional equations.

We hope that this Second English Edition will be useful to the mathematical community as well as to various categories of scientists, including Engineers, Physicists, and Statisticians.

CONSTANTIN CORDUNEANU

PREFACE TO THE ENGLISH EDITION

This new edition has been prepared with the substantial help of my colleagues N. Gheorghiu and V. Barbu from A. I. Cuza University, Jassy. The former edited Chapter VI, and the latter revised and completed the material included in the Rumanian edition which refers to almost periodic functions on groups. Thus arose Chapter VII of the present edition. The other chapters have not been changed, with the exception of Chapter V, to which there has been added a section regarding the almost periodicity of the solutions of parabolic equations. The List of References was also extended with more than 200 new titles.

I wish to express my warm appreciation to my colleagues N. Gheorghiu and V. Barbu for their collaboration in preparing this edition.

I also express my thanks to Mr. E. Tomer from Columbia University and Miss Gitta Berstein from the University of Massachusetts for their interest for this book, and for their amiability in assuming the responsibility of performing the translation.

I also want to express my thanks to Prof. L. Bers and to Interscience Publishers, a Division of John Wiley and Sons, for their interest in this book.

C. Corduneanu

PREFACE TO ROMANIAN EDITION

The theory of almost periodic functions is a relatively young branch of mathematics. It was started between 1923 and 1925 by the Danish mathematician Harald Bohr (1887–1951). As H. Bohr's predecessors one should mention the names of P. Bohl and E. Esclangon. In their papers concerning the generalization of the notion of periodic function, P. Bohl and E. Esclangon considered a class of functions called quasi-periodic functions, which form a special case of almost periodic functions. Some of the methods of P. Bohl and E. Esclangon have been used by Bohr to elaborate his theory.

The papers of Bohr quickly attracted the interest of a number of researchers who made many important contributions to the development of the theory of almost periodic functions.

Then in 1925, V. V. Stepanov generalized the class of almost periodic functions in the sense of Bohr without using the hypothesis of continuity. Thus a series of studies was begun and then continued by H. Weyl, A. S. Besicovitch, A. S. Kovanko and others.

In 1933, S. Bochner defined and studied the almost periodic functions with values in Banach spaces. He showed that these functions include certain earlier generalizations of the notion of almost periodic function. In a short time, S. Bochner and J. Von Neumann pointed out the usefulness of these functions in certain problems of mathematical physics. In 1945, S. L. Sobolev obtained conclusive results in this direction, establishing the almost periodicity of the solutions of the wave equation.

Another step in developing the theory of almost periodic functions was made in 1934 by J. Von Neumann.

By using a result of S. Bochner, Von Neumann elaborated the theory of almost periodic functions on groups. This was later simplified by W. Maak.

In connection with his studies in non-linear mechanics, N. N. Bogoliubov has made a profound analysis of Bohr's theory, and has succeeded in giving a new proof of the approximation theorem. This allows a constructive presentation of the theory of numerical almost periodic functions, which has been our purpose in the first chapter of the present book.

It would be too complicated to mention all the applications of the theory of almost periodic functions. We remark only that there are applications to the theory of functions of a complex variable, ordinary and partial differential equations, and to the theory of numbers and statistics.

vii

Of course not all the aspects of this theory can be covered in our exposition. We try to bring to the reader the basic results of the theory and to orient him toward some of the applications. In some places we include a few of the recent results of special character which might arouse the interest of researchers in this field.

We also tried to give a complete list of references. It was necessary to leave out certain papers concerned with some of the applications of the theory of almost periodic functions, and to leave out those papers in which the connection with the theory is accidental. We make an attempt with these references to help the interested reader to extend his knowledge in certain problems in the theory of almost periodic functions.

C. CORDUNEANU

CONTENTS

INTRODUCTION

Periodic functions. Fourier series

Before developing the theory of almost periodic functions, we shall recall a few of the problems connected with the periodic functions and the Fourier series associated with them. These problems will permit us to follow more easily the parallelism between the theory of periodic functions and the theory of almost periodic functions. They will also be useful for certain proofs in Chapter I.

Assume that $f(x)$ is a complex-valued function, defined and continuous for $-\infty < x < +\infty$. We say that $f(x)$ has period 2ω, if $f(x + 2\omega) = f(x)$, $-\infty < x < +\infty$. The smallest number $2\omega > 0$ for which the above property is true is called the fundamental period of the function $f(x)$. The unique periods of $f(x)$ are the numbers $2m\omega$, where m is an integer.

If the function $f(x)$ is not constant and has at least one period, then this function has a fundamental period.

Let us assert that $f(x)$ has at least one period 2ω. We can consider that $2\omega > 0$, since from $f(x + 2\omega) = f(x)$ it follows that $f(x') = f(x' - 2\omega)$. Thus if 2ω is a period, then -2ω is also a period. If there exists no smallest positive period, then we can determine a sequence of positive numbers $\{k_n\}$, monotonely decreasing towards zero, such that $f(x + k_n) = f(x)$, $-\infty < x < +\infty$. This also implies $f(x + mk_n) = f(x)$, for any real x and for any integers m and n, $n \geqslant 1$.

Let x_0 be any real number. For every n there exists a largest m denoted by m_n, such that $m_n k_n \leqslant x_0$. Since $|x_0 - m_n k_n| < k_n$, from the continuity of $f(x)$, we find that $f(x + x_0) = f(x)$. For a fixed x and an arbitrary x_0, we obtain $f = \text{const}$.

The study of periodic functions on the real line can be reduced to the study of these functions on bounded intervals of length equal to their period.

The following two properties of (continuous) periodic functions are obtained from properties of functions continuous on a finite interval:

(1) *Any periodic function is bounded on the real line.*

(2) *Any periodic function is uniformly continuous on the real line.*

We shall see that these properties belong also to the almost periodic functions.

1

It is well known that to any periodic function $f(x)$ of period 2ω corresponds a Fourier series

(1) $$f(x) \sim \sum_{k=-\infty}^{\infty} a_k e^{i\omega_k x},$$

where

(2) $$\omega_k = \frac{k\pi}{\omega}, \quad a_k = \frac{1}{2\omega} \int_0^{2\omega} f(x) e^{-i\omega_k x}\, dx,$$

$$k = 0, \pm 1, \pm 2, \ldots .$$

We cannot replace the sign $\sim$ in (1) by the sign $=$, since the Fourier series associated with a function does not converge in general to the given function. Every periodic and continuous function is completely defined by its Fourier series. More precisely, the following so-called uniqueness theorem is true:

The unique function, periodic and continuous with all Fourier coefficients zero, is the identically vanishing function.

We notice that without loss of generality one can assume that the function is real. If $f(x)$ is not real, and all the Fourier coefficients are zero, then the Fourier coefficients of $\bar{f}(x)$ are zero too. The functions

$$\frac{1}{2}[f(x) + \bar{f}(x)] \quad \text{and} \quad \frac{1}{2i}[f(x) - \bar{f}(x)]$$

are real and their Fourier coefficients are zero.

We also note that from

(3) $$\int_0^{2\omega} f(x) e^{-i\omega_k x}\, dx = 0, \quad k = 0, \pm 1, \pm 2, \ldots$$

it follows that

(4) $$\int_0^{2\omega} f(x) T(x)\, dx = 0,$$

for any $T(x)$ of form

(5) $$T(x) = \sum_k c_k e^{i\omega_k x},$$

$\sum$ being extended over a finite number of terms. In (5) c_k are arbitrary complex numbers.

The uniqueness theorem will be proved if we can succeed in constructing a $T(x)$ of the form (5) for which (4) is not valid, under the assumption that the function $f(x)$ does not vanish identically. We can assert without loss of generality that $f(\omega) > 0$. Indeed, since $f(x_0) \neq 0$, then for the function

$g(x) = f(x + \omega + x_0)$ we obtain $g(\omega) = f(2\omega + x_0) = f(x_0) \neq 0$. The functions $f(x)$ and $g(x)$ vanish or do not vanish simultaneously. Finally, if we consider the function $-g(x)$, the second condition $f(\omega) > 0$ will also be fulfilled.

Since $f(\omega) > 0$, there exist two positive numbers c and d, such that $f(x) > c > 0$, if $|x - \omega| \leqslant d < \omega$. Consider now the function

$$(6) \qquad T_1(x) = \cos \frac{\pi(x + \omega)}{\omega} + 1 - \cos \frac{\pi d}{\omega}.$$

Applying the Euler formulas, we notice that $T_1(x)$ is of the form (5). The function $T_1(x)$ has the following properties:

1) $T_1(\omega - d) = T_1(\omega + d) = 1$;

2) $|T_1(x)| < 1$ for $0 \leqslant x < \omega - d$ and $\omega + d < x \leqslant 2\omega$;

3) $|T_1(x)| > 1$ for $\omega - d < x < \omega + d$ and

$$T_1(x) > g > 1 \qquad \text{for} \qquad \omega - \delta < x < \omega + \delta, \qquad \delta < d.$$

Set now $T(x) = [T_1(x)]^n$. It is obvious that $T(x)$ is also of the form (5), and

$$\int_0^{2\omega} f(x)T(x)\,dx = \int_0^{\omega - d} + \int_{\omega - d}^{\omega - \delta} + \int_{\omega - \delta}^{\omega + \delta} + \int_{\omega + \delta}^{\omega + d} + \int_{\omega + d}^{2\omega} > -M(\omega - d)$$

$$+ 0 + 2\delta cg^n + 0 - M(\omega - d) > 2\delta cg^n - 2\omega M,$$

where $|f(x)| \leqslant M$. For a sufficiently large n, one obtains

$$\int_0^{2\omega} f(x)T(x)\,dx > 0,$$

which contradicts (4). Thus the uniqueness theorem is proved.

An immediate consequence of the uniqueness theorem which will be useful is the following statement:

If the Fourier series associated with a periodic function is uniformly convergent, then this series converges to the given function.

Consider

$$f(x) \sim \sum_{k=-\infty}^{\infty} a_k e^{i\omega_k x}.$$

If we denote by $S(x)$ the sum of the series on the right side, then $S(x)$ will be continuous and periodic, with the period 2ω:

$$(7) \qquad S(x) = \sum_{k=-\infty}^{\infty} a_k e^{i\omega_k x}.$$

Let us calculate the Fourier coefficients of $S(x)$. We use the following property:

$$\frac{1}{2\omega} \int_0^{2\omega} e^{i\omega_k x} e^{-i\omega_h x}\, dx = \begin{cases} 1 & k = h, \\ 0 & k \neq h. \end{cases}$$

Applying formulas (2) and integrating each term of series (7), we find

$$\frac{1}{2\omega} \int_0^{2\omega} S(x) e^{-i\omega_k x}\, dx = a_k, \qquad k = 0, \pm 1, \pm 2, \ldots .$$

Therefore, $S(x)$ has the same Fourier coefficients as $f(x)$. From the uniqueness theorem we obtain the equality $f(x) = S(x)$ which we desired to prove.

In particular, *if $\sum_{k=-\infty}^{+\infty} |a_k|$ is convergent, then*

$$(8) \qquad\qquad f(x) = \sum_{k=-\infty}^{\infty} a_k e^{i\omega_k x}.$$

Indeed, the series on the right is uniformly convergent since

$$|a_k e^{i\omega_k x}| = |a_k|$$

An elementary calculation shows that

$$\frac{1}{2\omega} \int_0^{2\omega} \left| f(x) - \sum_{k=-n}^{n} a_k e^{i\omega_k x} \right|^2 dx = \frac{1}{2\omega} \int_0^{2\omega} |f(x)|^2\, dx - \sum_{k=-n}^{n} |a_k|^2.$$

Since the left side of this equality is non-negative, it follows that

$$\sum_{k=-n}^{n} |a_k|^2 \leqslant \frac{1}{2\omega} \int_0^{2\omega} |f(x)|^2\, dx.$$

Since the equality holds for any n, we can write

$$(9) \qquad\qquad \sum_{k=-\infty}^{\infty} |a_k|^2 \leqslant \frac{1}{2\omega} \int_0^{2\omega} |f(x)|^2\, dx.$$

Inequality (9) is called *Bessel's inequality* and it is valid for any function for which the integral on the right exists. In particular, this inequality is true for any continuous and periodic function with period 2ω.

The inequality can be stated more precisely as follows: *The Parseval formula*

$$(10) \qquad\qquad \sum_{k=-\infty}^{\infty} |a_k|^2 = \frac{1}{2\omega} \int_0^{2\omega} |f(x)|^2\, dx$$

holds for any periodic function $f(x)$.

Let

(11)
$$Q(x) = \frac{1}{2\omega} \int_0^{2\omega} f(x + t)\bar{f}(t)\, dt.$$

By a simple calculation one may show that

$$Q(x) \sim \sum_{k=-\infty}^{\infty} |a_k|^2\, e^{i\omega kx}.$$

Now from (9) it follows that the series $\sum_{k=-\infty}^{+\infty} |a_k|^2$ is convergent. Therefore, the Fourier series of the function $Q(x)$ is uniformly convergent, and

(12)
$$Q(x) = \sum_{k=-\infty}^{\infty} |a_k|^2\, e^{i\omega kx}.$$

Setting $x = 0$ in (12) and using (11), we obtain equality (10).

A consequence of Parseval's formula (or even of Bessel's inequality) is

(13)
$$\lim_{n \to \infty} |a_n| = 0.$$

It is well known that not every periodic and continuous function is the sum of its Fourier series, *but any function* of this type can nevertheless be uniformly approximated by sums such as (5). More precisely, the following theorem of Weierstrass is true:

If $f(x)$ is a periodic and continuous function with the period 2ω, then for $\varepsilon > 0$ there exists a trigonometric polynomial

(14)
$$T_\varepsilon(x) = \sum_{k=-n}^{n} c_k\, e^{i\omega kx},$$

such that

(15)
$$|f(x) - T_\varepsilon(x)| < \varepsilon, \quad -\infty < x < +\infty.$$

We notice that if $0 \leqslant x \leqslant 2\omega$, then the point $z = e^{i\pi x/\omega}$ describes the circle $|z| = 1$. The correspondence is one-to-one except at the end points of the segment $0 \leqslant x \leqslant 2\omega$. The point $z = 1$ corresponds to both end points. If we take for $\log z$ the principal value, then we find $x = (\omega/i\pi) \log z$ for $0 \leqslant \arg z < 2\pi$. The function $F(z) = f(\omega/i\pi) \log z$ is continuous on the circle $|z| = 1$. The unique point at which $F(z)$ could be discontinuous is $z = 1$, since on the positive half-line the arguments of $\log z$ are discontinuous. If $\mathrm{Im}\, z > 0$, then $\lim_{z \to 1} F(z) = f(0)$. If $\mathrm{Im}\, z < 0$, then $\lim_{z \to 1} F(z) = f(\omega/i\pi) \cdot 2i\pi = f(2\omega) = f(0)$. Therefore, according to the Weierstrass theorem for functions of a complex variable (see Appendix), there exists a polynomial $P_\varepsilon(z, \bar{z})$ for any $\varepsilon > 0$, such that

(16)
$$|F(z) - P_\varepsilon(z, \bar{z})| < \varepsilon, \quad |z| = 1.$$

Set now

$$T_\varepsilon(x) = P_\varepsilon(e^{i\pi x/\omega}, e^{-i\pi x/\omega}).$$

From (16) and from the fact that $f(x) = F(z)$ for $z = e^{i\pi x/\omega}$, (15) follows.

Remark. The above theorem has a special significance in later considerations. It suggests the definition for almost periodic functions which will be given in Chapter I.

The Weierstrass approximation theorem has many consequences. We shall mention only one of these consequences.

Consider a continuous and periodic function with period 2ω, and let λ be a real number. Then

(17)
$$\lim_{T \to \infty} \frac{1}{T} \int_0^T f(x)e^{-i\lambda x}\, dx$$

can be different from zero only if $\lambda = \omega_k$:

Indeed we can write

(18)
$$f(x) = \sum_{k=-n}^{n} c_k e^{i\omega_k x} + r(x),$$

where $|r(x)| < \varepsilon$, $-\infty < x < +\infty$. Hence,

$$\frac{1}{T} \int_0^T f(x)e^{-i\lambda x}\, dx = \frac{1}{T} \int_0^T \left\{ \sum_{k=-n}^{n} c_k e^{i(\omega_k - \lambda)x} \right\} dx + \frac{1}{T} \int_0^T r(x)e^{-i\lambda x}\, dx.$$

But for $\lambda \neq \omega_k$

$$\left| \frac{1}{T} \int_0^T \left\{ \sum_{k=-n}^{n} c_k e^{i(\omega_k - \lambda)x} \right\} dx \right| \leqslant \frac{2}{T} \sum_{k=-n}^{n} \frac{|c_k|}{|\omega_k - \lambda|} \to 0 \qquad \text{as } T \to \infty.$$

Therefore,

(19)
$$\lim_{T \to \infty} \frac{1}{T} \int_0^T f(x)e^{-i\lambda x}\, dx = \lim_{T \to \infty} \frac{1}{T} \int_0^T r(x)e^{-i\lambda x}\, dx.$$

Since

$$\left| \frac{1}{T} \int_0^T r(x)e^{-i\lambda x}\, dx \right| \leqslant \varepsilon,$$

it follows from (19) that

(20)
$$\left| \lim_{T \to \infty} \frac{1}{T} \int_0^T f(x)e^{-i\lambda x}\, dx \right| \leqslant \varepsilon.$$

The above assertion is a consequence of (20) and of the fact that ε is arbitrary.

We shall see later that the limit exists.

Let us say a few words about the Fourier transform of an absolutely integrable function on the real line.

Consider a function absolutely integrable on the real line, i.e., such that

(21)
$$\int_{-\infty}^{+\infty} |f(x)|\, dx < +\infty.$$

Then we can write

(22)
$$F(u) = \frac{1}{2\pi} \int_{-\infty}^{+\infty} f(x)e^{iux}\, dx,$$

for any real number u. The function $F(u)$ is called the Fourier transform of the function $f(x)$. Since $|f(x)e^{iux}| = |f(x)|$, the integral on the right side of (22) is uniformly convergent with respect to u. This ensures the continuity of $F(u)$ on the real line.

If a continuous function $f(x)$ is absolutely integrable on the real line and has a bounded variation on any finite interval, then

(23)
$$f(x) = \int_{-\infty}^{+\infty} F(u)e^{-iux}\, du.$$

This is the inversion formula for the Fourier transform. We do not prove the above theorem, since the proof can be found in many books.*

The following property of the Fourier transform will be used in the sequel:

If $f(x)$ is absolutely integrable on the real line, and $F(u)$ is its Fourier transform, then

(24)
$$\lim_{|u| \to \infty} F(u) = 0.$$

If $\varepsilon > 0$, then

(25)
$$\int_{-\infty}^{-R} |f(x)|\, dx + \int_R^{\infty} |f(x)|\, dx < \pi\varepsilon,$$

for any $R = R(\varepsilon)$ sufficiently large.

---

*For example, S. Bochner, Lectures on Fourier Integrals, *Annals of Math Studies*, No. 42, 1959.

We now define a new function $g(x)$ such that

$$g(x) = \begin{cases} f(x) & \text{for} & |x| \leqslant R, \\ 0 & \text{for} & |x| > R. \end{cases}$$

The function $g(x)$ is absolutely integrable on the real line. If we denote its Fourier transform by $G(u)$ and consider (25), then for real u we obtain

$$(26) \qquad\qquad |F(u) - G(u)| < \frac{\varepsilon}{2}.$$

If $S = R + 1$ and $|u| > \pi$, then we may write

$$G(u) = \frac{1}{2\pi} \int_{-S}^{S} g(x)e^{iux}\, dx = \frac{1}{2\pi} \int_{-S+(\pi/u)}^{S+(\pi/u)} g(x)e^{iux}\, dx$$

$$= \frac{1}{2\pi} \int_{-S}^{S} g\left(y + \frac{\pi}{u}\right) e^{iu(y+\pi/u)}\, dy = -\frac{1}{2\pi} \int_{-S}^{S} g\left(x + \frac{\pi}{u}\right) e^{iux}\, dx,$$

whence it follows that

$$(27) \qquad\qquad G(u) = \frac{1}{4\pi} \int_{-S}^{S} \left[g(x) - g\left(x + \frac{\pi}{u}\right) \right] e^{iux}\, dx.$$

It follows from (27) that

$$(28) \qquad\qquad\qquad |G(u)| < \frac{\varepsilon}{2},$$

for $|u| > U(\varepsilon)$.

Inequalities (26) and (28) show that

$$(29) \qquad\qquad |F(u)| < \varepsilon \qquad \text{for} \qquad |u| > U(\varepsilon),$$

which proves the above assertion.

From (22) and (24) we find that

$$(30) \qquad\qquad \lim_{|u|\to\infty} \int_{-\infty}^{+\infty} f(x)\sin ux\, dx = 0,$$

if $f(x)$ is absolutely integrable on the real line. We have mentioned (30) because we shall use it later.

I

Numerical almost periodic functions

We shall prove in this chapter the fundamental theorems in the theory of almost periodic functions of one variable. We have chosen as definition of an almost periodic function the following property: it can be uniformly approximated on the real line by trigonometric polynomials. Starting from this definition it is easier to prove some of the properties of almost periodic functions. This has some advantages since it gives to readers with less background in mathematical analysis the possibility to learn about the theory of almost periodic functions just by reading only the first section of the present chapter.

We should remark that such an exposition became possible only after N. N. Bogoliuboff gave a constructive proof of the approximation theorem (theorem 1.11).

1. Definitions and fundamental properties

A function such as

(1.1) $$T(x) = \sum_{k=1}^{n} c_k e^{i\lambda_k x},$$

where c_k are complex numbers and λ_k real numbers, is called a complex trigonometric polynomial.

Set $c_k = a_k + ib_k$. Since $e^{i\lambda_k x} = \cos \lambda_k x + i \sin \lambda_k x$, we have

$$\text{Re}\{T(x)\} = \sum_{k=1}^{n} (a_k \cos \lambda_k x - b_k \sin \lambda_k x).$$

$$\text{Im}\{T(x)\} = \sum_{k=1}^{n} (b_k \cos \lambda_k x + a_k \sin \lambda_k x).$$

Thus the real and the imaginary parts of a complex trigonometric polynomial are real trigonometric polynomials.

A complex valued function $f(x)$ defined for $-\infty < x < +\infty$ is called almost periodic, if for any $\varepsilon > 0$ there exists a trigonometric polynomial $T_\varepsilon(x)$, such that

(1.2) $$|f(x) - T_\varepsilon(x)| < \varepsilon, \qquad -\infty < x < +\infty.$$

9

Thus almost periodic functions are those functions defined on the real line, which can be uniformly approximated by trigonometric polynomials.

From the definition of almost periodic functions it follows that any trigonometric polynomial is an almost periodic function.

From the theorem of approximation of periodic functions by trigonometric polynomials (see Introduction) it follows that *any periodic function is also almost periodic.*

Theorem 1.1. *There exist almost periodic functions which are not periodic.*

Proof. It is sufficient to show that there exists at least one trigonometric polynomial which is not a periodic function.

Set $f(x) = e^{ix} + e^{i\pi x}$ and let us assume the existence of a real number $\omega \neq 0$, such that $f(x + \omega) = f(x)$ for any x. This means that

$$(1.3) \qquad (e^{i\omega} - 1)e^{ix} + (e^{i\pi\omega} - 1)e^{i\pi x} \equiv 0.$$

But the functions e^{ix} and $e^{i\pi x}$ are linearly independent. Therefore, ω must satisfy the conditions $e^{i\omega} = e^{i\pi\omega} = 1$. Then we have $\omega = 2k\pi$, $\pi\omega = 2h\pi$, where k and h are integers. However these two equations are incompatible, so that the theorem is proved.

Almost periodic functions have many of the properties of periodic functions. Certain almost periodic functions have properties that the periodic functions do not have (see, for example, Theorem 1.25). We establish below some of the fundamental properties of almost periodic functions.

Theorem 1.2. *An almost periodic function is continuous and bounded on the real line.*

Proof. Let $f(x)$ be an almost periodic function and $T_n(x)$ a trigonometric polynomial such that

$$|f(x) - T_n(x)| < \frac{1}{n}, \qquad -\infty < x < \infty.$$

The sequence $T_n(x)$ converges to the function $f(x)$ on the whole real line. But the limit of a uniformly convergent sequence of continuous functions is a continuous function. Since the trigonometric polynomials are continuous functions, $f(x)$ is continuous at every point of the real line. Assume that $|T_1(x)| \leqslant M$, where $M > 0$. Then we have

$$|f(x)| \leqslant |f(x) - T_1(x)| + |T_1(x)| \leqslant M + 1, \qquad -\infty < x < \infty$$

which proves the theorem.

Theorem 1.3. *An almost periodic function is uniformly continuous on the real line.*

Proof. We first notice that any trigonometric polynomial is a uniformly continuous function on the real line. Any function $c_k e^{i\lambda_k x}$ is uniformly continuous and consequently, the sum of a finite number of such functions will also be uniformly continuous.

Consider an almost periodic function $f(x)$ and let $\varepsilon > 0$. There exists a trigonometric polynomial $T_\varepsilon(x)$ such that

$$(1.4) \qquad |f(x) - T_\varepsilon(x)| < \frac{\varepsilon}{3}, \qquad -\infty < x < +\infty.$$

From the uniform continuity of $T_\varepsilon(x)$ it follows the existence of a number $\delta(\varepsilon) > 0$, such that

$$(1.5) \qquad |T_\varepsilon(x_1) - T_\varepsilon(x_2)| < \frac{\varepsilon}{3}, \qquad |x_1 - x_2| < \delta.$$

Since

$$|f(x_1) - f(x_2)| \leqslant |f(x_1) - T_\varepsilon(x_1)| + |T_\varepsilon(x_1) - T_\varepsilon(x_2)| + |T_\varepsilon(x_2) - f(x_2)|,$$

from (1.4) and (1.5) it follows that

$$|f(x_1) - f(x_2)| < \varepsilon, \qquad |x_1 - x_2| < \delta.$$

Thus $f(x)$ is uniformly continuous.

Theorem 1.4. *If $f(x)$ is an almost periodic function, c a complex number, and a is a real number, the functions $\bar{f}(x)$, $cf(x)$, $f(x + a)$ and $f(ax)$ are almost periodic.*

The proof follows immediately from the fact that if $T(x)$ is a trigonometric polynomial, then $\bar{T}(x)$, $cT(x)$, $T(x + a)$ and $T(ax)$ are trigonometric polynomials.

Theorem 1.5. *If $f(x)$ and $g(x)$ are almost periodic functions, then $f(x) + g(x)$ and $f(x) \cdot g(x)$ are almost periodic functions.*

Proof. We omit the proof of the first statement of the theorem which is obvious.

Let ε be a number such that $0 < \varepsilon < 1$. There exist two trigonometric polynomials $S_\varepsilon(x)$ and $T_\varepsilon(x)$ such that

$$(1.6) \qquad |f(x) - S_\varepsilon(x)| < \frac{\varepsilon}{2(M + 1)}, \qquad |g(x) - T_\varepsilon(x)| < \frac{\varepsilon}{2(M + 1)},$$

where M is a number for which $|f(x)| \leqslant M$, $|g(z)| \leqslant M$ on the whole real line. Consequently,

$$(1.7) \qquad |S_\varepsilon(x)| \leqslant |S_\varepsilon(x) - f(x)| + |f(x)| < M + 1.$$

Since

$$|fg - S_\varepsilon T_\varepsilon| \leqslant |g| \cdot |f - S_\varepsilon| + |S_\varepsilon| \, |g - T_\varepsilon|,$$

from (1.6) and (1.7) it follows that

$$|f(x)g(x) - S_\varepsilon(x)T_\varepsilon(x)| < \varepsilon, \qquad -\infty < x < +\infty.$$

This proves the almost periodicity of the product $f(x) \cdot g(x)$, since $S_\varepsilon(x)T_\varepsilon(x)$ is also a trigonometric polynomial.

Corollary. Consider a polynomial $P(z_1, z_2, \ldots, z_n)$ and almost periodic functions $f_1(x), f_2(x), \ldots, f_n(x)$. Then the function $F(x) = P(f_1, \ldots, f_n)$ is almost periodic.

Theorem 1.6. *The limit of a uniformly convergent sequence of almost periodic functions is an almost periodic function.*

Proof. Let $\{f_n(x)\}$ be a sequence of almost periodic functions converging uniformly (on the real line) to the function $f(x)$. If $\varepsilon > 0$, then we have

$$(1.8) \qquad |f(x) - f_n(x)| < \frac{\varepsilon}{2}, \qquad -\infty < x < +\infty, \qquad n \geqslant N(\varepsilon).$$

Since $f_N(x)$ is almost periodic, there exists a trigonometric polynomial $T_\varepsilon(x)$ such that

$$(1.9) \qquad |f_N(x) - T_\varepsilon(x)| < \frac{\varepsilon}{2}, \qquad -\infty < x < +\infty.$$

From (1.8) and (1.9) we obtain

$$|f(x) - T_\varepsilon(x)| < \varepsilon, \qquad -\infty < x < +\infty,$$

which proves the theorem.

Remark. This theorem expresses the important fact that the set of almost periodic functions is closed with respect to uniform convergence. This can also be stated in the following way: any function that can be approximated uniformly by almost periodic functions, with any accuracy, is almost periodic.

Theorem 1.7. *Let* $\Phi(z_1, \ldots, z_n)$ *be a uniformly continuous function of* $(z_1, \ldots, z_n) \in \mathcal{M}$, *where* $\mathcal{M}$ *is a set in the n-dimensional complex space. If* $f_1(x), f_2(x), \ldots, f_n(x)$ *are almost periodic functions such that* $(f_1, f_2, \ldots, f_n) \in \mathcal{M}$ *for any real x, then the function*

$$F(x) = \Phi(f_1(x), f_2(x), \ldots, f_n(x))$$

is almost periodic.

Proof. Since the functions $f_1(x), \ldots, f_n(x)$ are bounded on the real line, the set of points $(f_1(x), \ldots, f_n(x))$ in the n-dimensional complex space is bounded. Thus, without loss of generality, we can assume that $\mathcal{M}$ is bounded and closed. According to the Weierstrass approximation theorem (see Appendix) for any $\varepsilon > 0$, there is a polynomial $P_\varepsilon(z_1, \ldots, z_n; \bar{z}_1, \ldots, \bar{z}_n)$, such that

(1.10) $|\Phi(z_1, \ldots, z_n) - P_\varepsilon(z_1, \ldots, z_n; \bar{z}_1, \ldots, \bar{z}_n)| < \varepsilon$

for any $(z_1, \ldots, z_n) \in \mathcal{M}$. Therefore we have

(1.11) $|F(x) - P_\varepsilon(f_1(x), \ldots, f_n(x); \bar{f}_1(x), \ldots, \bar{f}_n(x))| < \varepsilon$

for every real x. Inequality (1.11) proves that $F(x)$ can be approximated with any degree of accuracy by almost periodic functions.

Corollary. Let $f(x)$ and $g(x)$ be two almost periodic functions such that $0 < m \leqslant |g(x)|$. Then $f(x)/g(x)$ is an almost periodic function. According to Theorem 1.5 it is sufficient to show that $1/g(x)$ is an almost periodic function. Since $g(x)$ is bounded, we have $0 < m \leqslant |g(x)| \leqslant M$. The function $\Phi(z) = 1/z$ is uniformly continuous in $m \leqslant |z| \leqslant M$, and according to Theorem 1.7 the function $1/g(x)$ is almost periodic.

Theorem 1.8. *If* $f(x)$ *is almost periodic and* $f'(x)$ *is uniformly continuous on the real line, then* $f'(x)$ *is almost periodic.*

Proof. Let $f(x) = f_1(x) + if_2(x)$, $f_1(x)$ and $f_2(x)$ being real functions. Since $f'(x) = f'_1(x) + if'_2(x)$, it follows that $f'_1(x)$ and $f'_2(x)$ are uniformly continuous on the real line.

Consider the functions

$$\varphi_n(x) = n\left[f\left(x + \frac{1}{n}\right) - f(x)\right], \qquad n = 1, 2, \ldots.$$

Since

$$\varphi_n(x) = f'_1\left(x + \frac{\theta_n}{n}\right) + if'_2\left(x + \frac{\tau_n}{n}\right), \qquad 0 < \theta_n, \tau_n < 1,$$

this means that for any $\varepsilon > 0$, there is an $N(\varepsilon)$, such that

$$|\varphi_n(x) - f'(x)| < \varepsilon, \qquad -\infty < x < +\infty, \qquad n \geqslant N(\varepsilon).$$

The functions $\varphi_n(x)$ are almost periodic and consequently, $f'(x)$ is almost periodic.

Remark. In Chapter IV we shall study the almost periodicity of the primitive of an almost periodic function. We shall note that the result will generally be negative. Indeed, the constants are almost periodic whereas the indefinite integral of a nonzero constant is unbounded on the real line. Thus the primitive cannot be almost periodic.

2. Two characteristic properties of almost periodic functions

We shall assume in the following that the functions involved in the definitions and theorems are continuous on the whole real line and are complex valued.

The two characteristic properties of almost periodic functions are:

A. *From any sequence of the form* $\{f(x + h_n)\}$, *where* h_n *are real numbers, one can extract a subsequence converging uniformly on the real line.*

Briefly we shall say that the function $f(x)$ is *normal.*

B. *For any* $\varepsilon > 0$, *there exists a number* $l(\varepsilon) > 0$ *with the property that any interval of length* $l(\varepsilon)$ *of the real line contains at least one point with abscissa* ξ, *such that*

(1.12) $|f(x + \xi) - f(x)| < \varepsilon, \qquad -\infty < x < +\infty.$

The property B is the definition of almost periodic functions given by H. Bohr.

The number ξ is called *translation number* of $f(x)$ corresponding to ε, or an ε-translation number.

In order to establish the equivalence of the properties A and B with the definition of almost periodic functions, we proceed in the following manner: First, we show that any almost periodic function possesses property A; then, we prove that any function with property A also has property B; finally, we prove that a function with property B is almost periodic, i.e., it can be uniformly approximated on the entire real axis by trigonometric polynomials.

Theorem 1.9. *If $f(x)$ is an almost periodic function, then it possesses property* A.

Proof. Assume that $f(x) = e^{i\lambda x}$, λ being a real number. If $\{h_n\}$ is an arbitrary sequence of real numbers, then $\{f(x + h_n)\} = \{e^{ixh_n} \cdot e^{i\lambda x}\}$ contains a subsequence uniformly convergent on the real line. Indeed, since $|e^{i\lambda h_n}| = 1$, then according to Bolzano–Weierstrass criterion, we can extract a subsequence $\{h_{1n}\}$ of the sequence $\{h_n\}$, such that the numerical sequence $\{e^{i\lambda h_{1n}}\}$ will be convergent.

But

$$|f(x + h_{1n}) - f(x + h_{1m})| = |e^{i\lambda h_{1n}} - e^{i\lambda h_{1m}}|,$$

and according to Cauchy's criterion $\{f(x + h_{1n})\}$ is uniformly convergent on the real line. Obviously, any function of the form $ce^{i\lambda x}$, c being a complex number, also has property A.

Consider now a trigonometric polynomial

$$T(x) = \sum_{k=1}^{m} c_k e^{i\lambda_k x}$$

and a sequence of real numbers $\{h_n\}$. We can extract a subsequence $\{h_{1n}\}$, such that $\{c_1 e^{i\lambda_1(x + h_{1n})}\}$ will be uniformly convergent on the real line. Then from $\{h_{1n}\}$ we extract a subsequence $\{h_{2n}\}$, such that $\{c_2 e^{i\lambda_2(x + h_{2n})}\}$ will be uniformly convergent. Proceeding further in the same manner we reach the conclusion that there exists a subsequence $\{h_{mn}\}$ of the sequence $\{h_n\}$, such that all the sequences $\{c_k e^{i\lambda_k(x + h_{kn})}\}$, $k = 1, 2, \ldots, m$, converge uniformly. Thus, $\{T(x + h_{mn})\}$ is uniformly convergent on the real line, proving thereby the normality of any trigonometric polynomial.

Consider now an arbitrary almost periodic function $f(x)$ and a sequence of trigonometric polynomials $\{T_n(x)\}$ uniformly converging to $f(x)$ on the real line. If $\{h_n\}$ is a sequence of real numbers, then we can determine a subsequence $\{h_{1n}\}$ such that $\{T_1(x + h_{1n})\}$ will be uniformly convergent on the real line. Then, we extract from $\{h_{1n}\}$ a subsequence $\{h_{2n}\}$, such that $\{T_2(x + h_{2n})\}$ will be uniformly convergent on the real line and so on. There will exist a subsequence $\{h_{pn}\}$ for any positive integer p such that $\{T_q(x + h_{pn})\}$ is uniformly convergent on the real line for any $q \leqslant p$. Let us now construct the diagonal subsequence $\{h_{pp}\}$ which (with the exception of a finite number of terms) is a subsequence of any sequence $\{h_{qn}\}$. Therefore, $\{T_n(x + h_{pp})\}$ is uniformly convergent on the real line for any n. Let $\varepsilon > 0$ be sufficiently large so that

(1.13) $$|f(x) - T_n(x)| < \frac{\varepsilon}{3}, \qquad -\infty < x < +\infty.$$

There exists an $N(\varepsilon) > 0$, such that

$$(1.14) \qquad |T_n(x + h_{pp}) - T_n(x + h_{qq})| < \frac{\varepsilon}{3}, \qquad -\infty < x < +\infty.$$

if $p, q \geqslant N(\varepsilon)$. From (1.13) and (1.14) it follows that

$$|f(x + h_{pp}) - f(x + h_{qq})| \leqslant |f(x + h_{pp}) - T_n(x + h_{pp})|$$
$$+ |T_n(x + h_{pp}) - T_n(x + h_{qq})|$$
$$+ |T_n(x + h_{qq}) - f(x + h_{qq})| < \varepsilon,$$
$$-\infty < x < +\infty,$$

when $p, q \geqslant N(\varepsilon)$.

Thus, the sequence $\{f(x + h_{pp})\}$ is uniformly convergent on the real line, proving that $f(x)$ is a normal function.

Theorem 1.10. *If $f(x)$ has property* A, *then it also has property* B.

Proof. Assume on the contrary that $f(x)$ does not have property B. Then there exists at least one $\varepsilon > 0$, such that for any $l > 0$, we can determine an interval of length l which contains no ε-translation number of $f(x)$. Consider an arbitrary number h_1 and an interval (a_1, b_1) of the real line of length $> 2|h_1|$ which does not contain any ε-translation number of $f(x)$. If we denote by $h_2 = \frac{1}{2}(a_1 + b_1)$, then $h_2 - h_1 \in (a_1, b_1)$ and consequently, $h_2 - h_1$ cannot be an ε-translation number of $f(x)$. Then, there exists an interval (a_2, b_2) of the real line of length $> 2(|h_1| + |h_2|)$ which does not contain any ε-translation number of $f(x)$. Letting $h_3 = \frac{1}{2}(a_2 + b_2)$ we get that $h_3 - h_2, h_3 - h_1 \in (a_2, b_2)$ and thus $h_3 - h_1, h_3 - h_2$ are not ε-translation numbers of $f(x)$. Proceeding in a similar manner, we define the numbers $h_4, h_5, \ldots,$ such that none of the differences $h_i - h_j$ is an ε-translation number of the function $f(x)$. Therefore, for any i and j, $\sup |f(x + h_i) - f(x + h_j)| = \sup |f(x + h_i - h_j) - f(x)| \geqslant \varepsilon$, which proves that the sequence $\{f(x + h_n)\}$ cannot contain any uniformly convergent subsequence. This contradicts the fact that $f(x)$ is normal. Hence, the normality implies property B.

Theorem 1.11. *If $f(x)$ has property* B, *then it is an almost periodic function.*

Proof. The proof of this theorem is more complicated and therefore we shall divide it into several steps.

a. *A function with property* B *is bounded on the real line.* Let $\varepsilon = 1$ and let $l = l(1)$ be the corresponding length described in B. If x is an arbitrary real number, then there exists an 1-translation number of $f(x)$ for which $-x \leqslant \xi \leqslant -x + l$, or $0 \leqslant x + \xi \leqslant l$. Let $m = \sup |f(x)|$, $0 \leqslant x \leqslant l$. The number m exists because $f(x)$ is continuous. Since

$$|f(x)| \leqslant |f(x) - f(x + \xi)| + |f(x + \xi)|,$$

it follows that

$$|f(x)| < 1 + m = M.$$

b. *A function with property* B *is uniformly continuous on the real line.*

Let $\varepsilon > 0$ and let $l = l(\varepsilon/3)$ be the length corresponding to $\varepsilon/3$ according to property B. The function $f(x)$ is uniformly continuous on the interval $-1 \leqslant x \leqslant 1 + l$. Hence, there exists a number $\delta = \delta(\varepsilon/3)$, $0 < \delta < 1$, such that

$$|f(y_1) - f(y_2)| < \frac{\varepsilon}{3},$$

whenever

$$|y_1 - y_2| < \delta, \qquad -1 \leqslant y_1, y_2 \leqslant 1 + l.$$

Consider two real numbers x_1 and x_2 such that $|x_1 - x_2| < \delta$. There exists an $(\varepsilon/3)$-translation number of $f(x)$, say ξ, such that $-x_1 \leqslant \xi \leqslant -x_1 + l$ or $0 \leqslant x_1 + \xi \leqslant l$. Therefore we obtain $-1 \leqslant x_2 + \xi \leqslant 1 + l$. Thus

$$|f(x_1) - f(x_2)| \leqslant |f(x_1) - f(x_1 + \xi)| + |f(x_1 + \xi) - f(x_2 + \xi)|$$

$$+ |f(x_2 + \xi) - f(x_2)| < \varepsilon,$$

since $x_1 + \xi$ and $x_2 + \xi$ belong to the interval $[-1, 1 + l]$, and $|(x_1 + \xi) - (x_2 + \xi)| < \delta$.

c. *If* $f(x)$ *possesses property* B, *then for any* $\varepsilon > 0$ *there exist two positive numbers* $l(\varepsilon)$ *and* $\delta(\varepsilon)$ *with the property that any interval of length* l *of the real line contains a subinterval of length* δ *whose points are* ε-*translation numbers of* $f(x)$.

Assume $\varepsilon > 0$ and let $l_1 = l(\varepsilon/2)$ be the length corresponding to $\varepsilon/2$ according to B. As proved in b, there exists a positive number $\delta_1 = \delta(\varepsilon/2)$ for which

$$|f(x + h) - f(x)| < \frac{\varepsilon}{2}, \qquad -\infty < x < +\infty,$$

if $|h| < \delta_1$. Consider an $(\varepsilon/2)$-translation number, say ξ, of the function $f(x)$. It belongs to a certain interval $\alpha \leqslant \xi \leqslant \alpha + l_1$. We notice that for $|h| < \delta_1$, $\xi + h$ belongs to the interval $[\alpha - \delta_1, \alpha + l_1 + \delta_1]$. But

$$|f(x + \xi + h) - f(x)| \leqslant |f(x + \xi + h) - f(x + h)|$$

$$+ |f(x + h) - f(x)| < \varepsilon, \qquad -\infty < x < +\infty.$$

Therefore, $\xi + h$ is an ε-translation number of $f(x)$. Thus we can take $l = l_1 + 2\delta_1$ and $\delta = 2\delta_1$.

d. Below we shall assume that ε is a positive number. Let l and δ be the corresponding numbers whose meaning was established in c. Consider the intervals $I_k = [kl, (k + 1)l]$, $k = 0, \pm 1, \pm 2, \ldots$; in each interval I_k there exists a subinterval $\Delta_k = [\xi_k - \delta/2, \xi_k + \delta/2]$, such that any point from Δ_k is an ε-translation number of the function $f(x)$.

We define a function $K(s)$ on the whole real line, setting

$$K(s) = \begin{cases} \dfrac{3l}{\delta} & \text{for } s \in \bigcup_{k=-\infty}^{\infty} \Delta_k, \\[2mm] 0 & \text{for } s \bar{\in} \bigcup_{k=-\infty}^{\infty} \Delta_k. \end{cases}$$

If n is a natural number and

(1.15) $\qquad f_n(x) = \dfrac{1}{(6nl)^2} \int_{-nl}^{nl} \int_{-nl}^{nl} f(x + s + t)K(s)K(t)\, ds\, dt,$

then

(1.16) $\qquad |f(x) - f_n(x)| < 2\varepsilon, \qquad -\infty < x < +\infty.$

We notice that

$$\frac{1}{6nl} \int_{-nl}^{nl} K(s)\, ds = \frac{1}{6nl} \sum_{k=-n}^{n-1} \frac{3l}{\delta} \int_{\Delta_k} ds = \frac{1}{2n\delta} \sum_{k=-n}^{n-1} \int_{\Delta_k} ds$$

$$= \frac{1}{2n\delta} \cdot 2n\delta = 1,$$

and

$$\frac{1}{(6nl)^2} \int_{-nl}^{nl} \int_{-nl}^{nl} K(s)K(t)\, ds\, dt = \left\{ \frac{1}{6nl} \int_{-nl}^{nl} K(s)\, ds \right\}^2 = 1.$$

Therefore

$$f_n(x) - f(x) = \frac{1}{(6nl)^2} \int_{-nl}^{nl} \int_{-nl}^{nl} [f(x + s + t) - f(x)]K(s)K(t)\, ds\, dt.$$

From $K(s)K(t) \neq 0$ it follows that $s, t \in \bigcup_{k=-\infty}^{\infty} \Delta_k$; in other words s and t are ε-translation numbers for $f(x)$. Consequently

$$|f(x + s + t) - f(x)| \leqslant |f(x + s + t) - f(x + s)| +$$

$$|f(x + s) - f(x)| < 2\varepsilon, \qquad -\infty < x < +\infty.$$

Hence

$$|f(x) - f_n(x)| < 2\varepsilon, \qquad -\infty < x < +\infty,$$

for any natural number n.

e. For any natural number n we define two periodic functions $\bar{f}_n(x)$ and $K_n(x)$ with the period $6nl$, setting

$$\bar{f}_n(x) = f(x) \quad \text{for} \quad |x| \leqslant 3nl; \quad \bar{f}_n(x + 6nl) = \bar{f}_n(x);$$

$$K_n(x) = K(x) \quad \text{for} \quad |x| \leqslant nl,$$

$$K_n(x) = 0 \quad \text{for} \quad nl < |x| \leqslant 3nl, \quad K_n(x + 6nl) = K_n(x).$$

Let

$$\bar{f}_n(x) \sim \sum_{k=-\infty}^{\infty} A_{n,k}\, e^{i\omega_{n,k}x},$$

$$K_n(x) \sim \sum_{k=-\infty}^{\infty} a_{n,k}\, e^{i\omega_{n,k}x},$$

where

$$\omega_{n,k} = \frac{k\pi}{3nl}.$$

The Fourier coefficients $A_{n,k}$ and $a_{n,k}$ fulfill the following conditions

(1.17)
$$\sum_{k=-\infty}^{\infty} |A_{n,k}|^2 \leqslant M^2,$$

(1.18)
$$\sum_{k=-\infty}^{\infty} |a_{n,k}|^2 = \frac{3l}{\delta},$$

(1.19)
$$|a_{n,k}| \leqslant \min\left\{1, \frac{2}{\delta|\omega_{n,k}|}\right\},$$

where $M = \sup |f(x)|$, $-\infty < x < +\infty$.

Inequality (1.17) is an immediate consequence of Parseval's formula for the periodic function $\bar{f}_n(x)$. Equality (1.18) represents Parseval's formula applied to the periodic function $K_n(x)$ and can be immediately verified. Regarding inequality (1.19) we notice that

$$|a_{n,k}| = \left| \frac{1}{6nl} \int_{-3nl}^{3nl} K_n(x)\, e^{-i\omega_{n,k}x} dx \right| \leqslant \frac{1}{6nl} \int_{-nl}^{nl} K(x)\, dx$$

$$= \frac{1}{2n\delta} \sum_{j=-n}^{n-1} \int_{\Delta j} dx = 1,$$

$$|a_{n,k}| = \frac{1}{6nl} \left| \sum_{j=-n}^{n-1} \frac{3l}{\delta} \int_{\Delta j} e^{-i\omega_{n,k}x} dx \right| = \frac{1}{2n\delta} \left| \sum_{j=-n}^{n-1} \int_{\xi_j - \delta/2}^{\xi_j + \delta/2} e^{-i\omega_{n,k}x} dx \right|$$

$$= \frac{1}{2n\delta|\omega_{n,k}|} \left| \sum_{j=-n}^{n-1} \left\{ e^{-i\omega_{n,k}(\zeta_j + \delta/2)} - e^{-i\omega_{n,k}(\zeta_j - \delta/2)} \right\} \right| \leqslant \frac{4n}{2n\delta|\omega_{n,k}|}$$

$$= \frac{2}{\delta|\omega_{n,k}|}.$$

A consequence of (1.17) is that

$$|A_{n,k}| \leqslant M.$$

f. *If* $|x| < nl$, *then* $f_n(x)$ *can be expanded in the series*

(1.20) $$f_n(x) = \sum_{k=-\infty}^{\infty} A_{n,k} a_{n,-k}^2 e^{i\omega_{n,k}x},$$

the convergence being absolute and uniform.

Consider the auxiliary function

(1.21) $$f_n^*(x) = \frac{1}{(6nl)^2} \int_{-3nl}^{3nl} \int_{-3nl}^{3nl} \bar{f}_n(x + s + t) K_n(s) K_n(t) \, ds \, dt.$$

Obviously, this is a continuous periodic function with the period $6nl$. Since $K_n(s) = 0$, for $nl \leqslant |s| \leqslant 3nl$, it follows that

$$f_n^*(x) = \frac{1}{(6nl)^2} \int_{-nl}^{nl} \int_{-nl}^{nl} \bar{f}_n(x + s + t) K(s) K(t) \, ds \, dt.$$

But for $|x| < nl$, $|s| < nl$, $|t| < nl$ we have that $|x + s + t| < 3nl$ and thus, $\bar{f}_n(x + s + t) = f(x + s + t)$. Therefore, for $|x| < nl$,

(1.22) $$f_n^*(x) = \frac{1}{(6nl)^2} \int_{-nl}^{nl} \int_{-nl}^{nl} f(x + s + t) K(s) K(t) \, ds \, dt = f_n(x).$$

We now calculate the Fourier coefficients of the function $f_n^*(x)$. Let

$$f_n^*(x) \sim \sum_{k=-\infty}^{\infty} b_{n,k} e^{i\omega_{n,k}x}.$$

We have

$$b_{n,k} = \frac{1}{6nl} \int_{-3nl}^{3nl} f_n^*(x) e^{-i\omega_{,k}x} \, dx$$

$$= \frac{1}{(6nl)^2} \int_{-3nl}^{3nl} \int_{-3nl}^{3nl} K_n(s) K_n(t) \, ds \, dt \cdot \frac{1}{6nl} \int_{-3nl}^{3nl} \bar{f}_n(x + s + t) e^{-i\omega_{n,k}x} \, dx$$

$$= \frac{1}{(6nl)^2} \int_{-3nl}^{3nl} K_n(s) e^{i\omega_{n,k}s} \, ds \cdot \int_{-3nl}^{3nl} K_n(t) e^{i\omega_{n,k}t} \, dt$$

$$\times \frac{1}{6nl} \int_{-3nl}^{3nl} \bar{f}_n(u) e^{-i\omega_{n,k}u} \, du = A_{n,k} a_{n,-k}^2.$$

Since $|A_{n,k} a_{n,-k}^2| \leqslant |A_{n,k}| |a_{n,-k}| \leqslant \frac{1}{2}(|A_{n,k}|^2 + |a_{n,k}|^2)$, by (1.17) and (1.18) the series $\sum_{k=-\infty}^{\infty} |b_{n,k}|$ is convergent. Thus, the Fourier series of $f_n^*(x)$ is uniformly convergent and consequently, its sum is $f_n{}^*(x)$. According to (1.22), for $|x| < nl$, (1.20) holds true.

g. *There exists a positive number* $\lambda(\varepsilon)$ *independent of* n, *such that*

$$(1.23) \qquad \left| f_n(x) - \sum_{|\omega_{n,k}| \leqslant \lambda} A_{n,k} \, a_{n,-k}^2 \, e^{i\omega_{n,k}x} \right| < \varepsilon, \qquad |x| < nl.$$

First, let us show that we can determine a positive number $\lambda(\varepsilon)$, such that

$$(1.24) \qquad \sum_{|\omega_{n,k}| > \lambda} |A_{n,k}| \cdot |a_{n,-k}^2| < \varepsilon.$$

Since $|\omega_{n,k}| > \lambda$, it follows from (1.19) that $|a_{n,k}| < 2/\delta\lambda$. Moreover, since l and δ depend only on ε, and M is a constant, we have

$$\sum_{|\omega_{n,k}| > \lambda} |A_{n,k}| \cdot |a_{n,-k}^2| = \sum_{|\omega_{n,k}| > \lambda} |A_{n,k}| \cdot |a_{n,-k}| \cdot |a_{n,-k}|$$

$$< \frac{2}{\delta\lambda} \sum_{|\omega_{n,k}| > \lambda} |A_{n,k}| \cdot |a_{n,-k}|$$

$$\leqslant \frac{2}{\delta\lambda} \left(\sum_{k=-\infty}^{\infty} |A_{n,k}|^2 \right)^{1/2} \cdot \left(\sum_{k=-\infty}^{\infty} |a_{n,k}|^2 \right)^{1/2}$$

$$\leqslant \frac{2\sqrt{3}\,M\sqrt{l}}{\delta^{3/2}\lambda} < \varepsilon, \qquad \text{when} \qquad \lambda > \frac{2\sqrt{3}\,M\sqrt{l}}{\varepsilon\delta^{3/2}} = \lambda(\varepsilon).$$

From (1.20) it follows for $|x| < nl$ that

$$(1.25) \qquad f_n(x) - \sum_{|\omega_{n,k}| \leqslant \lambda} A_{n,k} \, a_{n,-k}^2 \, e^{i\omega_{n,k}x} = \sum_{|\omega_{n,k}| > \lambda} A_{n,k} \, a_{n,-k}^2 \, e^{i\omega_{n,k}x}.$$

Inequality (1.23) is now an immediate consequence of (1.24) and (1.25).

h. We now fix the number n and we consider the k's for which $|\omega_{n,k}| \leqslant \lambda$. Obviously, there exists only a finite number of values of k (integers) for which the preceding condition is satisfied. We arrange the numbers $A_{n,k}$ in the order of their decreasing modulus, and we denote them by $B_{n,1}, B_{n,2}, \ldots, B_{n,r_n} : |B_{n,q}| \geqslant |B_{n,q+1}|$. We denote by $\mu_{n,1}, \ldots, \mu_{n,r_n}$ the corresponding exponents $\omega_{n,k}$ in the established order, and by $\alpha_{n,1}, \ldots, \alpha_{n,r_n}$ the numbers $a_{n,-k}$. Set $C_{n,q} = B_{n,q} \, \alpha_{n,q}^2$.

There exists a positive integer $m = m(\varepsilon)$, *independent of* n, *such that*

$$(1.26) \qquad \left| f_n(x) - \sum_{q=1}^{m} C_{n,q} \, e^{i\mu_{n,q}x} \right| < 2\varepsilon, \qquad |x| < nl.$$

We shall show that one can select a number m, independent of n, such that

$$(1.27) \qquad \sum_{q=m}^{r_n} |C_{n,q}| < \varepsilon.$$

Consider a natural number $p \leqslant r_n$. Since $\sum_{q=1}^{p} |B_{n,q}|^2 \leqslant M^2$ and $B_{n,q}$ decreases in modulus, we have $M^2 \geqslant \sum_{q=1}^{p} |B_{n,q}|^2 \geqslant p|B_{n,p}|^2$.

Hence

$$|B_{n,p}| \leqslant \frac{M}{\sqrt{p}}.$$

But

$$\sum_{q=p}^{r_n} |C_{n,q}| = \sum_{q=p}^{r_n} |B_{n,q}| \cdot |\alpha_{n,q}^2| \leqslant \frac{M}{\sqrt{p}} \sum_{q=p}^{r_n} |\alpha_{n,q}^2| \leqslant \frac{M}{\sqrt{p}} \sum_{q=-\infty}^{\infty} |\alpha_{n,q}|^2 < \varepsilon,$$

if

$$p > m = \left[\frac{9l^2 M^2}{\delta^2 \varepsilon^2} \right],$$

where the brackets indicate the integer part.

Then we notice that (1.23) can also be written as

$$(1.28) \qquad \left| f_n(x) - \sum_{q=1}^{r_n} C_{n,q} e^{i\mu_{n,q}x} \right| < \varepsilon, \qquad |x| < nl.$$

From (1.27) and (1.28) immediately follows (1.26).

i. We observe from the proofs in d and h, that

$$(1.29) \qquad \left| f(x) - \sum_{q=1}^{m} C_{n,q} e^{i\mu_{n,q}x} \right| < 4\varepsilon, \qquad |x| < nl,$$

since $|C_{n,q}| \leqslant |B_{n,q}| \leqslant M$ and $|\mu_{n,q}| \leqslant \lambda(\varepsilon)$ ($B_{n,q}$ are $A_{n,k}$ and $\mu_{n,q}$ are $\omega_{n,k}$); by a successive application of the Bolzano–Weierstrass theorem we can extract a subsequence of indices $n_1 < n_2 < \cdots < n_k < \cdots$ such that

$$\lim_{k \to \infty} C_{n_k,q} = C_q, \qquad \lim_{k \to \infty} \mu_{n_k,q} = \mu_q.$$

Setting $n = n_k$ in (1.29) we obtain for $k \to \infty$

$$(1.30) \qquad \left| f(x) - \sum_{q=1}^{m} C_q e^{i\mu_q x} \right| \leqslant 4\varepsilon, \qquad -\infty < x < +\infty,$$

since for any x we can find a sufficiently large k such that $|x| < n_k l$.

In other words, given a function with property B and a number $\varepsilon > 0$, one can determine a trigonometric polynomial satisfying (1.30). This proves the almost periodicity of the function $f(x)$.

3. The Fourier series associated with an almost periodic function

In order to introduce the Fourier series of an almost periodic function, it is necessary to define the *mean value* of such a function.

Theorem 1.12. *If $f(x)$ is an almost periodic function, then*

$$(1.31) \qquad \lim_{T \to \infty} \frac{1}{T} \int_a^{a+T} f(x)\, dx = M\{f(x)\},$$

exists uniformly with respect to a.

$M\{f(x)\}$ *is independent of a and is called the mean value of the almost periodic function $f(x)$.*

Proof. We shall show first that the mean exists if $f(x)$ is a trigonometric polynomial. Let

$$S(x) = c_0 + \sum_{k=1}^{n} c_k e^{i\lambda_k x},$$

where $\lambda_k \neq 0$, $k = 1, 2, \ldots, n$. It follows without any difficulty that

$$\frac{1}{T} \int_a^{a+T} S(x)\, dx = c_0 + \sum_{k=1}^{n} c_k \frac{e^{i\lambda_k(a+T)} - e^{i\lambda_k a}}{i\lambda_k T}.$$

For $T > 0$, we obtain

$$\left| \frac{1}{T} \int_a^{a+T} S(x)\, dx - c_0 \right| \leqslant \frac{2}{T} \sum_{k=1}^{n} \left| \frac{c_k}{\lambda_k} \right|.$$

This implies that

$$\lim_{T \to \infty} \frac{1}{T} \int_a^{a+T} S(x)\, dx = c_0,$$

namely, the mean value exists for any trigonometric polynomial, uniformly, with respect to a. This mean value is independent of a.

Let $f(x)$ be an almost periodic function and $\varepsilon > 0$ an arbitrary number. One can determine a trigonometric polynomial $S(x)$, such that

$$(1.32) \qquad |f(x) - S(x)| < \frac{\varepsilon}{3}, \qquad -\infty < x < +\infty.$$

Since the mean of $S(x)$ exists, we can find a number $T(\varepsilon)$ such that

$$(1.33) \quad \left| \frac{1}{T_1} \int_a^{a+T_1} S(x)\,dx - \frac{1}{T_2} \int_a^{a+T_2} S(x)\,dx \right| < \frac{\varepsilon}{3}, \qquad T_1, T_2 \geqslant T(\varepsilon).$$

It follows from (1.32) and (1.33) that

$$\left| \frac{1}{T_1} \int_a^{a+T_1} f(x)\,dx - \frac{1}{T_2} \int_a^{a+T_2} f(x)\,dx \right|$$

$$\leqslant \frac{1}{T_1} \int_a^{a+T_1} |f(x) - S(x)|\,dx + \left| \frac{1}{T_1} \int_a^{a+T_1} S(x)\,dx - \frac{1}{T_2} \int_a^{a+T_2} S(x)\,dx \right|$$

$$+ \frac{1}{T_2} \int_a^{a+T_2} |S(x) - f(x)|\,dx < \varepsilon,$$

if $T_1, T_2 \geqslant T(\varepsilon)$.

Let us also show that

$$(1.34) \quad \lim_{T \to \infty} \frac{1}{T} \int_a^{a+T} f(x)\,dx = \lim_{T \to \infty} \frac{1}{T} \int_0^T f(x)\,dx,$$

whence it will follow that the mean value is independent of the real number a. Since $f(x)$ is bounded, there exists a number M such that $|f(x)| \leqslant M$, $-\infty < x < +\infty$. Assuming that $a > 0$ and $T > a$, we obtain that

$$\frac{1}{T} \left| \int_a^{a+T} f(x)\,dx - \int_0^T f(x)\,dx \right| = \frac{1}{T} \left| \int_T^{a+T} f(x)\,dx - \int_0^a f(x)\,dx \right| \leqslant \frac{2aM}{T}.$$

Therefore equality (1.34) is true. One proceeds in a similar way if $a < 0$.

Hence, there exists a mean value for any almost periodic function.

Remark. Let us assume that $f(x)$ is a periodic function with the period 2ω. For any real T, we can write $T = 2n\omega + \alpha_n$, α_n being a real number for which $0 \leqslant \alpha_n < 2\omega$. We notice that $T \to \infty$ if and only if $n \to \infty$. Let us calculate now $M\{f(x)\}$. We obtain:

$$M\{f(x)\} = \lim_{T \to \infty} \frac{1}{T} \int_0^T f(x)\,dx = \lim_{n \to \infty} \frac{1}{2n\omega + \alpha_n} \int_0^{2n\omega + \alpha_n} f(x)\,dx$$

$$= \lim_{n \to \infty} \frac{1}{2n\omega + \alpha_n} \left\{ n \int_0^{2\omega} f(x)\,dx + \int_0^{\alpha_n} f(x)\,dx \right\} = \frac{1}{2\omega} \int_0^{2\omega} f(x)\,dx.$$

In other words, the mean value defined for almost periodic functions coincides with the usual mean in case the function is periodic.

The mean value introduced above for almost periodic functions has some simple properties. We shall list them in the following.

Theorem 1.13. *Let $f(x)$ and $g(x)$ be two almost periodic functions, and let c be a complex number. Then*

$$(1.35) \qquad M\{\bar{f}(x)\} = \overline{M\{f(x)\}},$$

$$(1.36) \qquad M\{cf(x)\} = cM\{f(x)\},$$

$$(1.37) \qquad M\{f(x)\} \geqslant 0, \quad \text{if} \ \ f(x) \geqslant 0,$$

$$(1.38) \qquad M\{f(x) + g(x)\} = M\{f(x)\} + M\{g(x)\}.$$

If $\{f_n(x)\}$ is a uniformly convergent sequence of almost periodic functions which converges to $f(x)$, then

$$(1.39) \qquad \lim_{n \to \infty} M\{f_n(x)\} = M\{f(x)\}.$$

We shall not elaborate on the assertions made in Theorem 1.13.

Consider now a real number λ and let $f(x)$ be an almost periodic function. Since $f(x)e^{-i\lambda x}$ is also an almost periodic function, there exists the mean value of this function. We can write

$$(1.40) \qquad a(\lambda) = M\{f(x)e^{-i\lambda x}\}.$$

Let $\mu_1, \mu_2, \ldots, \mu_n$ be any real numbers and $c_1, c_2, \ldots, c_n$ be any complex numbers. Put

$$(1.41) \qquad \varphi(c_1, c_2, \ldots, c_n) = M\left\{\left|f(x) - \sum_{k=1}^{n} c_k e^{i\mu_k x}\right|^2\right\}.$$

Theorem 1.14. *The function $\varphi(c_1, c_2, \ldots, c_n)$ assumes a minimum value for $c_k = a(\mu_k), k = 1, 2, \ldots, n$. We have*

$$(1.42) \qquad \sum_{k=1}^{n} |a(\mu_k)|^2 \leqslant M\{|f(x)|^2\}.$$

Proof. Using the properties of the mean value listed in Theorem 1.13, we obtain

$$M\left\{\left|f(x) - \sum_{k=1}^{n} c_k e^{i\mu_k x}\right|^2\right\} = M\{|f(x)|^2\} - \sum_{k=1}^{n} \bar{c}_k M\{f(x)e^{-i\mu_k x}\}$$

$$- \sum_{k=1}^{n} c_k M\{\bar{f}(x)e^{i\mu_k x}\}$$

$$+ \sum_{k=1}^{n} \sum_{h=1}^{n} c_k \bar{c}_h M\{e^{i\mu_k x} \cdot e^{-i\mu_h x}\}.$$

An elementary calculation shows that

$$M\{e^{i\mu_k x} \cdot e^{-i\mu_h x}\} = \begin{cases} 1 & \text{if} \quad h = k, \\ 0 & \text{if} \quad h \neq k. \end{cases}$$

Therefore

$$M\left\{\left| f(x) - \sum_{k=1}^{n} c_k e^{i\mu_k x} \right|^2\right\} = M\{|f(x)|^2\} - \sum_{k=1}^{n} \bar{c}_k a(\mu_k)$$

$$- \sum_{k=1}^{n} c_k \bar{a}(\mu_k) + \sum_{k=1}^{n} |c_k|^2$$

$$= M\{|f(x)|^2\} + \sum_{k=1}^{n} |c_k - a(\mu_k)|^2 - \sum_{k=1}^{n} |a(\mu_k)|^2.$$

Consequently, $\varphi(c_1, c_2, \ldots, c_n)$ takes the minimum value for $c_k = a(\mu_k)$, and this minimum value is $M\{|f(x)|^2\} - \sum_{k=1}^{n} |a(\mu_k)|^2$. Since this minimum is non-negative we obtain (1.42).

The following statement is of special importance in the theory of almost periodic functions:

Theorem 1.15. *There exists at most a countable set of λ's for which*

(1.43) $a(\lambda) \neq 0.$

Proof. By (1.42) it follows that there can exist only a finite number of λ's for which $|a(\lambda)| \geq 1$. Similarly, for any n one establishes that there exists only a finite number of λ's for which

$$\frac{1}{n+1} \leq |a(\lambda)| < \frac{1}{n}.$$

Hence, the set of λ's for which $a(\lambda) \neq 0$ is a countable union of finite sets, and consequently it is countable.

The numbers $\lambda_1, \lambda_2, \ldots, \lambda_n, \ldots$ for which $a(\lambda_k) \neq 0$ are called the *Fourier exponents of the function $f(x)$*, and $a(\lambda_k)$ are the *Fourier coefficients of $f(x)$*

Let $A_k = a(\lambda_k)$, $k = 1, 2, \ldots$. The fact that λ_k are the Fourier exponents of $f(x)$ and A_k, $k = 1, 2, \ldots$, are the Fourier coefficients of $f(x)$ can briefly be written as

(1.44) $f(x) \sim \sum_{k=1}^{\infty} A_k e^{i\lambda_k x}.$

The series on the right side is called the *Fourier series associated with the function $f(x)$*.

If $f(x)$ is a periodic function, then the Fourier series reduces to the usual Fourier series with the meaning used in theory of period functions (see Introduction and the above remark).

Theorem 1.16. *If* $f(x) \sim \sum_{k=1}^{\infty} A_k e^{i\lambda_k x}$, *then*

$$\bar{f}(x) \sim \sum_{k=1}^{\infty} \bar{A}_k e^{-i\lambda_k x},$$

$$f(x + a) \sim \sum_{k=1}^{\infty} A_k e^{i\lambda_k a} \cdot e^{i\lambda_k x}, \qquad e^{i\lambda x} f(x) \sim \sum_{k=1}^{\infty} A_k e^{i(\lambda_k + \lambda)x},$$

where a *and* λ *are real numbers.*

The proof of this theorem is not difficult and therefore we shall omit it.

Theorem 1.17. *If the derivative (primitive) of an almost periodic function is almost periodic, then its Fourier series can be obtained by formal differentiation (integration).*

Proof. Consider an almost periodic function $f(x)$ with an almost periodic derivative. There exists then the mean value $M\{f'(x)e^{-i\lambda x}\}$. Since

$$\frac{1}{T}\int_a^{a+T} f'(x)e^{-i\lambda x}\, dx = \frac{1}{T}f(x)e^{-i\lambda x}\Big|_a^{a+T} + i\lambda \frac{1}{T}\int_a^{a+T} f(x)e^{-i\lambda x}\, dx,$$

then for $T \to \infty$, we obtain

(1.45) $$M\{f'(x)e^{-i\lambda x}\} = i\lambda M\{f(x)\}.$$

From (1.45) it follows that $f'(x)$ has the same Fourier exponents as $f(x)$, except for $\lambda = 0$, if it appears among the Fourier exponents of $f(x)$. If we denote by A'_k the Fourier coefficients of $f'(x)$, we obtain from (1.45) that

(1.46) $$A'_k = i\lambda_k A_k.$$

Therefore,

(1.47) $$f'(x) \sim \sum_{k=1}^{\infty} i\lambda_k A_k e^{i\lambda_k x}.$$

From (1.47) it follows that if the primitive $F(x)$ of the function $f(x)$ is almost periodic, we should have that

(1.48) $$F(x) = \int_0^x f(t)\, dt \sim C + \sum_{k=1}^{\infty} \frac{A_k}{i\lambda_k} e^{i\lambda_k x}.$$

We notice that the presence of the numbers λ_k in the denominator does not affect the validity of (1.48). Indeed, according to our above remarks,

$\lambda = 0$ cannot occur among the Fourier exponents of an almost periodic function which is the derivative of another almost periodic function.

In other words, for a primitive of an almost periodic function to be almost periodic it is necessary that among the Fourier exponents of this function $\lambda = 0$ does not occur (i.e., the Fourier series has no free term). However, there are examples showing that this condition is not sufficient.

Theorem 1.18. *If $f(x)$ is an almost periodic function such that $f(x) \sim \sum_{k=1}^{\infty} A_k e^{i\lambda_k x}$, then Parseval's equality*

$$(1.49) \qquad \sum_{k=1}^{\infty} |A_k|^2 = M\{|f(x)|^2\}.$$

is true.

Proof. Inequality (1.42) allows us to write

$$\sum_{k=1}^{n} |A_k|^2 \leqslant M\{|f(x)|^2\}.$$

Since the preceding inequality holds for any number n, this means that

$$(1.50) \qquad \sum_{k=1}^{\infty} |A_k|^2 \leqslant M\{|f(x)|^2\}.$$

Inequality (1.50) is known as the Bessel inequality.

Let $S(x)$ be any trigonometric polynomial. Set $S^*(x) \equiv 0$, if none of the exponents of $f(x)$ occurs among the Fourier exponents of $S(x)$, and $S^*(x) = \sum A_k c^{i\lambda_k x}$, the summation being extended over those k's for which λ_k is a Fourier exponent common to the functions $f(x)$ and $S(x)$.

Since the function $f(x)$ is almost periodic, there exists a sequence of trigonometric polynomials $\{S_n(x)\}$, such that

$$(1.51) \qquad |f(x) - S_n(x)| < \frac{1}{\sqrt{n}}, \qquad -\infty < x < +\infty.$$

From (1.51) we obtain

$$M\{|f(x) - S_n(x)|^2\} \leqslant \frac{1}{n}.$$

Applying Theorem 1.14, we obtain

$$M\{|f(x) - S_n^*(x)|^2\} \leqslant M\{|f(x) - S_n(x)|^2\} \leqslant \frac{1}{n}.$$

But

$$M\{|f(x) - S_n^*(x)|\}^2 = M\{|f(x)|^2\} - \sum_n |A_k|^2,$$

where the sum is extended over those k's for which λ_k is a Fourier exponent of $S_n(x)$. Hence

$$M\{|f(x)|^2\} \leqslant \sum_n |A_k|^2 + \frac{1}{n},$$

with the same conventions regarding the summation. Furthermore, we shall have that

$$M\{|f(x)|^2\} \leqslant \sum_{k=1}^{\infty} |A_k|^2 + \frac{1}{n},$$

and since n is arbitrarily large, we obtain

(1.52) $$M\{|f(x)|^2\} \leqslant \sum_{k=1}^{\infty} |A_k|^2.$$

Comparing (1.50) with (1.52) we obtain (1.49).

We are able to prove now the uniqueness theorem, which is one of the fundamental theorems in the theory of almost periodic functions.

Theorem 1.19. *Two distinct almost periodic functions have distinct Fourier series.*

Proof. If two distinct almost periodic functions had the same Fourier series, then from Parseval's equality applied to the function $f(x) - g(x)$ it would follow that

$$M\{|f(x) - g(x)|^2\} = 0,$$

whereas $f(x) - g(x) \neq 0$. It is sufficient to show that a non-negative and nonvanishing almost periodic function has a positive mean.

Let $\varphi(x) \geqslant 0$ be an almost periodic function such that $\varphi(x_0) = \alpha > 0$. We choose two numbers $l > 0$ and $\delta > 0$ such that any interval of length l of the real line will contain a subinterval of length 2δ whose points must all be $(\alpha/3)$-translation numbers of $\varphi(x)$, and $|x_1 - x_2| < \delta$ should imply that $|\varphi(x_1) - \varphi(x_2)| < \alpha/3$. This choice is possible as seen in c) in the proof of Theorem 1.11. Consider now any interval of length l, say $(a - \delta - x_0, a + l - \delta - x_0)$, a being a real number. Then there exists an $(\alpha/3)$-translation number ξ of $\varphi(x)$ which belongs to this interval. One establishes that $x_0 + \xi$

belongs to the interval $(a - \delta, a + l - \delta)$, and assuming that $|x - x_0| < \delta$, the number $x + \xi$ will range over an interval of length 2δ. We have

$$\varphi(x + \xi) = \varphi(x_0) + [\varphi(x) - \varphi(x_0)] + [\varphi(x + \xi) - \varphi(x)]$$

$$> \alpha - \frac{\alpha}{3} - \frac{\alpha}{3} = \frac{\alpha}{3},$$

which proves that any interval of length l on the real line contains a sub-interval of length 2δ at all points of which $\varphi(x) > \alpha/3$. This implies that

$$\frac{1}{nl} \int_0^{nl} \varphi(x)\, dx = \frac{1}{nl} \sum_{k=1}^{n} \int_{(k-1)l}^{kl} \varphi(x)\, dx > \frac{1}{nl}\, n \cdot 2\delta \cdot \frac{\alpha}{3} = \frac{2\alpha\delta}{3l}.$$

Letting $n \to +\infty$, we obtain

$$M\{\varphi(x)\} \geqslant \frac{2\alpha\delta}{3l} > 0,$$

which we wanted to show. Thus the uniqueness theorem is proved.

Remarks. Obviously, Theorem 1.19 can be stated as follows: the unique almost periodic function with zero Fourier coefficients is the identically vanishing function.

An important property of periodic functions transposed to almost periodic functions states:

Theorem 1.20. *If the Fourier series of an almost periodic function is uniformly convergent, then the sum of the series is the given function.*

Proof. Let

$$f(x) \sim \sum_{k=1}^{\infty} A_k e^{i\lambda_k x}, \qquad S(x) = \sum_{k=1}^{\infty} A_k e^{i\lambda_k x},$$

the convergence being uniform. If we set

$$S_n(x) = \sum_{k=1}^{n} A_k e^{i\lambda_k x},$$

then the sequence $\{S_n(x)\}$ converges uniformly to $S(x)$. Thus,

$$M\{S(x)e^{-i\lambda x}\} = \lim_{n \to \infty} \sum_{k=1}^{n} A_k M\{e^{i\lambda_k x} \cdot e^{-i\lambda x}\} = \begin{cases} A_k & \text{if} \quad \lambda = \lambda_k \\ 0 & \text{if} \quad \lambda \neq \lambda_k. \end{cases}$$

Therefore $S(x)$ has the same Fourier series as $f(x)$. According to the uniqueness theorem, we must have $f(x) \equiv S(x)$.

Corollary 1. There exist almost periodic functions whose Fourier exponents are the arbitrarily prescribed real numbers $\lambda_1, \lambda_2, \ldots, \lambda_n, \ldots$.

Indeed, if the complex numbers A_k, $k = 1, 2, \ldots$, are such that the series $\sum_{k=1}^{\infty} |A_k|$ is convergent, then the series

$$\sum_{k=1}^{\infty} A_k e^{i\lambda_k x}$$

is uniformly convergent on the real line. Thus the only Fourier exponents of the sum of this series are the numbers λ_k, $k = 1, 2, \ldots$, and no others.

We shall see in the next paragraph how a certain distribution of λ_k on the real line affects the behavior of the Fourier series.

Corollary 2. There exist almost periodic functions whose Fourier series does not have a free term, and whose primitive is not almost periodic.

This statement completes the remark made while proving Theorem 1.17. Consider the series

$$\sum_{k=1}^{\infty} \frac{1}{k^2} \exp\left(i \frac{x}{k^2}\right).$$

Acording to what was said above, the sum of this series is an almost periodic function. Let us denote it by $f(x)$. If the function

$$F(x) = \int_0^x f(t)\, dt$$

were almost periodic, then one should have

$$F(x) \sim C + \sum_{k=1}^{\infty} \frac{1}{i} \exp\left(i \frac{x}{k^2}\right).$$

But the right-hand side is not the Fourier series of an almost periodic function because it violates Parseval's equality.

4. Convergence of Fourier series

The convergence of the Fourier series associated with almost periodic functions is a difficult problem. There have been many results obtained in this direction, but nevertheless the problem is still far from a complete solution. From the numerous convergence criteria known so far, we shall establish in this paragraph only two. These two criteria are nearly immediate consequences of criteria known from the theory of Fourier series associated with periodic functions.

Let a and b be two real numbers such that $0 < a < b$. Consider also the function $\varphi_{a,b}(\lambda)$ of the real variable λ, defined by:

$$(1.53) \qquad \varphi_{a,b}(\lambda) = \begin{cases} 1 & \text{if} \quad |\lambda| \leqslant a, \\ \dfrac{1}{b-a}(b - |\lambda|) & \text{if} \quad a < |\lambda| < b, \\ 0 & \text{if} \quad |\lambda| \geqslant b. \end{cases}$$

Since $\varphi_{a,b}(\lambda)$ takes nonzero values only on the bounded interval $|\lambda| \leqslant b$ of the real line, being continuous on this interval, this function is absolutely integrable on the real line. Thus, there exists the Fourier transform

$$\psi_{a,b}(u) = \frac{1}{2\pi} \int_{-\infty}^{\infty} \varphi_{a,b}(\lambda) e^{-iu\lambda} \, d\lambda.$$

But

$$\psi_{a,b}(u) = \frac{1}{2\pi} \int_{-b}^{b} \varphi_{a,b}(\lambda) e^{-iu\lambda} \, d\lambda = \frac{1}{\pi} \int_{0}^{b} \varphi_{a,b}(\lambda) \cos \lambda u \, d\lambda$$

$$= \frac{1}{\pi} \int_{0}^{a} \cos \lambda u \, d\lambda + \frac{1}{\pi(b-a)} \int_{a}^{b} (b - \lambda) \cos \lambda u \, d\lambda$$

$$= \frac{2}{\pi(b-a)u^2} \sin \frac{b-a}{2} u \cdot \sin \frac{b+a}{2} u.$$

Therefore,

$$(1.54) \qquad \psi_{a,b}(u) = \frac{2}{\pi(b-a)u^2} \sin \frac{b-a}{2} u \cdot \sin \frac{b+a}{2} u.$$

Lemma 1.1. *The inequality*

$$(1.55) \qquad \int_{-\infty}^{\infty} |\psi_{a,b}(u)| \, du \leqslant \frac{2}{\pi}\left(2 + \ln \frac{b+a}{b-a}\right).$$

holds true.

Proof. Set $(b - a)u = 2v$ and $(b + a) = \eta(b - a)$. Then we have

$$\int_{-\infty}^{\infty} |\psi_{a,b}(u)| \, du = \frac{2}{\pi} \int_{0}^{\infty} \frac{|\sin v \cdot \sin \eta v|}{v^2} \, dv \leqslant \frac{2}{\pi} \int_{0}^{1} \frac{|\sin \eta v|}{v} \, dv + \frac{2}{\pi} \int_{1}^{\infty} \frac{dv}{v^2}$$

$$= \frac{2}{\pi} \int_{0}^{\eta} \frac{|\sin v|}{v} \, dv + \frac{2}{\pi}$$

$$= \frac{2}{\pi} \int_0^1 \frac{|\sin v|}{v} \, dv + \frac{2}{\pi} \int_1^\eta \frac{|\sin v|}{v} \, dv + \frac{2}{\pi} \leqslant \frac{2}{\pi} + \frac{2}{\pi} \ln \eta + \frac{2}{\pi}$$

$$= \frac{2}{\pi} (2 + \ln \eta) = \frac{2}{\pi} \left(2 + \ln \frac{b + a}{b - a} \right).$$

Consider now an almost periodic function

(1.56)
$$f(x) \sim \sum_{k=1}^\infty A_k e^{i\lambda_k x}$$

and the function

(1.57)
$$f_{a,b}(x) = \int_{-\infty}^\infty f(x + u) \psi_{a,b}(u) \, du.$$

The integral in (1.57) is obviously convergent, since $\psi_{a,b}(u)$ is absolutely integrable on the real line and $f(x)$ is bounded.

Lemma 1.2. *The function* $f_{a,b}(x)$ *is almost periodic, and we have*

(1.58)
$$f_{a,b}(x) \sim \sum_{k=1}^\infty \varphi_{a,b}(\lambda_k) A_k e^{i\lambda_k x}.$$

Proof. The continuity of $f_{a,b}(x)$ is an immediate consequence of (1.57), using (1.55). The almost periodicity of $f_{a,b}(x)$ follows from the inequality:

$$|f_{a,b}(x + \xi) - f_{a,b}(x)| \leqslant \int_{-\infty}^\infty |f(x + u + \xi) - f(x + u)| \cdot |\psi_{a,b}(u)| \, du$$

$$\leqslant \frac{2}{\pi} \left(2 + \ln \frac{b + a}{b - a} \right) \sup_x |f(x + \xi) - f(x)|,$$

$$-\infty < x < +\infty,$$

using the fact that property B can be taken as a definition of almost periodic functions.

We now calculate the Fourier series corresponding to $f_{a,b}(x)$. We observe that

$$\frac{1}{T} \int_0^T f_{a,b}(x) e^{-i\lambda x} \, dx = \int_{-\infty}^\infty \psi_{a,b}(u) \frac{1}{T} \int_0^T f(x + u) e^{-i\lambda x} \, dx \, du$$

$$= \int_{-\infty}^\infty \psi_{a,b}(u) e^{i\lambda u} \frac{1}{T} \int_u^{u+T} f(x) e^{-i\lambda x} \, dx \, du.$$

By Theorem 1.12 and Lemma 1.1, we obtain for $T \to +\infty$:

$$M\{f_{a,b}(x)e^{-i\lambda x}\} = M\{f(x)e^{-i\lambda x}\}\int_{-\infty}^{\infty} \psi_{a,b}(u)e^{i\lambda u}\,du = \varphi_{a,b}(\lambda)M\{f(x)e^{-i\lambda x}\}.$$

This equality shows that any Fourier exponent of $f(x)$ is also a Fourier exponent of $f_{a,b}(x)$, if $\varphi_{a,b}(\lambda_k) \neq 0$. Since for the subscripts k for which $\varphi_{a,b}(\lambda_k) = 0$, we have

$$M\{f_{a,b}(x)e^{-i\lambda_k x}\} = 0,$$

we can write

$$f_{a,b}(x) \sim \sum_{k=1}^{\infty} \varphi_{a,b}(\lambda_k)A_k\,e^{i\lambda_k x}.$$

Thus the lemma has been proved.

Lemma 1.3. *Suppose that $b \to \infty$, and $a < \theta b$, where $0 < \theta < 1$. Then*

$$\lim_{b \to \infty} f_{a,b}(x) = f(x),$$

the convergence being uniform on the whole real line.

Proof. From the inversion formula

$$\varphi_{a,b}(\lambda) = \int_{-\infty}^{\infty} \psi_{a,b}(u)e^{i\lambda u}\,du,$$

we obtain for $\lambda = 0$ that

$$\int_{-\infty}^{\infty} \psi_{a,b}(u)\,du = 1.$$

Let us now multiply both parts of this equality by $f(x)$ and subtract it term by term from (1.57). We obtain

$$f_{a,b}(x) - f(x) = \int_{-\infty}^{\infty} \{f(x + u) - f(x)\}\psi_{a,b}(u)\,du.$$

Making the substitution $(b - a)u = 2v$ and using (1.54), we find that

$$f_{a,b}(x) - f(x) = \frac{1}{\pi}\int_{-\infty}^{\infty} \left\{f\left(x + \frac{2v}{b - a}\right) - f(x)\right\}\frac{1}{v^2}\sin v \cdot \sin\frac{b + a}{b - a}v \cdot dv.$$

Consider now a positive number N. Obviously, we have

$$(1.59) \quad |f_{a,b}(x) - f(x)| \leqslant \frac{2M}{\pi}\int_{-\infty}^{-N}\frac{dv}{v^2} + \frac{2M}{\pi}\int_{N}^{\infty}\frac{dv}{v^2}$$

$$+ \frac{1}{\pi}\int_{-N}^{N}\left|f\left(x + \frac{2v}{b - a}\right) - f(x)\right|\frac{1}{v^2}\left|\sin v \cdot \sin\frac{b + a}{b - a}v\right|dv,$$

where $M = \sup |f(x)|$, $-\infty < x < +\infty$. Since

$$\int_{-\infty}^{-N} \frac{dv}{v^2} = \int_{N}^{\infty} \frac{dv}{v^2} = \frac{1}{N},$$

this means that

(1.60) $$\frac{2M}{\pi} \int_{-\infty}^{-N} \frac{dv}{v^2} + \frac{2M}{\pi} \int_{N}^{\infty} \frac{dv}{v^2} < \frac{\varepsilon}{2},$$

if

(1.61) $$N > \frac{8M}{\pi\varepsilon},$$

for any positive numbers a and b.

We now fix N so that (1.61) will be satisfied. Since $a < \theta b$, $0 < \theta < 1$, this means that $b - a < b - \theta b = b(1 - \theta) \to \infty$ as $b \to \infty$. If $|v| \leqslant N$, for b sufficiently large we shall obtain

$$\left| f\left(x + \frac{2v}{b-a}\right) - f(x) \right| < \frac{\varepsilon}{4\{2 + \ln [2/(1-\theta)]\}},$$

since $f(x)$ is uniformly continuous. Therefore

$$\frac{1}{\pi} \int_{-N}^{N} \left| f\left(x + \frac{2v}{b-a}\right) - f(x) \right| \frac{1}{v^2} \left| \sin v \cdot \sin \frac{b+a}{b-a} v \right| dv.$$

$$\leqslant \frac{\pi\varepsilon}{4\pi\{2 + \ln [2/(1-\theta)]\}} \int_{-\infty}^{\infty} \frac{1}{v^2} \left| \sin v \cdot \sin \frac{b+a}{b-a} v \right| dv,$$

$$\leqslant \frac{\varepsilon}{2} \cdot \frac{2 + \ln [(b+a)/(b-a)]}{2 + \ln [2/(1-\theta)]}.$$

However, $(b + a)/(b - a) < 2/(1 - \theta)$, so that for b sufficiently large we can write

(1.62) $$\frac{1}{\pi} \int_{-N}^{N} \left| f\left(x + \frac{2v}{b-a}\right) - f(x) \right| \frac{1}{v^2} \left| \sin v \cdot \sin \frac{b+a}{b-a} v \right| dv < \frac{\varepsilon}{2}.$$

Using (1.59), (1.60) and (1.62) we find that

$$|f_{a,b}(x) - f(x)| < \varepsilon, \qquad -\infty < x < +\infty,$$

when b is sufficiently large.

This proves Lemma 1.3.

We shall consider below only almost periodic functions with the property
that

(1.63)
$$\lim_{n \to \infty} |\lambda_n| = +\infty.$$

One can also say that the Fourier exponents have a unique limit point and
this is the point at infinity.

We observe that this class of almost periodic functions includes all the
periodic functions.

To give symmetry to the formulas, we agree to add to the Fourier exponents the numbers $-\lambda_n$, if they do not appear among the exponents.
Numbering them so that $\lambda_{-n} = -\lambda_n$, it is natural to write the Fourier
series in the form

(1.64)
$$f(x) \sim \sum_{k=-\infty}^{\infty} A_k e^{i\lambda_k x}.$$

For the exponents λ_n added according to the above convention, we shall
obviously have $A_n = 0$.

Theorem 1.21. *Assume that the almost periodic function $f(x)$ is such that*
$\lambda_{n+1} - \lambda_n \geq \alpha > 0$, $n = 1, 2, \ldots$, *so that the unique limit point of its*
Fourier exponents is the point at infinity. If x_0 is a point in the neighborhood
of which $f(x)$ has a bounded variation, then

$$\lim_{n \to \infty} \sum_{k=-n}^{n} A_k e^{i\lambda_k x_0} = f(x_0).$$

Proof. Setting $a = \lambda_n$, $b = \lambda_n + \alpha$, and using (1.58) we obtain

$$S_n(x) = \sum_{k=-n}^{n} A_k e^{i\lambda_k x} = f_{a,b}(x)$$

$$= \frac{2}{\pi\alpha} \int_{-\infty}^{\infty} f(x+u) \, \frac{\sin(\alpha u/2) \, \dfrac{\sin(\lambda_n + \alpha)}{2} u}{u^2} \, du.$$

If h is an arbitrary positive number, then we can write

(1.65)
$$S_n(x_0) = \frac{2}{\pi\alpha} \left\{ \int_{-\infty}^{-h} + \int_{-h}^{h} + \int_{h}^{\infty} \right\} = I_1 + I_2 + I_3.$$

The behavior of the integral

$$I_2 = \frac{2}{\pi\alpha} \int_{-h}^{h} f(x_0+u) \, \frac{\sin(\alpha u/2) \, \dfrac{\sin(\lambda_n + \alpha)}{2} u}{u^2} \, du$$

for $h \to 0$ is the same as that of the integral

$$(1.66) \qquad I'_2 = \frac{1}{\pi} \int_{-h}^{h} f(x_0 + u) \frac{\sin(\lambda_n + \alpha)}{2} u \frac{}{u} \, du.$$

Indeed,

$$I_2 - I'_2 = \frac{1}{\pi} \int_{-h}^{h} f(x_0 + u) \left[\frac{\sin(\lambda_n + \alpha)}{2} u \right] u \cdot \frac{2 \sin(\alpha u/2) - \alpha u}{\alpha u^2} \, du.$$

The function under the integral sign is bounded at the origin, uniformly with respect to n. Thus, given $\varepsilon > 0$, one can determine a $\delta(\varepsilon) > 0$, independent of n, such that

$$(1.67) \qquad |I_2 - I'_2| < \frac{\varepsilon}{3}, \qquad \text{if} \quad h < \delta.$$

Let us now assign a value to h such that (1.67) is satisfied. Since the function

$$\frac{f(x_0 + u)}{u^2} \sin \frac{\alpha u}{2}$$

is absolutely integrable on each of the intervals $(-\infty, -h)$ and (h, ∞), it follows (see Introduction) that

$$(1.68) \qquad |I_1| < \frac{\varepsilon}{3}, \qquad |I_3| < \frac{\varepsilon}{3}, \qquad \text{if} \quad n > N(\varepsilon).$$

Using (1.65), (1.66), (1.67), and (1.68), we can write

$$(1.69) \qquad |S_n(x_0) - I'_2| < \varepsilon, \qquad \text{if} \quad n > N(\varepsilon).$$

But

$$\lim_{n \to \infty} I'_2 = \lim_{n \to \infty} \frac{1}{\pi} \int_{-h}^{h} f(x_0 + u) \frac{\sin(\lambda_n + \alpha)}{2} u \frac{}{u} \, du$$

$$= \tfrac{1}{2}\{f(x_0 + 0) + f(x_0 - 0)\} = f(x_0).$$

Thus, considering (1.69) we find that

$$(1.70) \qquad |S_n(x_0) - f(x_0)| < \varepsilon, \qquad \text{if} \quad n > N_1(\varepsilon).$$

Inequality (1.70) proves the assertion of the theorem.

Theorem 1.21 has a local character in the sense that it concerns the behavior of the series at a point at which the function satisfies certain conditions. We shall now prove a theorem regarding the behavior of the series on the whole real line, i.e., having a global character.

Theorem 1.22. *Assume that the unique limiting point of the Fourier exponents of the almost periodic function* (1.64) *is the point at infinity. If there
exists a number θ such that $0 < \theta < 1$, and $\lambda_n \leqslant \theta \lambda_{n+1}$, $n = 1, 2, \ldots$, then*

$$(1.71) \qquad\qquad f(x) = \sum_{k=-\infty}^{\infty} A_k e^{i\lambda_k x},$$

the convergence being uniform on the whole real line.

Proof. The assertion follows easily from Lemma 1.3 if we set $a = \lambda_n$,
$b = \lambda_{n+1}$, $n > 0$. It suffices to remark that

$$(1.72) \qquad\qquad f_{a,b}(x) = \sum_{|\lambda_k| < \lambda_n} A_k e^{i\lambda_k x},$$

for the above chosen numbers a and b.

But (1.72) is an immediate consequence of (1.58) if we use the fact that
for $a = \lambda_n$, $b = \lambda_{n+1}$ we have $\varphi_{a,b}(\lambda_k) = 1$, if $|\lambda_k| < \lambda_n$.

5.   *Summability of Fourier series*

As we know, Fejér gave a simple summability method for the Fourier
series associated with a periodic function; instead of the sequence of partial
sums of the Fourier series, he considered the sequence of arithmetic means
of these sums. Bochner showed that this method can be extended to the
Fourier series associated with almost periodic functions. Before developing
this summability method, we must formulate some definitions.

Consider a family of functions continuous on the whole real line:
$\mathscr{F} = \{f(x)\}$.

*The functions belonging to $\mathscr{F}$ are called equi-continuous, if to any $\varepsilon > 0$
there corresponds a number $\delta(\varepsilon) > 0$, such that*

$$(1.73) \qquad\qquad |f(x_1) - f(x_2)| < \varepsilon, \qquad f \in \mathscr{F},$$

if $|x_1 - x_2| < \delta(\varepsilon)$.

*It is obvious that any function belonging to this family is uniformly con-
tinuous on the real line.*

*The functions belonging to the family $\mathscr{F}$ are called equi-almost periodic,
if to any $\varepsilon > 0$ there corresponds a number $l(\varepsilon) > 0$, such that any interval
of length $l(\varepsilon)$ contains at least one number ξ for which*

$$(1.74) \qquad |f(x + \xi) - f(x)| < \varepsilon, \qquad f \in \mathscr{F}, \qquad -\infty < x < +\infty.$$

A simple example of an equi-almost periodic family is the family
$\{f(x + h)\}$, where $f(x)$ is an almost periodic function and h is a real
parameter.

We also need another definition in order to state an important theorem concerning sequences of equi-continuous and equi-almost periodic functions.

A sequence of almost periodic functions $\{f_n(x)\}$ *is called convergent in the mean, if to any* $\varepsilon > 0$ *there corresponds an N such that*

$$(1.75) \qquad M\{|f_n(x) - f_m(x)|^2\} < \varepsilon,$$

provided that

$$n, m \geqslant N(\varepsilon).$$

Theorem 1.23. *If the sequence of equi-continuous and equi-almost periodic functions* $\{f_n(x)\}$ *converges in the mean, then it converges uniformly on the whole real line.*

Proof. Given a number $\varepsilon > 0$, there exists an $N(\varepsilon) > 0$, such that (1.75) holds for any $n, m \geqslant N(\varepsilon)$. We shall show that this implies that $|f_n(x) - f_m(x)| < \varepsilon$, if $n, m \geqslant N_1(\varepsilon)$. This will follow from the following statement: if at a point x_0 we have

$$|f_n(x_0) - f_m(x_0)| > c,$$

then there exists a number $c' > 0$, depending only on c, such that

$$M\{|f_n(x) - f_m(x)|^2\} > c'.$$

Since $\{f_n(x)\}$ is an equi-continuous family of functions, we shall obtain

$$(1.76) \qquad |f_n(x_1) - f_n(x_2)| \leqslant \frac{c}{8},$$

if $|x_1 - x_2| < \delta$. If at a point ξ we have

$$(1.77) \qquad |f_n(\xi) - f_m(\xi)| > \frac{c}{2},$$

then there exists a segment of length 2δ with the center at ξ, such that at any point x, $|x - \xi| < \delta$

$$(1.78) \qquad |f_n(x) - f_m(x)| > \frac{c}{4}.$$

Indeed, if at a point $\bar{x}$ of the interval $|x - \xi| < \delta$ inequality (1.78) is not valid, then

$$|f_n(\xi) - f_m(\xi)| \leqslant |f_n(\xi) - f_n(\bar{x})| + |f_n(\bar{x}) - f_m(\bar{x})| + |f_m(\bar{x}) - f_m(\xi)|$$

$$\leqslant \frac{c}{8} + \frac{c}{4} + \frac{c}{8} = \frac{c}{2},$$

which contradicts (1.77).

There exists a number $l_1 = l(c/4)$ such that any segment of length l_1 of the real line contains a translation number corresponding to $c/4$ common to all the functions of the sequence. Hence, any interval of length l_1 will contain a point x'_0 for which

$$|f_n(x'_0) - f_n(x_0)| \leqslant \frac{c}{4}.$$

Since

$$|f_n(x'_0) - f_m(x'_0)| \geqslant |f_n(x_0) - f_m(x_0)| - 2\frac{c}{4} > \frac{c}{2},$$

it means that (1.77) holds for $\xi = x'_0$. Therefore, any interval of length $l = l_1 + 2\delta$ of the real line contains a subinterval of length 2δ on which (1.78) is true. Consequently,

$$\frac{1}{nl}\int_0^{nl} |f_n - f_m|^2 \, dx = \frac{1}{nl}\sum_{k=0}^{n-1}\int_{kl}^{(k+1)l} |f_n - f_m|^2 \, dx \geqslant \frac{1}{nl}\left(\frac{c}{4}\right)^2 \cdot n \cdot 2\delta$$

$$= \frac{2\delta}{l}\left(\frac{c}{4}\right)^2.$$

This shows that

$$M\{|f_n(x) - f_m(x)|^2\} > \frac{\delta}{l}\left(\frac{c}{4}\right)^2 = c',$$

which contradicts the hypothesis. Therefore, Theorem 1.23 has been proved.

Consider now a countable set of real numbers $\{\beta_k\}$.

The numbers belonging to $\{\beta_k\}$ are called linearly independent if any relation of the form

$$r_1\beta_1 + r_2\beta_2 + \cdots + r_n\beta_n = 0, \qquad n = 1, 2, \ldots,$$

where $r_1, r_2, \ldots, r_n$ are rational numbers, implies that

$$r_1 = r_2 = \cdots = r_n = 0.$$

Let $\mathcal{M}$ be a set of real numbers.

We say that the set of linearly independent numbers $\{\beta_k\}$ constitutes a basis for the set $\mathcal{M}$, if any $\alpha \in \mathcal{M}$ can be represented as

$$\alpha = r_1\beta_1 + r_2\beta_2 + \cdots + r_n\beta_n,$$

where $r_1, r_2, \ldots, r_n$ are rational numbers depending on α.
A lemma of interest is the following:

Lemma 1.14. *Any countable set of real numbers admits a basis, whose elements belong to the given set.*

Proof. Let $\lambda_1, \lambda_2, \ldots$ belong to the given set. Set $\beta_1 = \lambda_1$ and leave out all the numbers λ_k of the form $r\beta_1$, where r is rational. Let β_2 be the first λ_k which is not of form $r\beta_1$. Then we eliminate all λ_k of the form $r_1\beta_1 + r_2\beta_2$ and so on. In this way we construct the sequence $\beta_1, \beta_2, \ldots, \beta_n, \ldots$, which is a basis. In some cases it is possible that the above indicated process will be terminated after a finite number of steps. Then we say that the basis is finite.
We can now prove the main result of this paragraph.

Theorem 1.24. *Given an almost periodic function*

$$(1.79) \qquad f(x) \sim \sum_{k=1}^{\infty} A_k e^{i\lambda_k x},$$

there exists a sequence of trigonometric polynomials

$$(1.80) \qquad \sigma_m(x) = \sum_{k=1}^{n} r_{k,m} A_k e^{i\lambda_k x}, \qquad n = n(m),$$

which converges uniformly to $f(x)$ on the whole real line as $m \to \infty$. The numbers $r_{k,m}$ are rational and depend on λ_k and m, but not on A_k.

Proof. We shall consider the Fejér kernel which is used in a similar problem concerning periodic functions:

$$(1.81) \qquad K_n(t) = \frac{1}{n} \frac{\sin^2(nt/2)}{\sin^2(t/2)}.$$

For $t = 2k\pi$, the values of K_n can be defined such that K_n is continuous everywhere.

One can prove without difficulty that

$$(1.82) \qquad K_n(t) = \sum_{v=-n}^{n} \left(1 - \frac{|v|}{n}\right) e^{-ivt}.$$

The following properties of $K_n(t)$ are obvious:

$$1)\ K_n(t) \geqslant 0; \qquad 2)\ M\{K_n(t)\} = 1.$$

Let $\beta_1, \beta_2, \ldots, \beta_n, \ldots$ be a basis of the Fourier exponents $\lambda_1, \lambda_2, \ldots, \lambda_n, \ldots$. Set

$$(1.83) \qquad \sigma_m(x) = M_t\left\{ f(x+t) K_{(m!)^2}\left(\frac{\beta_1 t}{m!}\right) \cdots K_{(m!)^2}\left(\frac{\beta_m t}{m!}\right)\right\},$$

where M_t indicates that the mean must be taken with respect to the variable t. This mean exists, since the function between the brackets (being a product of almost periodic functions) is almost periodic in t. If the basis is finite, then we agree to take $\beta_k = 0$ in (1.83), starting with a sufficiently large k.

If we consider (1.82), we obtain

$$\sigma_m(x) = \sum_{v_1=-(m!)^2}^{(m!)^2} \cdots \sum_{v_m=-(m!)^2}^{(m!)^2} \left(1 - \frac{|v_1|}{(m!)^2}\right) \cdots$$

$$\cdots \left(1 - \frac{|v_m|}{(m!)^2}\right) a\left(\frac{v_1\beta_1}{m!} + \cdots + \frac{v_m\beta_m}{m!}\right) \exp\left[i\left(\frac{v_1\beta_1}{m!} + \cdots + \frac{v_m\beta_m}{m!}\right)x\right],$$

where, as usual, we have

$$a(\lambda) = M\{f(x) e^{-i\lambda x}\}.$$

From the above expression for $\sigma_m(x)$, it follows that we can write

$$\sigma_m(x) = \sum_{k=1}^{n} r_{k,m} A_k e^{i\lambda_k x},$$

where we have put

$$(1.84) \qquad r_{k,m} = \left(1 - \frac{|v_1|}{(m!)^2}\right) \cdots \left(1 - \frac{|v_m|}{(m!)^2}\right),$$

the numbers $v_1, \ldots, v_m$ being such that

$$(1.85) \qquad \lambda_k = \frac{v_1\beta_1}{m!} + \cdots + \frac{v_m\beta_m}{m!}.$$

A representation of the form (1.85) is always possible for a sufficiently large m. Indeed, we know that

$$\lambda_k = r_1\beta_1 + \cdots + r_h\beta_h,$$

since $\{\beta_k\}$ is a basis for the Fourier exponents of $f(x)$. If m is sufficiently large (so that it exceeds h and the product of the denominators of $r_1, \ldots, r_h$), then λ_k can be represented as (1.85) with $v_i = r_i m!$ for $i \leqslant h$, and $v_i = 0$ for $i > h$.

If we consider that

$$\frac{|v_i|}{(m!)^2} = \frac{|r_i|}{m!},$$

then from (1.84) it follows that

(1.86)
$$\lim_{m \to \infty} r_{k,m} = 1.$$

Let us write now the Parseval equality for the function $f(x) - \sigma_m(x)$. We have

(1.87)
$$M\{|f(x) - \sigma_m(x)|^2\} = \sum_{k=1}^{\infty} (r_{k,m} - 1)^2 |A_k|^2.$$

Taking N sufficiently large so that

(1.88)
$$\sum_{k=N+1}^{\infty} |A_k|^2 \leqslant \frac{\varepsilon}{2},$$

where ε is a given positive number, we also obtain

(1.89)
$$\sum_{k=1}^{N} (r_{k,m} - 1)^2 |A_k|^2 \leqslant \frac{\varepsilon}{2},$$

for m sufficiently large, $m \geqslant N_1(\varepsilon)$. This follows from (1.86).

From (1.87), (1.88), and (1.89) we find that

(1.90)
$$M\{|f(x) - \sigma_m(x)|^2\} \leqslant \varepsilon, \qquad m \geqslant N_1(\varepsilon).$$

Inequality (1.90) implies that the sequence $\{\sigma_m(x)\}$ converges in the mean to $f(x)$. If we show that the functions $\sigma_m(x)$ are equi-continuous and equi-almost periodic, then the assertion which we wish to prove will follow from Theorem 1.23.

Let us refer now to the definition of the trigonometric polynomials $\sigma_m(x)$ by formulas (1.83), and notice that

$$M\left\{ K_{(m!)^2}\left(\frac{\beta_1 t}{m!}\right) \cdots K_{(m!)^2}\left(\frac{\beta_m t}{m!}\right) \right\} = 1.$$

This equality follows from the simple remark that the product of the kernels is a trigonometric polynomial whose free term is equal to 1. Consequently,

$$(1.91) \qquad \sup_x |\sigma_m(x + \xi) - \sigma_m(x)| \leqslant \sup_x |f(x + \xi) - f(x)|,$$

if we consider that the kernels are non-negative. But (1.91) shows that $\sigma_m(x)$ are equi-continuous on the real line (since $f(x)$ is uniformly continuous), and equi-almost periodic functions [any ε-translation number of $f(x)$ is an ε-translation number common to $\sigma_m(x)$].

Hence, Theorem 1.24 is proved.

It is a remarkable fact that all we said previously with regard to the summability method permits us to obtain convergence criteria for the Fourier series. We shall confine ourselves to only one criterion using this method.

Theorem 1.25. *If the Fourier exponents of an almost periodic function $f(x)$ are linearly independent, then the Fourier series associated with $f(x)$ is uniformly convergent.*

Proof. Let

$$f(x) \sim \sum_{k=1}^{\infty} A_k e^{i\lambda_k x}.$$

We can write $A_k = |A_k| e^{i\theta_k}$, $k = 1, 2, \ldots$, where θ_k is the argument of A_k. We construct the functions

$$\varphi_k(x) = 1 + \tfrac{1}{2} \left[e^{-i(\lambda_k x + \theta_k)} + e^{i(\lambda_k x + \theta_k)} \right]$$

$$\Phi_n(x) = \prod_{k=1}^{n} \varphi_k(x) = 1 + \frac{1}{2} \sum_{k=1}^{n} e^{-i(\lambda_k x + \theta_k)} + \sum_{h} c_h e^{i\mu_h x}.$$

The numbers μ_k are linear combinations of $\lambda_1, \lambda_2, \ldots, \lambda_n$ with integral coefficients (at least one of which is distinct from 0). Therefore none of μ_k can vanish and none can be equal to λ_m. We have

$$M\{\Phi_n(x)\} = 1, \qquad M\{f(x)\Phi_n(x)\} = \frac{1}{2} \sum_{k=1}^{n} |A_k|.$$

Since $\Phi_n(x) \geqslant 0$, it follows that

$$|M\{f(x)\Phi_n(x)\}| \leqslant \sup_x |f(x)| \cdot M\{\Phi_n(x)\} = \sup_x |f(x)|.$$

Therefore

$$\sum_{k=1}^{n} |A_k| \leqslant 2 \sup_{x} |f(x)|,$$

for any integer n. Consequently, we find

$$\sum_{k=1}^{\infty} |A_k| \leqslant 2 \sup_{x} |f(x)|.$$

This inequality shows that the Fourier series associated with the function $f(x)$ is absolutely and uniformly convergent. According to Theorem 1.20, the statement of Theorem 1.25 has been established.

6. Numerical almost periodic sequences and their connection with almost periodic functions

The numerical sequences considered in this paragraph will be regarded as complex-valued functions of an integer variable: $\{a_n\} = \{f(n)\}$, $n = 0$, ± 1, ± 2, ... In the following both notations will be used

A function of an integer variable $f(n)$ is called almost periodic, if to any $\varepsilon > 0$ there corresponds an integer $N(\varepsilon)$, such that among any N consecutive integers there exists an integer p with the property

(1.92) $|f(n + p) - f(n)| < \varepsilon, \qquad n = 0, \pm 1, \pm 2, \dots .$

Exactly as in the case of the almost periodic functions of a real variable, p is called an ε-translation number of $f(n)$.

To characterize the almost periodic real-valued functions of an integer variable we need another definition.

A function of an integer variable $f(n)$ is called *normal*, if from any sequence $\{f(n + m_k)\}$ one can extract a subsequence which converges uniformly with respect to n; $\{m_k\}$ is an arbitrary sequence of integers.

Theorem 1.26. *A necessary and sufficient condition for a function of an integer variable to be almost periodic is that it be normal.*

Proof. Assume that $f(n)$ is an almost periodic function of the integer variable n. Let $\{f(n + m_k)\}$ be a sequence of translations of $f(n)$ and let $\varepsilon > 0$. By the definition of almost periodicity, there exists an integer $N(\varepsilon)$, such that in any interval $(m_k - N, m_k]$ there occurs an ε-translation number p. Now, $m_k - N < p_k \leqslant m_k$ leads to $0 \leqslant m_k - p_k < N$. Set $q_k = m_k - p_k$.

From the above considerations we obtain $q_k = q = \text{const.}$ for an infinity of values of k. Since

$$|f(n + m_k) - f(n + q_k)| = |f(n + m_k) - f(n + m_k - p_k)|$$

$$= |f[(n + m_k - p_k) + p_k] - f(n + m_k - p_k)| < \varepsilon,$$

we have for a convenient subsequence $\{m_{1k}\} \subset \{m_k\}$

(1.93) $|f(n + m_{1k}) - f(n + q)| < \varepsilon, \qquad k = 1, 2, \dots .$

In other words, given the sequence $\{f(n + m_k)\}$ and the number $\varepsilon > 0$, there exists a subsequence $\{m_{1k}\}$ of the sequence $\{m_k\}$ and an integer $q = q(\varepsilon)$ such that (1.93) is true.

Consider now a sequence of positive numbers $\varepsilon_1 > \varepsilon_2 > \cdots > \varepsilon_s > \cdots$ converging to zero. From the sequence $\{f(n + m_k)\}$ we extract the subsequence $\{f(n + m_{1k})\}$ which satisfies (1.93) for $\varepsilon = \varepsilon_1$. From this sequence we extract a subsequence $\{f(n + m_{2k})\}$ for which an equality analogous to (1.93) is valid. Of course, q will not be the same, but will depend on the subsequence. We proceed further in the same way and then we form the diagonal sequence $\{f(n + m_{rr})\}$, $r = 1, 2, \dots .$ Let us show that this sequence is uniformly convergent with respect to n.

Assume that $\varepsilon > 0$ and that R is sufficiently large so that $2\varepsilon_R < \varepsilon$. As a result we obtain

$$|f(n + m_{rr}) - f(n + m_{ss})| \leqslant |f(n + m_{rr}) - f(n + \bar{q})|$$

$$+ |f(n + \bar{q}) - f(n + m_{ss})| < \varepsilon,$$

for $r, s \geqslant R$; $\bar{q}$ is the number $q(\varepsilon_R)$ corresponding to the sequence $\{f(n + m_{Rk})\}$. Hence, the sequence $\{f(n + m_{rr})\}$ is uniformly convergent with respect to n.

Conversely, if $f(n)$ is a normal function, then it is almost periodic. The proof of this fact is entirely similar to the proof of Theorem 1.10, so that we do not reproduce it. We have to consider only that the argument of our functions is no longer a real number, but an integer. Thus the proof of the theorem has been completed.

We notice that this theorem permits the easy establishment of certain properties of almost periodic sequences, such as: if $\{a_n\}$ and $\{b_n\}$ are two almost periodic sequences, then $\{a_n + b_n\}$ and $\{a_n b_n\}$ are also almost periodic.

We can now give a characterization of almost periodic sequences, using the almost periodic functions of a real variable.

Theorem 1.27. *A necessary and sufficient condition for a sequence* $\{a_n\}$ *to be almost periodic is the existence of an almost periodic function* $f(x)$ *such that* $a_n = f(n)$, $n = 0, \pm 1, \pm 2, \ldots$.

Proof. The condition is sufficient because if such a function exists, then from any sequences $\{f(x + m_k)\}$, m_k being integers, one can extract a subsequence $\{f(x + m_{1k})\}$ converging uniformly on the real line. Consequently, the sequence $\{f(n + m_{1k})\}$ will be uniformly convergent with respect to n as $k \to \infty$. This shows that the function of an integer variable $f(n)$ is normal, and this proves the almost periodicity of the sequence $\{a_n\}$.

Conversely, assume that $\{a_n\}$ is an almost periodic sequence. We define a function $f(x)$ of a real variable by the following relations:

$$(1.94) \quad f(x) = a_n + (x - n)(a_{n+1} - a_n), \qquad n \leqslant x < n + 1,$$

$$n = 0, \pm 1, \pm 2, \ldots .$$

The graph of this function for real a_n is obtained by joining successively the points (n, a_n) by segments of straight lines.

We notice that $f(n) = a_n$. It remains to show only that $f(x)$ defined by (1.94) is an almost periodic function.

Since $f(n)$ is an almost periodic function of an integer variable, then to any $\varepsilon > 0$ there corresponds a number N with the property that among N consecutive integers, there exists at least one $(\varepsilon/3)$-translation number of $f(n)$. Consider an $(\varepsilon/3)$-translation number p. If $n \leqslant x < n + 1$, then $n + p \leqslant x + p < n + p + 1$. Since

$$f(x + p) - f(x) = f(n + p) - f(n)$$
$$+ (x - n)\{[f(n + p + 1) - f(n + 1)]$$
$$- [f(n + p) - f(n)]\},$$

considering that $0 \leqslant x - n < 1$, we obtain

$$|f(x + p) - f(x)| < \varepsilon.$$

Thus, any $(\varepsilon/3)$-translation number of the sequence $\{a_n\}$ is an ε-translation number of the function $f(x)$.

Remark. Since $f(x\delta)$ is almost periodic, whenever $f(x)$ is almost periodic, this means that the sequence $f(n\delta)$ will be almost periodic whenever $f(x)$ is an almost periodic function.

The preceding theorem allows us to prove in a simple way the existence of the mean value for almost periodic sequences, using the existence of the mean value for functions of a real variable.

Theorem 1.28. *If $\{a_n\}$ is an almost periodic sequence, then the limit*

$$(1.95) \qquad \lim_{k \to \infty} \frac{a_{n+1} + a_{n+2} + \cdots + a_{n+k}}{k} = M,$$

exists uniformly with respect to n. The number M is independent of n and it is called the mean value of the sequence $\{a_n\}$.

Proof. Consider an almost periodic function $f(x)$ of a real variable defined by (1.94). According to §3 the limit

$$(1.96) \qquad \lim_{k \to \infty} \frac{1}{k} \int_n^{n+k} f(x)\, dx = M = M\{f(x)\},$$

exists and the convergence is uniform with respect to n.

From (1.94) we find by a simple calculation that

$$(1.97) \qquad \frac{1}{k} \int_n^{n+k} f(x)\, dx = \frac{a_{n+1} + a_{n+2} + \cdots + a_{n+k}}{k} + \frac{a_n - a_{n+k}}{2k}.$$

Since $\{a_n\}$ is a bounded sequence, the second term on the right side of (1.97) converges to zero uniformly with respect to n, as $k \to \infty$. The relation (1.95) follows now from (1.96) and (1.97).

We shall now establish another theorem which shows how almost periodic functions of a real variable can be characterized by almost periodic numerical sequences.

Theorem 1.29. *Necessary and sufficient conditions for a function $f(x)$ to be almost periodic is that $f(x)$ be uniformly continuous on the real line, and that there exists a sequence of positive numbers $\{\delta_k\}$, monotonely decreasing to zero, such that $\{f(n\delta_k)\}$, $k = 1, 2, \ldots$, are almost periodic sequences.*

Proof. The conditions are necessary since if $f(x)$ is almost periodic, then it is uniformly continuous on the real line and the sequences $\{f(n\delta_k)\}$, $k = 1, 2, \ldots$, are almost periodic (as seen above).

Conversely, let us assume that $f(x)$ is uniformly continuous on the real line, and the sequences $\{f(n\delta_k)\}$, $k = 1, 2, \ldots$ are almost periodic.

For any $\varepsilon > 0$, there exists $\delta > 0$ such that

$$|f(x') - f(x'')| < \frac{\varepsilon}{3}, \qquad \text{for} \quad |x' - x''| < \delta.$$

Then there exists a δ_k such that $\delta_k < \delta$. The sequence $\{f(n\delta_k)\}$, for k fixed in this way is almost periodic and therefore, we can determine an integer N such that among any N consecutive integers there exists an $(\varepsilon/3)$-translation number of $\{f(n\,\delta_k)\}$. If p is an $(\varepsilon/3)$-translation number of $\{f(n\delta_k)\}$, then we shall show that $\xi = p\delta_k$ is an ε-translation number of $f(x)$. For any x there exists an n such that $n\delta_k \leqslant x < (n+1)\delta_k$. If we consider that $|x - n\delta_k| < \delta$, $|f(n\delta_k + p\delta_k) - f(n\delta_k)| < \varepsilon/3$, we obtain as a result

$$|f(x + \xi) - f(x)| \leqslant |f(x + \xi) - f(n\delta_k + \xi) +$$
$$+ |f(n\delta_k + \xi) - f(n\delta_k)| + |f(n\delta_k) - f(x)| < \varepsilon.$$

It is obvious that we can take $l(\varepsilon) = (N+1)\delta_k$, since an interval of length $(N+1)\delta_k$ indeed contains N consecutive integral multiples of δ_k, and consequently, as seen above, this interval will contain at least one ε-translation number of $f(x)$.

To conclude this paragraph, we shall establish two more results on almost periodic sequences. These results have many applications in the theory of discrete dynamical processes described by difference equations.

Theorem 1.30. *Let $\{a_n\}$ be an almost periodic sequence, and let $\{A_n\}$ be the sequence constructed as follows:*

(1.98) *A_0 is chosen arbitrarily, and $A_{n+1} - A_n = a_n$ for any $n \in Z$.*

A necessary and sufficient condition for $\{A_n\}$ to be almost periodic is that it be bounded.

Proof. Given $\{a_n\}$, let us consider the almost periodic function $f(x)$ defined by (1.94). It is an elementary matter to show that the following relation holds true for any real x:

(1.99) $$\int_0^x f(t)\,dt = A_{[x]+1} - \int_{[x]}^x f(t)\,dt - \frac{1}{2}(a_0 + a_{[x]}).$$

By $[x]$ we have denoted, as usual, the greatest integer in x.

The last two terms in the right-hand side of (1.99) are bounded with respect to x. Consequently, the first term in the right-hand side of (1.99) and the left-hand side of (1.99) are bounded or not bounded simultaneously. Therefore, the integral in the left-hand side of (1.99) is an almost periodic function (see Theorem 4.1) in case and only in case the sequence $\{A_n\}$ is bounded. Letting $x = n$ in (1.99), we obtain

(1.100) $$\int_0^n f(t)\,dt = A_{n+1} - \frac{1}{2}(a_0 + a_n).$$

It remains only to apply Theorem 1.27 in order to obtain the almost periodicity of the sequence $\{A_n\}$.

Theorem 1.31. *Let* $\{k_n\}$ *be a summable sequence, i.e., let it be such that* $K = \Sigma_{n \in Z} |k_n| < \infty$. *Then, for any almost periodic sequence* $\{a_n\}$, *the sequence* $\{b_n\}$ *defined by*

$$(1.101) \qquad b_n = \sum_{m \in Z} k_m a_{n-m}, \qquad n \in Z,$$

is also almost periodic.

The proof of Theorem 1.31 follows immediately from the elementary inequality

$$\sup |b_{n+p} - b_n| \le K \sup |a_{n+p} - a_n|,$$

where the supremum is taken with respect to $n \in Z$. This easily follows from the definition of $\{b_n\}$.

Bibliographical notes

The results of this chapter can be found in the memoirs of H. Bohr (99, 100) or in the monographs of H. Bohr (121), J. Favard (247), A. Besicovitch (49), B. M. Levitan (465), W. Maak (489).

The problems developed in §6 represent a synthesis of the results of A. Walther (649), I. Seynsche (586), and N. Gheorghiu (289), results obtained in connection with almost periodic sequences. In this respect we mention the paper of Ky Fan (241), which considers almost periodic sequences of elements of a metric space, applying it to the ergodic theory.

H. Bohr (116, 120) showed that the argument of an almost periodic function $f(x)$, which satisfies the condition $|f(x)| \geqslant k > 0$, is of the form $\arg f(x) = cx + g(x)$, where c is constant and $g(x)$ is an almost periodic function.

If one begins with H. Bohr's definition of an almost periodic function, then the theorem of approximation by trigonometric polynomials (which shows the equivalence with our definition) is one of the fundamental theorems of this theory. We considered the proofs given by N. N. Bogoliubov (91) and S. Bochner (61). Other proofs of this theorem were given by N. Wiener (667), H. Weyl (661), and N. N. Bogoliubov (87, 88).

The quasiperiodic functions are a special class of almost periodic functions and were studied a long time ago by P. Bohl and E. Esclangon (see Bibliography). These authors also showed the applications of these functions to the theory of differential equations. The following definition of these functions is in the form given by Bogoliubov: Let $\omega_1, \omega_2, \ldots, \omega_p$ be linearly independent real numbers. We say that $f(x)$ is a quasi-periodic function with fundamental frequencies $\omega_1, \omega_2, \ldots, \omega_p$ if to any $\varepsilon > 0$ corresponds a trigonometric polynomial $T(x) = \sum_{k=1}^{n} A_k e^{i\lambda_k x}$, such that λ_k are linear combinations of ω_j with integral coefficients, and $|f(x) - T(x)| < \varepsilon$, $-\infty < x < +\infty$.

Another important class of almost periodic functions is the class of limit-periodic functions. To this class belong the limits of sequences of periodic continuous functions (having in general different periods). H. Bohr (100) showed that these functions are characterized by the fact that their Fourier exponents are of the form $r_k \beta$, β being a real number and r_k being rational numbers.

B. M. Levitan (447, 450, 451, 454) introduced and studied a class of functions called N-almost periodic, which generalizes the class of almost periodic functions in the sense of Bohr. We now define these functions.

The number $\xi = \xi(\varepsilon, N)$ is called an (ε, N)-translation number of a function $f(x)$, if for $|x| < N$ we have $|f(x \pm \xi) - f(x)| < \varepsilon$.

A function $f(x)$ continuous on the real line is called N-almost periodic, if the following conditions are fulfilled: (1) for any $\varepsilon > 0$ and $N > 0$, there exists an $l(\varepsilon, N) > 0$, such that any interval of the real line contains at least one (ε, N)-translation number of $f(x)$; (2) there exists a function $\delta(\varepsilon_1, \varepsilon_2, N) > 0$ such that $\delta \to 0$ as $\varepsilon_1, \varepsilon_2 \to 0$, and N is constant, with the property that $\xi(\varepsilon_1, N) \pm \xi(\varepsilon_2, N) = \xi(\delta, N)$. The last relation expresses the fact that the sum or the difference of any (ε_1, N)-translation number and any (ε_2, N)-translation number is a (δ, N)-translation number.

Obviously, an almost periodic function is also an N-almost periodic function. One can take for example $\delta(\varepsilon_1, \varepsilon_2, N) = \varepsilon_1 + \varepsilon_2$.

Besides the papers of B. M. Levitan mentioned above, the N-almost periodic functions are studied in the papers of B. Ja. Levin (441, 442), in the papers of B. Ja. Levin and B. M. Levitan (445, 446), and in the papers of V. A. Marčenko (498, 501).

M. Fréchet (279, 281) defined and studied the asymptotically almost periodic functions. We say that a function $f(x)$ continuous on the semi-axis $x \geqslant a$ is asymptotically almost periodic, if it can be represented as $f(x) = p(x) + r(x)$, where $p(x)$ is an almost periodic function, and $r(x) \to 0$ as $x \to +\infty$. One finds that this representation is unique. The mean of such a function exists, and many other properties of almost periodic functions can be extended to the asymptotically almost periodic functions.

In the papers of M. Fréchet (282, 283) and Ky Fan (241) we find interesting applications of these functions to ergodic theory. The asymptotically almost periodic functions have many applications in the theory of differential equations.

Other generalizations of almost periodic functions will be studied in Chapter VI and VII.

Besides the definition of almost periodic functions adopted in this chapter, and the equivalent properties A and B, one knows other characteristic properties of almost periodic functions. In his survey paper (729), A. M. Fink presents three more properties equivalent to the definition. Another characteristic property has been given by L. H. Loomis (478), and it is known as the *spectral characterization.* For the spectral property and its consequences see B. M. Levitan and V. V. Zhikov (745), or Y. Katznelson (743).

An interesting recent contribution which provides a new characterization of almost periodicity (not necessarily for numerical functions) is due to A. Haraux (736). Namely, if $f(t)$ is continuous and bounded on the real axis, then the compactness of the family $\{f(t + h)\}$, with h belonging to a relatively dense set on R, implies the almost periodicity of $f(t)$. We recall that a set $E \subset R$ is called relatively dense if there exists a length l, such that every interval of the real axis of length l contains at least one element of E. This shows that the normality condition can be weakened, which has many implications in the applications.

In connection with the theory of discrete dynamical processes, the significance of almost periodic sequences (or, of almost periodic functions defined on the most common group, Z) has considerably increased during the last years. There are important applications of this concept in the Systems Engineering (see, for instance, the papers by T. A. C. M. Claasen and W. F. G. Mecklenbrauker (712), A. M. Davis (723), as well as the books by A. Halanay and D. Wexler (731) and by C. J. Harris and J. F. Miles (738)). For applications to Statistics see Y. Isokawa (739). Almost periodicity results for discrete dynamical processes can be also found in C. Corduneanu (716). For various generalizations, see A. Precupanu (759). See also J. L. Mauclaire (746).

A new approach to approximation problems related to almost periodic functions, can be found in A. M. Fink's book (727).

For further contributions concerning the theory of numerical almost periodic functions see the bibliographies due to A. M. Fink (730) and G. H. Meisters (747). These bibliographies also contain papers dedicated to various applications of the theory of almost periodic functions.

Several papers of C. A. Desoer and his co-workers (F. M. Callier, J. F. Barman, and W. Chan) have been published in the IEEE Transactions

(AC or CAS Series), during the 1970's and early 1980's, containing significant applications of the theory of almost periodic functions to the Engineering Sciences.

A. M. Fink (728) deals with almost automorphic sequences and their extensions.

II

Almost periodic functions depending on parameters

We shall study in this chapter two classes of almost periodic functions depending on parameters:* the almost periodic functions depending uniformly on parameters and the functions almost periodic in the mean. The first class is encountered especially in connection with the existence of almost periodic solutions of nonlinear ordinary differential equations. An example of functions of the first category is the analytic almost periodic functions which will be studied in Chapter III. The functions almost periodic in the mean appear in connection with the almost periodicity of solutions of partial differential equations.

1. Almost periodic functions depending uniformly on parameters

Let Ω be a set of the n-dimensional complex space and let $Z = (z_1, \ldots, z_n)$ be the point with coordinates $z_1, \ldots, z_n$.

We shall study in this paragraph only complex-valued functions defined on the set $\Omega \times R$, where R is the real line. We shall assume that any function appearing in theorems or in definitions is continuous on the set $\Omega \times R$.

A function $f(Z, x)$ is called almost periodic in x, uniformly with respect to $Z \in \Omega$, if to any $\varepsilon > 0$ corresponds a number $l(\varepsilon)$ such that any interval of the real line of length $l(\varepsilon)$ contains at least one number ξ for which

$$(2.1) \qquad |f(Z, x + \xi) - f(Z, x)| < \varepsilon, \quad Z \in \Omega, \quad x \in R.$$

The number ξ is called an ε-translation number of $f(Z, x)$. The uniform dependence on parameters follows from the fact that $l(\varepsilon)$ and ξ are independent of Z.

Most of the properties of almost periodic functions can easily be extended to almost periodic functions depending on parameters.

Theorem 2.1. *Consider a function $f(Z, x)$ which is almost periodic in x uniformly with respect to $Z \in \Omega$. If c is a complex number and a is real, then $cf(Z, x), f(Z, x + a), \bar{f}(Z, x), |f(Z, x)|$ are almost periodic functions of x, uniformly with respect to $Z \in \Omega$.*

If $f(z, x) \geq m > 0$, then the function $1/f(Z, x)$ is almost periodic in x, uniformly with respect to $Z \in \Omega$.

---

*A third class of almost periodic functions to be investigated in this chapter is a relatively new one: *random functions almost periodic in probability*. In this case, the parameter is a stochastic variable and a few concepts from the theory of probability spaces are necessary.

The proof of the assertions made in the theorem is immediate. For instance, in order to prove the last assertion it is sufficient to note that

$$\left| \frac{1}{f(Z, x + \xi)} - \frac{1}{f(Z, x)} \right| \leqslant \frac{1}{m^2} |f(Z, x + \xi) - f(Z, x)|.$$

This inequality shows that any $m^2\varepsilon$-translation number of $f(Z, x)$ is an ε-translation number for $1/f(Z, x)$.

Theorem 2.2. *If Ω is a closed, bounded set, then the function $f(Z, x)$, almost periodic in x uniformly with respect to Z, is bounded on $\Omega \times R$.*

Proof. Let $\varepsilon = 1$, and let $l = l(1)$ be the corresponding length described in the definition of almost periodic functions depending uniformly on parameters. The function $f(Z, x)$ is bounded on the set $\Omega \times [0, l]$ since it is continuous. Set $m = \sup |f(Z, x)|$, $Z \in \Omega$, $x \in [0, l]$. Consider now any real number x and a 1-translation number ξ of $f(Z, x)$, which belongs to the interval $[-x, -x + l]$. It follows that $0 \leqslant x + \xi \leqslant l$. But

$$|f(Z, x)| \leqslant |f(Z, x) - f(Z, x + \xi)| + |f(Z, x + \xi)| < 1 + m = M,$$

which proves the theorem.

Corollary. *If the conditions of Theorem 2.2 are fulfilled, then $f^2(Z, x)$ is almost periodic in x, uniformly with respect to $Z \in \Omega$.*

This fact follows from the inequality

$$|f^2(Z, x + \xi) - f^2(Z, x)| \leqslant 2M|f(Z, x + \xi) - f(Z, x)|,$$

where M has the same meaning as in the proof of Theorem 2.2.

Before stating other theorems, let us recall that the distance between the points $Z = (z_1, \ldots, z_n)$ and $Z' = (z'_1, \ldots, z'_n)$ of the complex n-dimensional space is given by the formula

$$(2.2) \qquad |Z - Z'| = \sqrt{\sum_{i=1}^{n} |z_i - z'_i|^2}.$$

Theorem 2.3. *Under the same hypotheses as in Theorem 2.2, it follows that $f(Z, x)$ is uniformly continuous on the set $\Omega \times R$.*

Proof. Let ε be a positive number, and $l = l(\varepsilon/3)$ the length associated with $\varepsilon/3$ according to the definition of an almost periodic function. The function $f(Z, x)$ is continuous on the set $\Omega \times [-1, 1 + l]$ and therefore is uniformly continuous. Thus we can determine a positive number $\delta = \delta(\varepsilon/3) < 1$, such that

(2.3) $$|f(Z_2, y_2) - f(Z_1, y_1)| < \frac{\varepsilon}{3},$$

provided that

$$|Z_2 - Z_1| < \delta, |y_2 - y_1| < \delta, (Z_2, y_2), (Z_1, y_1) \in \Omega \times [-1, 1 + l].$$

Consider two points (Z_2, x_2) and (Z_1, x_1) of $\Omega \times R$ such that

(2.4) $$|Z_2 - Z_1| < \delta, \quad |x_2 - x_1| < \delta.$$

If ξ is an $(\varepsilon/3)$-translation number of $f(Z, x)$ belonging to the interval $[-x_1, -x_1 + l]$, then we have $0 \leqslant x_1 + \xi \leqslant l, -1 \leqslant x_2 + \xi \leqslant 1 + l$. The last double inequality follows from the first one, according to (2.4) and from the fact that $\delta < 1$. Therefore, $(Z_2, x_2 + \xi), (Z_1, x_1 + \xi) \in \Omega \times [-1, 1 + l]$ and consequently,

$$\begin{aligned}
|f(Z_2, x_2) - f(Z_1, x_1)| &\leqslant |f(Z_2, x_2) - f(Z_2, x_2 + \xi)| \\
&+ |f(Z_2, x_2 + \xi) - f(Z_1, x_1 + \xi)| \\
&+ |f(Z_1, x_1 + \xi) - f(Z_1, x_1)| \\
&< \frac{\varepsilon}{3} + \frac{\varepsilon}{3} + \frac{\varepsilon}{3} = \varepsilon,
\end{aligned}$$

if inequalities (2.4) are satisfied.

Theorem 2.4. *If a sequence $\{f_n(Z, x)\}$ of almost periodic functions uniformly depending on parameters is uniformly convergent on $\Omega \times R$ to $f(Z, x)$, then the function $f(Z, x)$ is almost periodic in x, uniformly with respect to $Z \in \Omega$.*

Proof. Consider a positive number ε, and let n be a sufficiently large number so that

(2.5) $$|f(Z, x) - f_n(Z, x)| < \frac{\varepsilon}{3}, \quad (Z, x) \in \Omega \times R.$$

Let ξ be an $(\varepsilon/3)$-translation number of $f_n(Z, x)$. We have

$$\begin{aligned}
|f(Z, x + \xi) - f(Z, x)| &\leqslant |f(Z, x + \xi) - f_n(Z, x + \xi)| \\
&+ |f_n(Z, x + \xi) - f_n(Z, x)| + |f_n(Z, x) - f(Z, x)| \\
&< \frac{\varepsilon}{3} + \frac{\varepsilon}{3} + \frac{\varepsilon}{3} = \varepsilon.
\end{aligned}$$

This inequality proves the assertion of the theorem.

A function $f(Z, x)$ is called Ω-uniformly continuous on the real line, if to any $\varepsilon > 0$ corresponds a $\delta(\varepsilon) > 0$, such that

(2.6) $$|f(Z, x_2) - f(Z, x_1)| < \varepsilon,$$

provided that

$$|x_2 - x_1| < \delta, \qquad x_1, x_2 \in R, \qquad Z \in \Omega.$$

Obviously, any function uniformly continuous on $\Omega \times R$ is Ω-uniformly continuous on the real line.

Theorem 2.5. *If $f(Z, x)$ is almost periodic in x, uniformly with respect to $Z \in \Omega$, and $f_x(Z, x)$ is Ω-uniformly continuous on the real line, then $f_x(Z, x)$ is almost periodic in x, uniformly with respect to $Z \in \Omega$.*

The proof is the same as in Chapter I, §1, but this time one makes use of Theorem 2.4. One shows that the sequence of functions

$$\varphi_n(Z, x) = n\left\{ f\left(Z, x + \frac{1}{n}\right) - f(Z, x) \right\}$$

is uniformly convergent on $\Omega \times R$ to $f_x(Z, x)$. But $\varphi_n(Z, x)$ are almost periodic functions of x, uniformly with respect to $Z \in \Omega$.

This theorem is similar to Theorem 1.8 regarding the almost periodicity of the derivative of an almost periodic function.

As shown in Chapter I, §2, normality is a property equivalent to almost periodicity. This equivalence occurs also in the case of almost periodic functions which depend uniformly on parameters, and are defined (as functions of parameters) on bounded and closed sets Ω.

Theorem 2.6. *A necessary and sufficient condition for a function $f(Z, x)$ to be almost periodic in x, uniformly with respect to $Z \in \Omega$, where Ω is a bounded and closed set, is that the family $\{f(Z, x + h)\}$ be normal on $\Omega \times R$.*

Proof. Let us prove the necessity of this condition. Let $f(Z, x)$ be almost periodic in x, uniformly with respect to $Z \in \Omega$. Consider a sequence of real numbers $\{h_k\}$. We must prove that from the sequence $\{f(Z, x + h_k)\}$ one can extract a subsequence which converges uniformly on the set $\Omega \times R$.

Let (Z_k, x_k) be a countable set of points everywhere dense in $\Omega \times R$. Since $f(Z, x)$ is bounded, then from the numerical sequence $\{f(Z_1, x_1 + h_k)\}$ we can choose a convergent subsequence $\{f(Z_1, x_1 + h_{1k})\}$. Then, from the sequence $\{f(Z_2, x_2 + h_{1k})\}$ we can choose a convergent subsequence $\{f(z_2, x_2 + h_{2k})\}$ and so on. Consider the diagonal sequence $\{f(Z, x + h_{kk})\}$.

Since the numerical sequence $\{h_{kk}\}$ is, except for a finite number of terms, a subsequence of any sequence $\{h_{mk}\}$, this means that $\{f(Z, x + h_{kk})\}$ converges at all the points (Z_k, x_k). Let us show that $\{f(Z, x + h_{kk})\}$ is uniformly convergent on $\Omega \times R$. Let (Z_0, x_0) be any point from $\Omega \times R$, an ε a positive number. Consider the numbers $l = l(\varepsilon/5)$ and $\delta = \delta(\varepsilon/5)$ corresponding to $\varepsilon/5$, by the definition of almost periodic functions depending uniformly on parameters and by Theorem 2.3. The set $\Omega \times [0, l]$ is bounded and closed. Therefore, there exist certain points $(Z_1, y_1), \ldots,$ (Z_p, y_p) which belong to this set, such that for any point $(Z, y) \in \Omega \times [0, l]$ there exists a subscript i, $1 \leqslant i \leqslant p$ for which $|Z - Z_i| < \delta$, $|y - y_i| < \delta$. On the other hand, there exists an integer $N(\varepsilon)$ depending only on ε, such that

$$(2.7) \qquad |f(Z_i, y_i + h_{rr}) - f(Z_i, y_i + h_{ss})| < \frac{\varepsilon}{5}, \qquad i = 1, 2, \ldots, p,$$

whenever $r, s \geqslant N(\varepsilon)$.

Consider an $(\varepsilon/5)$-translation number ξ of $f(Z, x)$ belonging to the interval $[-x_0, -x_0 + l]$. Then the number $y_0 = x_0 + \xi$ belongs to the interval $[0, l]$ and consequently, $(Z_0, y_0) \in \Omega \times [0, l]$. There exists an i, $1 \leqslant i \leqslant p$, for which $|Z_i - Z_0| < \delta$, $|y_i - y_0| < \delta$. This means that

$$(2.8) \qquad |f(Z_i, y_i + h_{kk}) - f(Z_0, y_0 + h_{kk})| < \frac{\varepsilon}{5}, \qquad k = 1, 2, \ldots.$$

From (2.7) and (2.8) one finds

$$\begin{aligned}
|f(Z_0, x_0 + h_{rr}) - f(Z_0, x_0 + h_{ss})| &\leqslant |f(Z_0, x_0 + h_{rr}) - f(Z_0, y_0 + h_{rr})| \\
&\quad + |f(Z_0, y_0 + h_{rr}) - f(Z_i, y_i + h_{rr})| \\
&\quad + |f(Z_i, y_i + h_{rr}) - f(Z_i, y_i + h_{ss})| \\
&\quad + |f(Z_i, y_i + h_{ss}) - f(Z_0, y_0 + h_{ss})| \\
&\quad + |f(Z_0, y_0 + h_{ss}) - f(Z_0, x_0 + h_{ss})| \\
&< \varepsilon,
\end{aligned}$$

if $r, s \geqslant N$. The absolute values of the five differences are smaller than $\varepsilon/5$ for the following reasons: the first and the last absolute values are less than $\varepsilon/5$ due to the almost periodicity of $f(Z, x)$, the second and the fourth—due to the uniform continuity of $f(Z, x)$, and the third due to the condition (2.7).

Since ε is arbitrary and N depends only on ε, the condition is necessary.

The sufficiency of the condition is proved in the same way as Theorem 1.10. One should always bear in mind that $f(Z, x)$ is normal on $\Omega \times R$.

Using Theorem 2.6 one can show easily that the arithmetic operations applied to almost periodic functions depending uniformly on parameters lead to the same kind of functions.

Theorem 2.7. *If Ω is a bounded closed set, and $f(Z, x)$ and $g(Z, x)$ are almost periodic in x, uniformly with respect to $Z \in \Omega$, then $f(Z, x) + g(Z, x)$ and $f(Z, x) \cdot g(Z, x)$ are almost periodic in x, uniformly with respect to $Z \in \Omega$. If $|g(Z, x)| \geqslant m > 0$, then the function $f(Z, x)/g(Z, x)$ is almost periodic in x, uniformly with respect to $Z \in \Omega$.*

Proof. To prove the almost periodicity of the sum, it suffices to show that from each sequence $\{f(Z, x + h_k) + g(Z x + h_k)\}$ one can extract a subsequence converging uniformly on $\Omega \times R$. We shall extract from $\{h_k\}$ a subsequence $\{h_{1k}\}$ such that the sequence $\{g(Z, x + h_{1k})\}$ will be uniformly convergent on $\Omega \times R$. Then, from $\{h_{1k}\}$ we extract a subsequence $\{h_{2k}\}$ such that the sequence $\{f(Z, x + h_{2k})\}$ will be uniformly convergent on $\Omega \times R$. The sequence $\{f(Z, x + h_{2k}) + g(Z, x + h_{2k})\}$ is uniformly convergent on $\Omega \times R$, which proves the assertion made in the theorem, regarding the sum. One proceeds similarly for the product. It is more convenient to use the corollary of Theorem 2.2 and the identity

$$fg = \tfrac{1}{4}\{(f + g)^2 - (f - g)^2\}.$$

For the quotient we shall consider Theorem 2.1 and the relation $f/g = f \times (1/g)$.

Remark. It may appear at the first glance that the hypothesis that Ω be bounded and closed is unnecessary for the validity of Theorem 2.7. An example however will show that the theorem is not generally true if we omit this hypothesis.

Consider functions which depend only on one complex parameter z. The functions $f(z, x) = z$ and $g(z, x) = e^{ix}$ are almost periodic in x, uniformly with respect to z, where z belongs to the complex plane. The first function is constant with respect to x, and the second one is independent of z. But $fg = ze^{ix}$ is almost periodic in x, uniformly with respect to z, where z belongs to a bounded set of the complex plane; this function is not almost periodic in x, uniformly with respect to z, when z belongs to the whole complex plane.

Before stating a new theorem, we shall define the vector-almost periodic functions.

We say that $Z = Z(x)$ is an almost periodic function, if all the components $z_i = z_i(x)$, $i = 1, 2, \ldots, n$ are almost periodic.

From this definition follows without any difficulty that $Z = Z(x)$ is almost periodic if and only if the family $\{Z(x + h)\}$ is normal.

Theorem 2.8. *If $f(Z, x)$ is almost periodic in x, uniformly with respect to $Z \in \Omega$, if $Z = Z(x)$ is almost periodic and $Z(x) \in \Omega$ when $x \in R$, then $f(Z(x), x)$ is almost periodic.*

Proof. Without loss of generality, we can assume that Ω is bounded and closed due to the fact that the set of points $Z(x)$, $x \in R$ is bounded. Let $\{h_k\}$ be an arbitrary sequence of real numbers. We must show that the sequence $\{f(Z(x + h_k), x + h_k)\}$ contains a subsequence converging uniformly on the real line. Since $\{Z(x + h)\}$ is normal, it follows from Theorem 2.6 that there exists a subsequence $\{h'_k\}$ of the sequence $\{h_k\}$ such that $\{Z(x + h'_k)\}$ is uniformly convergent on the real line, and $\{f(Z, x + h'_k)\}$ is uniformly convergent on $\Omega \times R$. Let $\varepsilon > 0$ be an arbitrary number. There exists a number $\delta(\varepsilon) > 0$ such that

$$(2.9) \qquad |f(Z_1, x) - f(Z_2, x)| < \frac{\varepsilon}{2},$$

provided that

$$(2.10) \qquad |Z_1 - Z_2| < \delta, \qquad Z_1, Z_2 \in \Omega, \qquad x \in R.$$

But now we have

$$(2.11) \qquad |f(Z, x + h'_r) - f(Z, x + h'_s)| < \frac{\varepsilon}{2},$$

if $r, s \geq N_1(\varepsilon)$, and also

$$(2.12) \qquad |Z(x + h'_r) - Z(x + h'_s)| < \delta,$$

for $r, s \geq N_2(\varepsilon)$. Set $N = \max\{N_1, N_2\}$. It follows from (2.9)–(2.12) that

$$|f(Z(x + h'_r), x + h'_r) - f(Z(x + h'_s), x + h'_s)|$$
$$\leq |f(Z(x + h'_r), x + h'_r) - f(Z(x + h'_r), x + h'_s)|$$
$$+ |f(Z(x + h'_r), x + h'_s) - f(Z(x + h'_s), x + h'_s)|$$
$$< \frac{\varepsilon}{2} + \frac{\varepsilon}{2} = \varepsilon,$$

for

$$r, s \geq N(\varepsilon).$$

Thus Theorem 2.8 has been proved.

Theorem 2.8 is an extension of Theorem 1.7 to almost periodic functions depending on parameters. This theorem will be especially useful in Chapter IV, where we shall discuss almost periodic solutions of differential equations.

The presentation in this paragraph implies that the properties A and B of almost periodic functions hold for almost periodic functions depending uniformly on parameters. In this case they are also characteristic properties (one of them can be taken as definition).

The question which arises naturally is whether in the case of almost periodic functions depending on parameters there occurs a property similar to the definition of almost periodic functions of Chapter I; in other words, whether an almost periodic function depending uniformly on parameters can be approximated uniformly by functions such as:

$$(2.13) \qquad T(Z, x) = \sum_{k=1}^{n} c_k(Z)e^{i\lambda_k x},$$

where $c_k(Z)$ are continuous functions for $Z \in \Omega$, and λ_k are real numbers.

The answer to this question is, under certain conditions, positive. This will follow from certain considerations in Chapter VI, when we shall study the almost periodic functions with values in a Banach space.

We note that any function such as (2.13) is almost periodic in x, uniformly with respect to $Z \in \Omega$, if Ω is a bounded and closed set.

Finally, we observe that all the facts discussed in the present paragraph remain valid if the parameters on which a function depends are real.

2.   Functions almost periodic in the mean

The functions almost periodic in the mean form a class of almost periodic functions depending on parameters. These functions were introduced and studied by C. F. Muckenhoupt in 1929 in connection with the equation of the vibrating string.

Since the applications which will be considered concern functions almost periodic in the mean depending only on one real parameter, we confine our exposition to this case. There is no difficulty, except for notation, if we consider several parameters instead of one.

The functions $f(x, t)$ considered in this paragraph will be defined for any real t; for a given t they will be measurable with respect to x on a finite interval $[a, b]$, the same throughout the paragraph.

We shall say that $f(x, t)$ is an almost periodic function in the mean, if for any $\varepsilon > 0$ there exists a trigonometric polynomial

$$(2.14) \qquad T_\varepsilon(x, t) = \sum_{k=1}^{n} c_k(x)e^{i\lambda_k t},$$

where $c_k(x)$ are continuous functions on the interval $[a, b]$ and λ_k are real numbers, such that

$$(2.15) \qquad \int_a^b |f(x, t) - T_\varepsilon(x, t)|^2 \, dx < \varepsilon, \qquad t \in R.$$

It is obvious that the variable t gives to $f(x, t)$ the character of almost periodicity.

To avoid the complication of the language we shall not mention every time that $f(x, t)$ is almost periodic in t, in the mean with respect to x.

This definition is similar to that given in Chapter I, §1. It states that the functions almost periodic in the mean are those which can be approximated in the mean with respect to the parameter, and uniformly with respect to the real variable, by trigonometric polynomials such as (2.14).

It is not true that the assumption concerning the continuity of the coefficients c_k reduces the generality, as it would appear at first glance. If we would have assumed that $c_k(x)$ are square summable functions on the interval $[a, b]$, then we would obtain the same results. This is a consequence of the fact that the set of functions continuous on $[a, b]$ is everywhere dense in the space $L_2[a, b]$ of square summable functions on the interval $[a, b]$.

Let us establish now certain fundamental properties of almost periodic functions in the mean.

Theorem 2.9. *If $f(x, t)$ is almost periodic in the mean, then this function is bounded in the mean. That is, there exists a number M, such that*

$$(2.16) \qquad \int_a^b |f(x, t)|^2 \, dx \leqslant M, \qquad t \in R.$$

Proof. Consider $\varepsilon > 0$ and let $T_\varepsilon(x, t)$ be a trigonometric polynomial satisfying (2.15). Obviously, there exists a number $A > 0$ such that

$$(2.17) \qquad \int_a^b |T_\varepsilon(x, t)|^2 \, dx \leqslant A, \qquad t \in R.$$

But $f(x, t) = [f(x, t) - T_\varepsilon(x, t)] + T_\varepsilon(x, t)$. By a very well-known inequality of integration theory and by (2.17), we obtain

$$\int_a^b |f(x, t)|^2 \, dx \leqslant 2 \int_a^b |f(x, t) - T_\varepsilon(x, t)|^2 \, dx$$

$$+ 2 \int_a^b |T_\varepsilon(x, t)|^2 \, dx < 2(A + \varepsilon) = M.$$

Remark. From the above considerations it follows that for any real t, $f(x, t)$ is a square summable function on $[a, b]$.

Theorem 2.10. *Any function almost periodic in the mean is uniformly continuous in the mean. That is, for any $\varepsilon > 0$, there exists a $\delta(\varepsilon) > 0$ such that*

(2.18)
$$\int_a^b |f(x, t_1) - f(x, t_2)|^2 \, dx < \varepsilon,$$

if

$$|t_1 - t_2| < \delta, \qquad t_1, t_2 \in R.$$

Proof. The statement is obvious for $f(x, t) = c(x)e^{i\lambda t}$. One finds easily that it is true if $f(x, t)$ is a trigonometric polynomial such as (2.14).

Consider an arbitrary number $\varepsilon > 0$. One can determine a trigonometric polynomial $T(x, t) = \sum_{k=1}^n c_k(x)e^{i\lambda_k t}$, such that

(2.19)
$$\int_a^b |f(x, t) - T(x,t)|^2 \, dx < \frac{\varepsilon}{9}, \qquad t \in R.$$

Since $T(x, t)$ is uniformly continuous in the mean, there exists a number $\delta(\varepsilon) > 0$, such that

(2.20)
$$\int_a^b |T(x, t_1) - T(x, t_2)|^2 \, dx < \frac{\varepsilon}{9}$$

for

$$|t_1 - t_2| < \delta.$$

From the inequality $|A + B + C|^2 \leqslant 3(|A|^2 + |B|^2 + |C|^2)$, and using (2.19) and (2.20), we find

$$\int_a^b |f(x, t_1) - f(x, t_2)|^2 \, dx \leqslant 3\int_a^b |f(x, t_1) - T(x, t_1)|^2 \, dx$$

$$+ 3\int_a^b |T(x, t_1) - T(x, t_2)|^2 \, dx$$

$$+ 3\int_a^b |T(x, t_2) - f(x, t_2)|^2 \, dx < \varepsilon,$$

if $|t_1 - t_2| < \delta$.

Theorem 2.11. *If $f(x, t)$ and $g(x, t)$ are almost periodic in the mean, then $f(x, t) + g(x, t)$ is almost periodic in the mean.*

Proof. If $\varepsilon > 0$ is an arbitrary number, then there exist two trigono-metric polynomials of the form (2.14) such that

$$\int_a^b |f(x, t) - S(x, t)|^2 \, dx < \frac{\varepsilon}{4}, \qquad \int_a^b |g(x, t) - T(x, t)|^2 \, dx < \frac{\varepsilon}{4}$$

for any real t. Since $|A + B|^2 \leqslant 2(|A|^2 + |B|^2)$, one obtains

$$\int_a^b |f(x, t) + g(x, t) - S(x, t) - T(x, t)|^2 \, dx$$

$$\leqslant 2 \int_a^b |f(x, t) - S(x, t)|^2 \, dx + 2 \int_a^b |g(x, t) - T(x, t)|^2 \, dx < \varepsilon,$$

for any $t \in R$. Since $S(x, t) + T(x, t)$ is also a trigonometric polynomial of the form (2.14), the theorem is proved.

Remark. The product of two functions almost periodic in the mean is not necessarily almost periodic in the mean.

On the other hand, if $f(x, t)$ and $g(x, t)$ are almost periodic in the mean, and $h = fg$, then for any $\varepsilon > 0$ there exists a trigonometric polynomial $T(x, t)$, such that

$$(2.21) \qquad \int_a^b |h(x, t) - T(x, t)| \, dx < \varepsilon, \qquad t \in R.$$

This follows easily if one considers Schwarz-Buniakovski's integral in-equality.

The product of a function almost periodic in the mean with a function almost periodic in the usual sense (Chapter I, §1,) is a function almost periodic in the mean.

Theorem 2.12. *If the sequence $\{f_n(x, t)\}$ of almost periodic functions in the mean converges in the mean to $f(x, t)$, uniformly with respect to $t \in R$, then $f(x, t)$ is almost periodic in the mean.*

Proof. The fact that $\{f_n(x, t)\}$ converges in the mean to $f(x, t)$, uniformly with respect to $t \in R$, means that for any $\varepsilon > 0$, there exists an $N(\varepsilon)$ such that

$$(2.22) \qquad \int_a^b |f_n(x, t) - f(x, t)|^2 \, dx < \varepsilon, \qquad t \in R,$$

provided that

$$n \geqslant N(\varepsilon).$$

Let us take an n for which (2.22) is true. There exists a trigonometric polynomial of form (2.14) such that

(2.23) $\int_a^b |f_n(x, t) - T(x, t)|^2 \, dx < \frac{\varepsilon}{4}, \qquad t \in R.$

From (2.22) and (2.23) we find

$$\int_a^b |f(x, t) - T(x, t)|^2 \, dx \leqslant 2 \int_a^b |f(x, t) - f_n(x, t)|^2 \, dx$$

$$+ 2 \int_a^b |f_n(x, t) - T(x, t)|^2 \, dx < \varepsilon, \qquad t \in R,$$

which proves that $f(x, t)$ is almost periodic in the mean.

Corollary. If the functions $f_n(x, t)$ are almost periodic in the mean and the sequence $\{f_n(x, t)\}$ converges uniformly on $[a, b] \times R$ to $f(x, t)$, then $f(x, t)$ is a function almost periodic in the mean.

This is a consequence of the fact that a sequence which converges uniformly on $[a, b] \times R$, also converges in the mean on $[a, b]$, uniformly with respect to $t \in R$.

Theorem 2.13. *If $f(x, t)$ is almost periodic in the mean, and $f_t(x, t)$ is uniformly continuous in the mean, then $f_t(x, t)$ is almost periodic in the mean.*

Proof. It is obvious that we may confine ourselves to the real-valued function $f(x, t)$. Consider the functions

$$\varphi_n(x, t) = n \left[f\left(x, t + \frac{1}{n}\right) - f(x, t) \right].$$

Since $f(x, t + 1/n)$ is almost periodic in the mean, the same can be said about $\varphi_n(x, t)$, $n = 1, 2, \ldots$. But we have

$$\varphi_n(x, t) - f_t(x, t) = f_t\left(x, t + \frac{\theta_n}{n}\right) - f_t(x, t), \qquad 0 < \theta_n < 1.$$

Therefore,

$$\int_a^b |\varphi_n(x, t) - f_t(x, t)|^2 \, dx = \int_a^b \left| f_t\left(x, t + \frac{\theta_n}{n}\right) - f_t(x, t) \right|^2 dx \to 0,$$

as $n \to \infty$, uniformly with respect to $t \in R$.

Theorem 2.12 shows that $f_t(x, t)$ is almost periodic in the mean.

To establish other properties of functions almost periodic in the mean, the notion of a *function normal in the mean* is useful.

We shall say that the function $f(x, t)$ is normal in the mean, if from any sequence $\{f(x, t + h_n)\}$ one can extract a subsequence $\{f(x, t + h_{1n})\}$ which will be convergent in the mean, uniformly with respect to $t \in R$.

Theorem 2.14. *If $f(x, t)$ is a function almost periodic in the mean, then it is normal in the mean.*

Let us assume for the moment that $f(x, t) = c(x)e^{i\lambda t}$, where $c(x)$ is a continuous function on $[a, b]$ and λ is a real number. For any sequence of real numbers $\{h_n\}$ one can determine a subsequence $\{h_{1n}\}$ such that the sequence $\{e^{i\lambda(t+h_{1n})}\}$ is uniformly convergent on the real line. But

$$\int_a^b |f(x, t + h_{1m}) - f(x, t + h_{1n})|^2 \, dx = |e^{i\lambda h_{1m}} - e^{i\lambda h_{1n}}|^2 \cdot \int_a^b |c(x)|^2 \, dx.$$

From this relation follows that $\{f(x, t + h_{1n})\}$ is a convergent sequence in the mean, uniformly with respect to $t \in R$. Therefore, the functions $f(x, t)$, of the form considered above, are normal in the mean. A simple argument (see Theorem 1.9) shows that any trigonometric polynomial of the form (2.14) is a function normal in the mean.

Assume now that $f(x, t)$ is an arbitrary almost periodic function in the mean and that $\{T_n(x, t)\}$ is a sequence of trigonometric polynomials of the form (2.14), which converges in the mean to $f(x, t)$, uniformly with respect to $t \in R$. If $\{h_n\}$ is an arbitrary sequence of real numbers, then one can determine a subsequence $\{h_{1n}\}$ so that the sequence $\{T_1(x, t + h_{1n})\}$ is convergent in the mean, uniformly with respect to $t \in R$. Then, from $\{h_{1n}\}$ can be extracted a subsequence $\{h_{2n}\}$ such that $\{T_2(x, t + h_{2n})\}$ is convergent in the mean, uniformly with respect to $t \in R$. Proceeding in the same way we reach the conclusion that for any number p there exists a subsequence $\{h_{pn}\}$ of the sequence $\{h_n\}$, such that the sequence $\{T_q(x, t + h_{pn})\}$, $q \leqslant p$, is convergent in the mean, uniformly with respect to $t \in R$.

Consider now the diagonal sequence $\{h_{pp}\}$. Except for a finite number of terms, this sequence is a subsequence of each of the sequences $\{h_{qn}\}$. Therefore, for any q the sequence $\{T_q(x, t + h_{pp})\}$ is convergent in the mean, uniformly with respect to $t \in R$.

Let $\varepsilon > 0$ be an arbitrary number, and n sufficiently large, such that

$$(2.24) \qquad \int_a^b |f(x, t) - T_n(x, t)|^2 \, dx < \frac{\varepsilon}{9}, \qquad t \in R.$$

If we choose n such that (2.24) is satisfied, we can determine an $N(\varepsilon) > 0$, such that

(2.25) $\displaystyle\int_a^b |T_n(x, t + h_{pp}) - T_n(x, t + h_{qq})|^2 \, dx < \frac{\varepsilon}{9},$ $t \in R,$

if $p, q \geqslant N(\varepsilon)$. According to (2.24) and (2.25) we obtain

$$\int_a^b |f(x, t + h_{pp}) - f(x, t + h_{qq})|^2 \, dx$$

$$\leqslant 3 \int_a^b |f(x, t + h_{pp}) - T_n(x, t + h_{pp})|^2 \, dx$$

$$+ 3 \int_a^b |T_n(x, t + h_{pp}) - T_n(x, t + h_{qq})|^2 \, dx$$

$$+ 3 \int_a^b |T_n(x, t + h_{qq}) - f(x, t + h_{qq})|^2 \, dx < \varepsilon, \qquad t \in R,$$

if $p, q \geqslant N(\varepsilon)$.

It follows that $\{f(x, t + h_{pp})\}$ is a sequence convergent in the mean, uniformly with respect to $t \in R$, and this proves that the function $f(x, t)$ is normal.

Theorem 2.15. *If the function $f(x, t)$ is almost periodic in the mean, then for any $\varepsilon > 0$, there exists a number $l(\varepsilon) > 0$ with the property that any interval of the real line of length $l(\varepsilon)$ contains at least one number τ for which*

(2.26) $\displaystyle\int_a^b |f(x, t + \tau) - f(x, t)|^2 \, dx < \varepsilon,$ $t \in R.$

The proof of this theorem is the same as that of Theorem 1.10. The only property of $f(x, t)$ which we have to consider is the normality in the mean. The uniform convergence is now replaced by convergence in the mean, uniformly with respect to $t \in R$.

Remark. The number τ is called an ε-translation number of $f(x, t)$ or a translation number corresponding to ε.

As in the case of almost periodic functions in the Bohr sense, the normality in the mean and the property mentioned in Theorem 2.15 are characteristic properties of functions almost periodic in the mean. This follows from

Theorem 2.16. *Let* $f(x, t)$ *be a function satisfying the following conditions:*
a. f *is defined for* $t \in R$ *and almost anywhere for* $x \in [a, b]$*;*
b. *for any* $t \in R$ *we have*

$$\int_a^b |f(x, t)|^2 \, dx < +\infty;$$

c. f *is continuous in the mean, i.e., for any* $\varepsilon > 0$ *there exists a* $\delta(\varepsilon, t) > 0$*, such that*

$$\int_a^b |f(x, t') - f(x, t)|^2 \, dx < \varepsilon, \qquad for \quad |t' - t| < \delta;$$

d. *for any* $\varepsilon > 0$ *there exists a number* $l(\varepsilon) > 0$*, such that any interval of length* l *of the real line contains at least one number* τ *for which*

$$\int_a^b |f(x, t + \tau) - f(x, t)|^2 \, dx < \varepsilon, \qquad t \in R.$$

Under these assumptions $f(x, t)$ *is almost periodic in the mean.*

The proof of this theorem assumes the construction of a sequence of trigonometric polynomials of the form (2.14), which converges in the mean to $f(x, t)$, uniformly with respect to $t \in R$. Since we shall not use the theorem in the sequel, and its proof is complicated, we shall omit it.

We observe that the functions almost periodic in the mean can be considered almost periodic functions of a real variable t, with values in the space $L_2[a, b]$. Indeed, for any $t \in R$, $f(x, t)$ is square summable on $[a, b]$. Since $L_2[a, b]$ is a Banach space (even a Hilbert space), the theory of functions almost periodic in the mean is a special case of the theory of almost periodic functions with values in a Banach space, which will be developed in Chapter VI.

3. *Random functions almost periodic in probability*

Let (Ω, K, P) be an arbitrary probability space, and denote by $\mathcal{L}(\Omega, K, P)$ the set of all random variables with complex values. For the reader's convenience, we provide the definition of a *random variable:* the map $f: \Omega \to C$ (= the complex field) is called a random variable if the inverse image of any Borel set in the complex plane is in K. In other words, f is measurable with respect to the probability measure P. Of course, two random variables are considered identical if they coincide at all points, excepting perhaps at the points of a set of P-measure zero.

A *random function* is a map $x: S \times \Omega \to C$, such that for every $s \in S$ (= a topological space), $x(s, \omega)$ is a random variable. We shall be primarily concerned with the case when $S = R$ (= the real line). Such a random function is sometimes called a *random process*.

The definition of *continuity in probability* can be formulated as follows: Let $f(t; \omega)$ be a random function defined in a neighborhood of a point $t_0 \in R$; $f(t; \omega)$ is continuous in probability at t_0 if for any $\epsilon > 0$, $\eta > 0$, there exists $\delta(\epsilon, \eta) > 0$ such that $P(\omega; |f(t, \omega) - f(t_0, \omega)| \geq \epsilon) \leq \eta$, provided $|t - t_0| \leq \delta(\epsilon, \eta)$.

A function which is continuous in probability at each point of a given set is called continuous in probability on that set. The definition of uniform continuity in probability can be easily formulated and we leave the task to the reader.

We shall now formulate the definition of *almost periodicity in probability* for a random function. Namely, a random function continuous on R is called almost periodic in probability, if for any $\epsilon > 0$, $\eta > 0$, there corresponds a number $l(\epsilon, \eta) > 0$ with the property that every interval of length l of the real axis contains a number τ for which $P(\omega; |f(t + \tau, \omega) - f(t, \omega)| \geq \epsilon) < \eta$, for all t in R.

This definition of almost periodicity in probability is similar to the definition given by H. Bohr for almost periodic functions. As we shall see later in this paragraph, this definition is equivalent to those based on normality (compactness) or on the approximation property.

Theorem 2.17. *Let $f(t, \omega)$ be a random function, almost periodic in probability. Then $f(t, \omega)$ is uniformly continuous in probability on the whole real axis R.*

Proof. Let $\epsilon > 0$, $\eta > 0$ be two arbitrary numbers, and consider also the number $l(\epsilon/3, \eta/3)$, which exists, according to the definition of almost periodic functions in probability. Since $f(t, \omega)$ is uniformly continuous in probability on any compact (bounded and closed) interval of the real axis, it means that we can find $\delta = \delta(\epsilon/3, \eta/3)$, $0 < \delta < 1$, such that $|t - s| < \delta$, $t, s \in [-1, -1 + l]$ imply $P(\omega; |f(t, \omega) - f(s, \omega)| \geq \epsilon/3) < \eta/3$. Consider now two real numbers t_1 and t_2, such that $|t_1 - t_2| < \delta$, and let t be an arbitrary real number. Since $f(t, \omega)$ is almost periodic in probability, there exists $\tau \in [-t_1, -t_1 + l]$, such that $P(\omega; |f(t + \tau, \omega) - f(t, \omega)| \geq \epsilon/3) < \eta/3$. Since $t_1 + \tau \in [0, l]$, we have $t_2 + \tau \in [-1, 1 + l]$. On the other hand, the set $\{\omega; |f(t_2, \omega) - f(t_1, \omega)| \geq \epsilon\}$ belongs to the union of the following three sets (in Ω) : $\{\omega; |f(t_2, \omega) - f(t_2 + \tau, \omega)| \geq \epsilon/3\}$, $\{\omega; |f(t_2 + \tau, \omega) - f(t_1 + \tau, \omega)| \geq \epsilon/3\}$, and $\{\omega; |f(t_1 + \tau, \omega) - f(t_1, \omega)| \geq \epsilon/3\}$. Consequently, the following inequality must hold: $P(\omega; |f(t_2, \omega) - f(t_1, \omega)| \geq \epsilon) < \eta$, for t_1, t_2 with $|t_2 - t_1| < \eta$. This shows the uniform continuity in probability of the random function $f(t, \omega)$.

The next result provides a characterization of almost periodicity in probability for random functions.

Theorem 2.18. *Let $f(t, \omega)$ be a random function continuous in probability on the real axis. Then $f(t, \omega)$ is almost periodic in probability if and only if for any $\epsilon > 0$, $\eta > 0$, there corresponds a finite set of real numbers $A = \{a_1, a_2, \ldots, a_n\}$, such that for every real a, one can get an $a_j \in A$, with the property*

$$P(\omega; |f(t + a, \omega) - f(t + a_j, \omega)| \geq \epsilon) < \eta, \qquad \text{for all } t \in R.$$

Proof. We will first prove the necessity. Let $\epsilon > 0$, $\eta > 0$ be given, and consider the number $l(\epsilon/2, \eta/2)$ which corresponds to $\epsilon/2$ and $\eta/2$ in accordance with the definition of almost periodicity in probability. As Theorem 2.17 states, $f(t, \omega)$ is uniformly continuous in probability on the whole real axis. In the interval $[0, l]$, we can find a finite set of points $A = \{a_1, a_2, \ldots, a_n\}$ such that for every $t \in [0, l]$, there exists $a_j \in A$ with the property

$$P(\omega; |f(t + a, \omega) - f(t + a_j, \omega)| \geq \epsilon/2) < \eta/2, \qquad \text{for all } t \in R.$$

Now let a be an arbitrary real number, and consider $\tau \in [-a, -a + l]$, with l as required by the definition of almost periodicity in probability. We obtain

$$P(\omega; |f(t + a + \tau, \omega) - f(t + a, \omega)| \geq \epsilon/2) < \eta/2, \qquad \text{for all } t \in R.$$

But $a + \tau \in [0, l]$ implies the existence of an $a_j \in A$ such that

$$P(\omega; |f(t + a + \tau, \omega) - f(t + a_j, \omega)| \geq \epsilon/2) < \eta/2, \qquad \text{for all } t \in R.$$

If we take into account the fact that the set $\{\omega; |f(t + a, \omega) - f(t + a_j, \omega)| \geq \epsilon\}$ is contained in the union of the sets $\{\omega; |f(t + a, \omega) - f(t + a + \tau, \omega)| \geq \epsilon/2\}$ and $\{\omega; |f(t + a + \tau, \omega) - f(t + a_j, \omega)| \geq \epsilon/2\}$ for all $t \in R$, then we obtain from the above inequalities

$$P(\omega; |f(t + a, \omega) - f(t + a_j, \omega)| \geq \epsilon/2) < \eta, \qquad \text{for all } t \in R.$$

This completes the proof of the necessity of the condition given in Theorem 2.18.

In order to prove the sufficiency of the condition, let us choose some numbers $\epsilon > 0$, $\eta > 0$, and consider the corresponding set A. It is obvious that the inequality in the statement of Theorem 2.18 can be rewritten in the form

$$P(\omega; |f(t, \omega) - f(t + a - a_j, \omega)| \geq \epsilon) < \eta, \qquad \text{for all } t \in R.$$

Let $l = \max \{|a_j|; a_j \in A\}$. Then $a - a_j \in [a - l, a + l]$. Consequently, each interval of the real axis of length $2l$ contains a number which is an almost period for $f(t; \omega)$, corresponding to ϵ and η, as seen from the last inequality above. This completes the proof of Theorem 2.18.

Before we can establish new properties of random functions almost periodic in probability, we need to discuss the concept of uniform convergence in probability.

Let $\{f_n(t, \omega)\}$ be a sequence of random functions defined on some interval $I \subset R$. We shall say that the sequence is *uniformly convergent* (on I) in *probability* to the random function $f(t, \omega)$, if for any $\epsilon > 0$, $\eta > 0$, there corresponds a natural member $N(\epsilon, \eta)$ with the property $P(\omega; |f_n(t, \omega) - f(t, \omega)| \geqslant \epsilon) < \eta$ for $n \geqslant N(\epsilon, \eta)$, and all $t \in I$.

The following result shows the significance of the concept of uniform convergence in probability, in the class of random functions almost periodic in probability.

Theorem 2.19. *Let* $\{f_n(t, \omega)\}$ *be a sequence of random functions almost periodic in probability. If* $f(t, \omega)$ *is the limit random function, then* $f(t, \omega)$ *is also almost periodic in probability.*

Proof. It is known from Probability Theory that $f(t, \omega)$ is continuous in probability on the entire real axis. From the uniform convergence in probability of the sequence $\{f_n(t, \omega)\}$ it results that for any $\epsilon > 0$, $\eta > 0$ we can find $N(\epsilon, \eta)$ with the property $P(\omega; |f_n(t, \omega) - f(t, \omega)| \geqslant \epsilon/3) < \eta/3$, as soon as $n \geqslant N(\epsilon, \eta)$, for all $t \in R$. Let us fix now a natural n for which the inequality holds true. From the almost periodicity in probability of the random function $f_n(t, \omega)$, there follows the existence of a positive $l(\epsilon, \eta)$ with the property that every interval of the real axis of length l contains a number τ such that $P(\omega; |f_n(t + \tau, \omega) - f_n(t, \omega)| \geqslant \epsilon/3) < \eta/3$, for all $t \in R$. If we take now into account the inequality

$$|f(t + \tau, \omega) - f(t, \omega)| \leqslant |f(t + \tau, \omega) - f_n(t + \tau, \omega)|$$
$$+ |f_n(t + \tau, \omega) - f_n(t, \omega)| + |f_n(t, \omega) - f(t, \omega)|,$$

which holds true for all $t \in R$ and $\omega \in \Omega$, as well as the fact that the set $\{\omega; |f(t + \tau, \omega) - f(t, \omega)| \geqslant \epsilon\}$ belongs to the union of three sets: $\{\omega; |f(t + \tau, \omega) - f_n(t + \tau, \omega)| \geqslant \epsilon/3\}$, $\{\omega; |f_n(t + \tau, \omega) - f_n(t, \omega)| \geqslant \epsilon/3\}$ and $\{\omega; |f_n(t, \omega) - f(t, \omega)| \geqslant \epsilon/3\}$, then we obtain

$$P(\omega; |f(t + \tau, \omega) - f(t, \omega)| \geqslant \epsilon) < \eta, \qquad \text{for all } t \in R.$$

This inequality proves the almost periodicity in probability of the random function $f(t, \omega)$, and Theorem 2.19 is thereby proved.

The next definition is necessary in order to provide one more characterization of almost periodic functions in probability, similar to property A in Chap. I.

We shall say that the random function $f(t, \omega)$ is *normal in probability* if from every sequence $\{t_n\}$ of real numbers one can extract a subsequence $\{t_{n_k}\}$ with the property that $\{f(t + t_{n_k}, \omega)\}$ is uniformly convergent in probability on the real axis.

Theorem 2.20. *Let $f(t, \omega)$ be a random function continuous in probability on R. A necessary and sufficient condition for the almost periodicity in probability of the random function $f(t, \omega)$ is its normality in probability.*

Proof. The sufficiency of the condition, i.e., normality in probability implies almost periodicity in probability, can be obtained by paraphrasing the proof of the similar property in the case of numerical almost periodic functions (see the proof of Theorem 1.10).

The necessity of the condition follows immediately from Theorem 2.18. Indeed, if $\{t_n\}$ is an arbitrary sequence of real numbers then, according to Theorem 2.18, for any $\epsilon > 0$, $\eta > 0$, there exists a finite set $A = \{a_1, a_2, \ldots, a_m\}$ such that for any real a one can find an a_j with the property $P(\omega; |f(t + a, \omega) - f(t + a_j, \omega)| \geqslant \epsilon) < \eta$, for all $t \in R$. If we successively choose $a = t_n$ in this inequality, and denote by $a_{j(n)}$ the element of A that corresponds to t_n, then we can write the property above in the following form: $P(\omega; |f(t + t_n, \omega) - f(t + a_{j(n)}, \omega)| \geqslant \epsilon) < \eta$, for all $t \in R$, and $n = 1, 2, \ldots$. But $j(n)$ can take only m distinct values, and consequently there exists a value for $j(n) = j_0$, $1 \leqslant j_0 \leqslant m$, such that $j(n) = j_0$ for infinitely many n, say $n = n_k$, $k = 1, 2, \ldots$. Therefore, we can write the above inequality in the form

$$P(\omega; |f(t + t_{n_k}, \omega) - f(t + a_{j_0}, \omega)| \geqslant \epsilon) < \eta$$

for all $t \in R$, and all $k \geqslant 1$. This completes the proof of Theorem 2.20.

Remark. Theorem 2.18 actually provides the normality in probability of the random function $f(t, \omega)$, and shows its equivalence with the definition of almost periodicity in probability. Indeed, since the topology of $\mathcal{L}(\Omega, K, P)$ is a topology of a metric space (as noted in books on Probability), Theorem 2.18 expresses exactly the fact that the family $\{f(t + h, \omega); h \in R\}$ is compact with respect to uniform convergence in probability.

Theorem 2.20 allows us to prove without difficulty the almost periodicity in probability of the sum of two random functions which are almost periodic in probability, as well as many other properties.

Let us note that most of the properties of almost periodic numerical functions can be adapted to the case of almost periodicity in probability.

Before we can state another basic result regarding random functions almost periodic in probability, let us define the concept of *random trigonometric polynomial*.

A random function that can be represented in the form

$$T_n(t, \omega) = \sum_{k=1}^{n} c_k(\omega) e^{i\lambda_k t}, \quad \lambda_k = \text{real},$$

where $c_k(\omega)$ are random variables, is called a random trigonometric polynomial.

The following characteristic property of random functions almost periodic in probability can be formulated.

Theorem 2.21. *Let $f(t, \omega)$ be a random function defined on the real axis R. A necessary and sufficient condition for the almost periodicity in probability of $f(t, \omega)$ is the following: for any $\epsilon > 0, \eta > 0$, there exists a random trigonometric polynomial $T_n(t, \omega)$ such that*

$$P(\omega; |f(t, \omega) - T_n(t, \omega)| \geq \epsilon) < \eta, \quad \text{for all } t \in R.$$

The sufficiency of the condition can be easily obtained if we note that any random trigonometric polynomial is almost periodic in probability. It suffices to consider only one random function of the form $c(\omega) \exp(i\lambda t)$, which can be shown to be almost periodic in probability by using the normality criterion. The necessity of the condition is more complicated, and we refer the reader to such sources as (752).

The definition of random function almost periodic in probability was given by O. Onicescu and V. Istratescu in their joint paper (753), where they adopted the definition based on the property of approximation in probability by means of random trigonometric polynomials. Contributions to the theory of random functions almost periodic in probability have been made by G. Cenusa (710), A. M. Precupanu (758), (760), G. Cenusa and I. Sacuiu (711). A book by O. Onicescu, G. Cenusa and I. Sacuiu (752) has been recently published on the subject under consideration. There are some applications to stochastic differential equations in (752), as well as many other almost periodicity problems related to random processes. For instance, the existence of the mean value is discussed, together with the Fourier series attached to random functions. Other concepts of almost periodicity for random processes are also discussed. Finally, the book (752) contains complete references on the subject up to 1983.

III

Analytic almost periodic functions

The analytic almost periodic functions are a class of almost periodic functions which depend on a real parameter. We did not discuss them in the preceding chapter because they have a number of properties related to the theory of analytic functions, and the methods used in this theory are different from those used in that chapter.

Before entering into the contents of this chapter, we shall make a remark which shows that the definition of the almost periodic analytic functions is the only definition of interest.

Indeed, one may believe that in the case of functions of a complex variable, the definition of almost periodicity ought to be formulated as follows: a function $f(z)$, defined for all complex values of z is called almost periodic, if to any $\varepsilon > 0$ corresponds an $l(\varepsilon)$ such that in any square of side l in the complex plane whose sides are parallel to the coordinate axes, there exists at least one point ζ for which

$$|f(z + \zeta) - f(z)| < \varepsilon,$$

for any complex number z.

If we assume that $f(z)$ is analytic, from the above definition follows that $f(z)$ is bounded in the whole plane. According to a well-known Liouville theorem, we see that $f(z)$ reduces to a constant.

By *analyticity* we shall understand *holomorphism*.

If we assume only the continuity of $f(z)$, the above definition leads to a class of almost periodic functions completely similar to the class of almost periodic functions of a real variable. To be more precise, any function $f(z)$ represents a pair of functions of two variables $u(x, y)$ and $v(x, y)$, almost periodic with respect to both variables.

1. Preliminary theorems

In order to develop the theory of analytic almost periodic functions, it is necessary to give some theorems regarding analytic functions in a half-plane or in a strip. Some of these theorems are frequently found in text-books treating the theory of functions, so that we omit them here. Other theorems can rarely be found and these we shall prove them in the present paragraph together with those consequences useful later.

76

The statements used for proving the theorems of this paragraph are found in many books*.

Theorem 3.1. *Consider a function $f(z)$ analytic in the half-plane* Re $z > 0$, *such that* Re $f(z) > 0$, $f(1) = 1$. *Then we have*

(3.1)
$$\left| \frac{f(z) - 1}{f(z) + 1} \right| \leqslant \left| \frac{z - 1}{z + 1} \right|.$$

Proof. The transformation

$$Z = \frac{z - 1}{z + 1}$$

maps the half-plane Re $z > 0$ into the disc $|Z| < 1$. Set

$$F(Z) = \frac{f(z) - 1}{f(z) + 1},$$

where

$$z = \frac{1 + Z}{1 - Z}$$

is the inverse of the above transformation. Consequently, $F(Z)$ is analytic for $|Z| < 1$, $F(0) = 0$, and $|F(Z)| \leqslant 1$ for $|Z| < 1$. Applying Schwarz's lemma to $F(Z)$, we find

$$|F(Z)| \leqslant |Z|,$$

i.e., (3.1).

Theorem 3.2. *Consider a function $f(z)$ analytic in the half-plane* Re $z > x_1$, *such that the following conditions are satisfied:*
1) *there exist two numbers $c > 0$ and $k > 0$ for which*

$$|f(z)| > ce^{kx_1}$$

in the considered half-plane;
2) *on the straight line* Re $z = x_2 > x_1$ *we have*

$$|f(z)| \leqslant ce^{kx_2}.$$

Then we have

(3.2)
$$|f(z)| \leqslant ce^{kx},$$

in the half-plane Re $z > x_2$.

---

* For instance, Einar Hille, *Analytic function theory*, Vol. II.

Proof. As usual, $x = \text{Re } z$, $y = \text{Im } z$. Without loss of generality, we can assume that $x_1 = 0$, $x_2 = 1$, $c = 1$. We reach this case by the substitution $\zeta = (z - x_1)/(x_2 - x_1)$, $\varphi = (1/c)f$. The conditions of the theorem become:

$$(3.3) \qquad\qquad |f(z)| > 1, \qquad \text{if} \quad \text{Re } z > 0,$$

$$(3.4) \qquad\qquad |f(z)| < e^k, \qquad \text{if} \quad \text{Re } z = 1.$$

We must prove that

$$(3.5) \qquad\qquad |f(z)| \leqslant e^{kx}, \qquad \text{if} \quad x > 1.$$

It is obvious that (3.5) needs to be proved only for real z. Setting $f_y(x) = f(x + iy)$, where y is a real parameter, the function $f_y(x)$ will satisfy (3.3) and (3.4), even if x is complex. If (3.5) is proved to be true for real x, then we have

$$|f_y(x)| = |f(z)| < e^{kx}, \qquad \text{if} \quad x > 1.$$

Since y is arbitrary, this is equivalent to (3.5), z being a complex number with $\text{Re } z > 1$.

Consider any branch of $F(z) = 1/k \, \log f(z)$. According to (3.3), $F(z)$ is analytic for $\text{Re } z > 0$, and we have

$$(3.6) \qquad\qquad \text{Re } F(z) > 0 \qquad \text{if} \quad \text{Re } z > 0,$$

$$(3.7) \qquad\qquad \text{Re } F(z) \leqslant 1 \qquad \text{if} \quad \text{Re } z = 1.$$

Setting $F(1) = \alpha + i\beta$, according to (3.6) and (3.7) we obtain that $0 < \alpha \leqslant 1$. Consider the function

$$\varphi(z) = \frac{1}{\alpha}[F(z) - i\beta]$$

and apply Theorem 3.1. For $z = x > 1$, we find

$$\left| \frac{\varphi(x) - 1}{\varphi(x) + 1} \right| \leqslant \frac{x - 1}{x + 1}.$$

Now $|\varphi(x) - 1| \geqslant |\varphi(x)| - 1$, $|\varphi(x) + 1| \leqslant |\varphi(x)| + 1$. Hence,

$$\frac{|\varphi(x)| - 1}{|\varphi(x)| + 1} \leqslant \left| \frac{\varphi(x) - 1}{\varphi(x) + 1} \right| \leqslant \frac{x - 1}{x + 1},$$

$$(|\varphi(x)| - 1)(x + 1) \leqslant (|\varphi(x)| + 1)(x - 1), \qquad |\varphi(x)| \leqslant x,$$

from which it follows that

$$(3.8) \qquad\qquad \text{Re } \varphi(x) \leqslant x.$$

But we have

(3.9) $$\operatorname{Re} \varphi(x) = \frac{1}{k\alpha} \log |fx)| \geqslant \frac{1}{k} \log |f(x)|.$$

From (3.8) and (3.9) it follows that

$$|f(x)| \leqslant e^{kx},$$

which we wanted to prove.

We shall consider below functions analytic in a strip so that $a \leqslant \operatorname{Re} z \leqslant b$ or $a < \operatorname{Re} z < b$. We shall say briefly that such a function is analytic in the strip $[a, b]$ or in the strip (a, b).

Theorem 3.3. *If $f(z)$ is analytic and bounded in the strip $[a, b]$, and*

(3.10) $$|f(z)| \leqslant K,$$

for $\operatorname{Re} z = a$, $\operatorname{Re} z = b$, then (3.10) is valid in the whole strip $[a, b]$.

Proof. Let $c > 0$. For $\operatorname{Im} z = \pm l$, we shall obtain

$$|f(z)e^{cz^2}| \leqslant K' \max\{e^{c(a^2 - l^2)}, e^{c(b^2 - l^2)}\},$$

where K' is such that $|f(z)| \leqslant K'$ in the strip $[a, b]$.

Let us apply the maximum modulus principle in a rectangle $a \leqslant \operatorname{Re} z \leqslant b$, $|\operatorname{Im} z| < l$ to the function $f(z)e^{cz^2}$, where l is sufficiently large so that the maximum will be reached either on the straight line $\operatorname{Re} z = a$ or on $\operatorname{Re} z = b$. We obtain

$$|f(z)e^{cz^2}| \leqslant K \max\{e^{ca^2}, e^{cb^2}\}.$$

Since $c > 0$ is arbitrary, for $c \to 0$ we obtain (3.10) in the whole strip $[a, b]$; this is true since for any z in this strip, we shall have $|\operatorname{Im} z| < l$, for a sufficiently large l.

Corollary. If the sequence of exponential polynomials

$$P_k(z) = \sum_{n=1}^{n_k} a_{n,k} e^{\lambda_{n,k} z},$$

where $a_{n,k}$ are complex numbers and $\lambda_{n,k}$ are real, converges uniformly on the straight lines $\operatorname{Re} z = a$ and $\operatorname{Re} z = b$, then this sequence converges uniformly in the whole strip $[a, b]$. Indeed, if $\varepsilon > 0$, then there exists an $N(\varepsilon)$ such that for $\operatorname{Re} z = a$, $\operatorname{Re} z = b$ we find

(3.11) $$|P_k(z) - P_h(z)| < \varepsilon,$$

if $k, h \geqslant N(\varepsilon)$. According to Theorem 3.3 inequality (3.11) is true in the whole strip $[a, b]$, if $k, h \geqslant N(\varepsilon)$, thus proving the corollary.

Theorem 3.4. *Let $f(z)$ be a bounded function, analytic in the strip $[a, b]$. If we set*

$$(3.12) \qquad M(x) = \sup_{y} |f(x + iy)|, \qquad -\infty < y < +\infty,$$

then for any $x_1, x, x_2, a \leqslant x_1 < x, < x_2 \leqslant b$, the following inequality is true:

$$(3.13) \qquad M(x) \leqslant \{M(x_1)\}^{\frac{x_2 - x}{x_2 - x_1}} \cdot \{M(x_2)\}^{\frac{x - x_1}{x_2 - x_1}}.$$

Proof. Let α be the real root of the equation $M(x_1)e^{\alpha x_1} = M(x_2)e^{\alpha x_2}$. Then

$$(3.14) \qquad e^{\alpha} = \{M(x_1)\}^{1/(x_2 - x_1)} \cdot \{M(x_2)\}^{1/(x_1 - x_2)}.$$

Applying Theorem 3.2 to the function $f(z)e^{\alpha z}$, we find

$$(3.15) \qquad M(x)e^{\alpha x} \leqslant M(x_1)e^{\alpha x_1}.$$

Inequality (3.13) is an immediate consequence of (3.14) and (3.15).

Note. This theorem is known as Doetsch's theorem of three straight lines.

Corollary. For an exponential polynomial $\sum_{n=1}^{N} a_n e^{\lambda_n z}$, where $\lambda_n < 0$, $n = 1, 2, \ldots, N$, the function $M(x)$ is decreasing on the whole real line.

Indeed, $\lim_{x \to \infty} M(x) = 0$. Thus $\lim_{x \to \infty} \log M(x) = -\infty$. For any real numbers x_1, x, x_2, such that $x_1 < x < x_2$, we have

$$(3.16) \qquad \log M(x) \leqslant \frac{(x_2 - x) \log M(x_1) + (x - x_1) \log M(x_2)}{x_2 - x_1}.$$

This inequality shows that the function $\log M(x)$ is convex. From the geometric point of view, (3.16) expresses the following fact: if one takes the points of abscissa x_1 and x_2 on the graph of $\log M(x)$ and joins them by a segment, then the graph of $\log M(x)$ in (x_1, x_2) will be situated below this segment. Since $\lim_{x \to \infty} \log M(x) = -\infty$, $\log M(x)$ must decrease. Otherwise, one could determine two numbers $x_1 < x$, such that $\log M(x_1) \leqslant \log M(x)$. There exists a real number $x_2, x < x_2$, such that $\log M(x)_1 > \log M(x_2)$. For the numbers x_1, x, x_2, (3.16) would not be true, which is impossible.

Theorem 3.5. *Let $f(z)$ be a function bounded and analytic in the strip $[a, b]$, and consider three numbers a', x_0, b' such that $a < a' < x_0 < b' < b$.*

If $\varepsilon > 0$, there exists a $\delta > 0$ such that

(3.17) $$|f(z)| < \varepsilon \quad \text{in} \quad [a', b'],$$

provided that

(3.18) $$M(x_0) < \delta.$$

Proof. Let $K > 0$ be such that $|f(z)| \leqslant K$ in the strip $[a, b]$. According to Theorem 3.4 it suffices to choose δ such that the following inequalties are satisfied

$$K^{\frac{x_0 - b'}{x_0 - a}} \cdot \delta^{\frac{a' - a}{x_0 - a}} < \varepsilon,$$

$$K^{\frac{b' - x_0}{b - x_0}} \cdot \delta^{\frac{b - b'}{b - x_0}} < \varepsilon.$$

These inequalities ensure the fact that for $\operatorname{Re} z = a'$ and $\operatorname{Re} z = b'$ we have $|f(z)| \leqslant \varepsilon_1 < \varepsilon$. A consequence of Theorem 3.3 will be that $|f(z)| < \varepsilon$ in the strip $[a', b']$.

Theorem 3.6. *Assume that the function $f(z)$ satisfies the following conditions:*

1) *f is analytic in the strip (a, b);*
2) *f is bounded in any strip $[a_1, b_1] \subset (a, b)$;*
3) *on the straight line $\operatorname{Re} z = x_0$, $a < x_0 < b$, there exists a sequence of numbers $x_0 + iy_n$, such that*

(3.19) $$\lim_{n \to \infty} f(x_0 + iy_n) = 0;$$

4) *there exist two positive numbers d and l, such that any interval of length l on the straight line $\operatorname{Re} z = x_0$ contains a point $x_0 + iy$ for which*

(3.20) $$|f(x_0 + iy)| > d.$$

Under these conditions the function $f(z)$ vanishes in the strip $(x_0 - \delta, x_0 + \delta)$ for any $\delta > 0$.

Proof. We construct a sequence of functions analytic in the strip $[a, b]$, setting $f_n(z) = f(z + iy_n)$, $n = 1, 2, \ldots$. In any rectangle $x_0 - \delta \leqslant \operatorname{Re} z \leqslant x_0 + \delta$, $\delta < \min\{x_0 - a, b - x_0\}$, $-l \leqslant \operatorname{Im} z \leqslant l$, this sequence will be bounded. By Montel's theorem we can extract a subsequence $\{f_{n_k}(z)\}$ converging uniformly in $x_0 - \delta \leqslant \operatorname{Re} z \leqslant x_0 + \delta$, $-l \leqslant \operatorname{Im} z \leqslant +l$ to an analytic function $g(z)$. From (3.19) it follows that $g(x_0) = 0$, and from (3.20) that $g(z) \not\equiv 0$. Since the zeros of an analytic function form an isolated set, there exists a positive number $r < \delta$ such that $g(z) \neq 0$ for

$|z - x_0| = r$. But the sequence $\{f_{n_k}(z)\}$ converges uniformly to $g(z)$ on the circle $|z - x_0| = r$; since on this circle $\inf |g(z)| > 0$, one can determine a k_0 with the property that

$$|f_{n_k}(z) - g(z)| < |g(z)|, \qquad k > k_0,$$

for $|z - x_0| = r$. A consequence of Rouché's theorem is that the function $f_{n_k}(z) = g(z) + \{f_{n_k}(z) - g(z)\}$ has in the domain $|z - x_0| < r$ the same number of zeros as $g(z)$, if $k > k_0$. Therefore, $f(z + iy_{n_k})$ has at least one zero in the strip $(x_0 - \delta, x_0 + \delta)$, if k is sufficiently large. This means that $f(z)$ vanishes in the strip mentioned above.

Theorem 3.7. *Consider a function $f(z)$ analytic in the strip (a, b), bounded in any strip $[a_1, b_1] \subset (a, b)$. Then $f(z)$ and all its derivatives are uniformly continuous in any strip $[a_1, b_1] \subset (a, b)$.*

Proof. Consider any strip $[a_1, b_1] \subset (a, b)$. If $\delta < \min(a_1 - a, b - b_1)$, then $[a_1 - \delta, b_1 + \delta] \subset (a, b)$ and consequently, there exists an $M > 0$ such that $|f(z)| \leqslant M$ in the strip $[a_1 - \delta, b_1 + \delta]$. If z is an arbitrary point of the strip $[a_1, b_1]$, then the circle of radius δ and center z is fully contained in the strip $[a_1 - \delta, b_1 + \delta]$. Let us denote by γ_z the circumference of this circle. According to Cauchy's formula we have

$$(3.21) \qquad f^{(n)}(z) = \frac{n!}{2\pi i} \int_{\gamma_z} \frac{f(\zeta)}{(\zeta - z)^{n+1}} \, d\zeta.$$

As a consequence

$$(3.22) \qquad |f^{(n)}(z)| \leqslant \frac{n!\,M}{\delta^n},$$

for any z in the strip $[a_1, b_1]$.

Let z_1 and z_2 be any two points in $[a_1, b_1]$. Since

$$(3.23) \qquad f^{(n-1)}(z_2) - f^{(n-1)}(z_1) = \int_{z_1}^{z_2} f^{(n)}(\zeta) \, d\zeta,$$

where the integral is taken on the segment connecting z_1 with z_2, from (3.22) and (3.23) we obtain

$$(3.24) \qquad |f^{(n-1)}(z_2) - f^{(n-1)}(z_1)| \leqslant \frac{n!\,M}{\delta^n} |z_2 - z_1|.$$

Inequality (3.24) proves that all the derivatives of $f(z)$, including $f(z)$ considered as its derivative of order zero, are uniformly continuous in the strip $[a_1, b_1]$.

Remark. Inequality (3.22) shows that all the derivatives of the function $f(z)$ are bounded in any strip $[a_1, b_1] \subset (a, b)$.

2. *Elementary properties of analytic almost periodic functions*

As mentioned at the beginning of this chapter, the almost periodic functions defined in the whole complex plane reduce to constants if we impose the condition of analyticity. On the contrary, the functions almost periodic in a strip are a class of functions with interesting properties.

Let us formulate the definition of functions almost periodic in a strip.

We say that a function $f(z)$ continuous in the strip $[a, b]$, is almost periodic in this strip, if to any $\varepsilon > 0$ corresponds a number $l(\varepsilon) > 0$ such that any interval of length l on the imaginary axis contains at least one point $i\eta$ for which

$$|f(z + i\eta) - f(z)| < \varepsilon,$$

for any z in the considered strip.

In other words, the function $\varphi(x, y) = f(x + iy)$ is almost periodic in y, uniformly with respect to x, $a \leqslant x \leqslant b$.

This permits us to state the following properties of analytic functions almost periodic in a strip $[a, b]$.

Theorem 3.8. *If $f(z)$ is an analytic function almost periodic in the strip $[a, b]$, then it is bounded in this strip.*

Theorem 3.9. *Any analytic function almost periodic in the strip $[a, b]$ is uniformly continuous in this strip.*

This theorem represents a particular case of Theorem 2.3 regarding almost periodic functions which depend uniformly on parameters. It follows also from Theorems 3.7 and 3.8.

Theorem 3.10. *If $f(z)$ and $g(z)$ are analytic almost periodic functions in the strip $[a, b]$, then $f(z) + g(z)$ and $f(z) \cdot g(z)$ are also analytic almost periodic in the same strip.*

If $|g(z)| \geqslant m > 0$, then $f(z)/g(z)$ is an analytic almost periodic function in the strip $[a, b]$.

Corollary. Every exponential polynomial

$$P_n(z) = \sum_{k=1}^{n} a_k e^{\lambda_k z}$$

where λ_k are real numbers, is an analytic almost periodic function in any strip $[a, b]$.

From the theorems proved for almost periodic functions uniformly depending on parameters, one can also obtain other properties of analytic almost periodic functions. We are to prove now a theorem which will allow us to establish in a simple way certain properties of almost periodic analytic functions, using the corresponding properties of almost periodic functions of a real variable.

Theorem 3.11. *Let $f(z)$ be an analytic function in the strip (a, b), bounded in any strip $[a_1, b_1] \subset (a, b)$. If the function $f(x_0 + iy)$ is almost periodic in y on the line $\mathrm{Re}\, z = x_0$, $a < x_0 < b$, then $f(z)$ is almost periodic in any strip $[a_1, b_1] \subset (a, b)$.*

Proof. Consider two numbers a' and b' such that $a < a' < a_1 < b_1 < b' < b$, $a' < x_0 < b'$. Consider also $M = \sup |f(z)|$, z belonging to $[a', b']$. Then, the analytic function $\varphi(z) = f(z + i\eta) - f(z)$ is bounded for any η: $|\varphi(z)| \leqslant 2M$, in the strip $[a', b']$. According to Theorem 3.5 we shall have $|\varphi(z)| < \varepsilon$ in the whole strip $[a_1, b_1]$, if $|(\varphi)|z < \delta$ on the straight line $\mathrm{Re}\, z = x_0$. Thus,

$$|f(z + i\eta) - f(z)| < \varepsilon$$

in the strip $[a_1, b_1]$ if η is a δ-translation number of $f(x_0 + iy)$. This shows that $f(z)$ is almost periodic in the strip $[a_1, b_1]$.

Note. In Theorem 3.5 the conditions are of the form $a' < a_1 < x_0 < b_1 < b'$. We did not assume that $a_1 < x_0 < b_1$. If we would have assumed that $x_0 \leqslant a_1 < b_1$, the conclusion would have nevertheless been the same. The same is true for $a_1 < b_1 \leqslant x_0$.

From Theorem 3.11 we can obtain Theorem 3.10 without difficulty. We propose that the reader prove Theorem 3.10 in this new way.

We shall now derive from Theorem 3.11 other properties of analytic almost periodic functions.

Theorem 3.12. *Consider a sequence $\{f_n(z)\}$ of analytic functions in the strip (a, b), almost periodic in any strip $[a_1, b_1] \subset (a, b)$. If this sequence converges uniformly to $f(z)$ in any strip $[a_1, b_1] \subset (a, b)$, then $f(z)$ is an analytic almost periodic function in any strip $[a_1, b_1] \subset (a, b)$.*

Proof. Let x_0 be such that $a < x_0 < b$. On the straight line $\mathrm{Re}\, z = x_0$ all the functions $f_n(z)$ considered as functions of y are almost periodic. Thus, $f(z)$ is almost periodic on this line. According to Theorem 3.11, $f(z)$ is almost periodic in any strip $[a_1, b_1] \subset (a, b)$.

Corollary. If λ_k are real numbers and the series

$$\sum_{k=1}^{\infty} a_k e^{\lambda_k z}$$

is uniformly convergent in $[a, b]$, then its sum is an analytic almost periodic function in $[a, b]$.

Theorem 3.13. *If the function $f(z)$ is analytic almost periodic in any strip $[a_1, b_1] \subset (a, b)$, then its derivatives are analytic almost periodic in any strip $[a_1, b_1] \subset (a, b)$.*

Proof. It is sufficient to prove that $f'(z)$ is almost periodic in any strip $[a_1, b_1] \subset (a, b)$. From this will follow that $f''(z)$ has the same property and so on.

We consider a straight line $\operatorname{Re} z = x_0, a < x_0 < b$. On this line $f(z) = f(x_0 + iy)$ is almost periodic in y and $f'(z) = if'(x_0 + iy)$. By Theorems 3.7 and 1.8, $f'(z)$ is almost periodic on the straight line $\operatorname{Re} z = x_0$. According to Theorem 3.11, $f'(z)$ is almost periodic in any strip $[a_1, b_1] \subset (a, b)$.

Before concluding this paragraph we shall prove another theorem. More precisely, we shall show that the theorem regarding the almost periodicity of the quotient is valid under the single hypothesis that the denominator does not vanish [in Theorem 3.10 we assumed that $|g(z)| \geqslant m > 0$]. Let us first prove

Lemma 3.1. *Let $f(z)$ be an analytic almost periodic function in the strip $[a, b]$, and let G be the set of values assumed by this function on the straight line $\operatorname{Re} z = x_0, 0 < x_0 < b$. Then, for any $\delta < \min\{x_0 - a, b - x_0\}$, $f(z)$ takes in the strip $(x_0 - \delta, x_0 + \delta)$ any value from the derived set G'.*

Proof. Consider $w_0 \in G'$. The case when $f(z) - w_0 \neq 0$ for $\operatorname{Re} z = x_0$ is of interest. From the definition of the derived set it follows that there exists a sequence of points $x_0 + iy_n$, such that $f(x_0 + iy_n) - w_0 \to 0$ as $n \to \infty$. If $x_0 + iy'$ is any point on the straight line $\operatorname{Re} z = x_0$, then $|f(x_0 + iy') - w_0| = 2d > 0$. There exists a number $l > 0$ with the property that any interval of length l on the imaginary axis contains at least one d-translation number of $f(z)$. Hence, any interval of length l of the straight line $\operatorname{Re} z = x_0$ contains at least one point y at which $|f(x_0 + iy) - w_0| > d > 0$. But this means that $f(z) - w_0$ satisfies the conditions of Theorem 3.6. Therefore this function vanishes in any strip $(x_0 - \delta, x_0 + \delta)$, $\delta < \min\{x_0 - a, b - x_0\}$. Therefore $f(z)$ takes in the strip $(x_0 - \delta, x_0 + \delta)$ any value from the derived set G'.

Theorem 3.14. *If $f(z)$ and $g(z)$ are analytic in the strip (a, b) and almost periodic in any strip $[a_1, b_1] \subset (a, b)$, and $g(z) \neq 0$ in (a, b), then $f(z)/g(z)$ is an analytic almost periodic function in any strip $[a_1, b_1] \subset (a, b)$.*

Proof. Using Theorem 3.11, it will be sufficient to prove that $f(z)/g(z)$ is almost periodic on the straight line $\operatorname{Re} z = x_0$, $a < x_0 < b$. This will follow from the fact that $|g(z)| \geqslant m > 0$ on such a straight line. Indeed, if there were not an $m > 0$ such that $|g(z)| \geqslant m > 0$, for $\operatorname{Re} z = x_0$, then 0 would be element of the set G' (see Lemma 3.1). Therefore, $g(z)$ would vanish in any strip $(x_0 - \delta, x_0 + \delta)$, $\delta < \min\{x_0 - a, b - x_0\}$. This contradicts the assumption that $g(z) \neq 0$ in the strip (a, b).

Remark. Theorem 3.14 can be considered as an improvement of the last part of Theorem 3.10.

Finally, let us remark that we could have considered analytic almost periodic functions in a strip such as $a < \operatorname{Re} z < b$. The definition can be stated in the same way as in the case of strips like $a \leqslant \operatorname{Re} z \leqslant b$. It is obvious that an analytic almost periodic function in a strip (a, b) is almost periodic in any strip $[a_1, b_1] \subset (a, b)$. This remark helps us to transpose some of the properties of analytic almost periodic functions in a closed strip to almost periodic functions in an open strip.

For example, from Theorem 3.8 it follows that an analytic almost periodic function in the strip (a, b) is bounded in any strip $[a_1, b_1] \subset (a, b)$.

Similar statements can be derived from Theorems 3.9 and 3.10.

Generally speaking, the properties mentioned in Theorems 3.8, 3.9, and 3.10 are not valid in the case of analytic almost periodic functions in an open strip.

Let us illustrate by an example that Theorem 3.8 does not apply if we replace the closed strip by an open one.

Consider the function $f(z) = \sinh z = \frac{1}{2}(e^z - e^{-z})$. As can be seen, this function is periodic with period $2\pi i$ and the only points at which it vanishes are the points $z_k = k\pi i$, where k is an integer.

In any strip $(0, a)$, $a > 0$, the function

$$\varphi(z) = \frac{1}{f(z)} = \frac{2}{e^z - e^{-z}}$$

is almost periodic in y, being periodic with respect to this variable. Therefore $\varphi(z)$ is almost periodic and analytic in any strip $(0, a)$.

Since

$$\lim_{z \to z_k} \varphi(z) = \infty,$$

this means that $\varphi(z)$ is not bounded in $(0, a)$.

One can give examples which show that neither Theorem 3.9 nor 3.10 remains valid.

3. The Dirichlet series associated with an analytic almost periodic function

In the first chapter we showed that with any almost periodic function of a real variable can be associated a Fourier series. An almost periodic function of a real variable is completely determined by its Fourier series, as follows from the uniqueness theorem. We indicated also the Fejér–Bochner summation method which permits us to construct an almost periodic function, starting with the Fourier series associated with this function.

All this can be extended naturally to functions analytic and almost periodic in a strip or even in a half-plane, except that this time it is natural to associate with every function a series of the form

$$\sum_{k=1}^{\infty} A_k e^{\lambda_k z},$$

which is called the Dirichlet series of the given function.

Theorem 3.15. Let $f(z)$ be an analytic almost periodic function in the strip (a, b). The Fourier series of the function $f(x + iy)$, considered as function of y is

$$(3.25) \qquad f(x + iy) \sim \sum_{k=1}^{\infty} A_k e^{\lambda_k x} e^{i\lambda_k y}$$

where the numbers A_k and λ_k are independent of x and y.

Proof. Let us now keep x constant and prove that the real values of λ for which

$$(3.26) \qquad M_y\{f(x + iy)e^{-i\lambda y}\} \neq 0$$

are independent of x. Let $x_1 < x_2$ be two numbers in the strip (a, b). Applying Cauchy's theorem to the function $f(z)e^{-\lambda z}$ in a rectangle $x_1 \leqslant x \leqslant x_2, 0 \leqslant y \leqslant T$, we find

$$\int_{x_1}^{x_1+iT} f(z)e^{-\lambda z}\, dz + \int_{x_1+iT}^{x_2+iT} f(z)e^{-\lambda z}\, dz + \int_{x_2+iT}^{x_2} f(z)e^{-\lambda z}\, dz$$

$$+ \int_{x_2}^{x_1} f(z)e^{-\lambda z}\, dz = 0.$$

Dividing both sides by T and then making $T \to +\infty$ we see that the second and fourth terms approach zero. Since the limits of the other two terms exist as $T \to +\infty$, we can write

$$\lim_{T \to +\infty} \frac{1}{T} \int_{x_1}^{x_1 + iT} f(z)e^{-\lambda z} \, dz = \lim_{T \to +\infty} \frac{1}{T} \int_{x_2}^{x_2 + iT} f(z)e^{-\lambda z} \, dz,$$

or

$$e^{-\lambda x_1} \lim_{T \to +\infty} \frac{1}{T} \int_0^T f(x_1 + iy)e^{-i\lambda y} \, dy = e^{-\lambda x_2} \lim_{T \to +\infty} \frac{1}{T} \int_0^T f(x_2 + iy)e^{-i\lambda y} \, dy.$$

The last equality shows that $M_y\{f(x + iy)\}$ is simultaneously zero or nonzero for all $x \in (a, b)$. This equality can also be written as

$$(3.27) \qquad e^{-\lambda x_1} M_y\{f(x_1 + iy)e^{-i\lambda y}\} = e^{-\lambda x_2} M_y\{f(x_2 + iy)e^{-i\lambda y}\}.$$

Denoting as usual the Fourier exponents of $f(x + iy)$ by $\lambda_1, \lambda_2, \ldots$ (as we have seen they are the same for all x), we can write

$$(3.28) \qquad f(x + iy) \sim \sum_{k=1}^{\infty} A_k(x)e^{i\lambda_k y},$$

where

$$(3.29) \qquad A_k(x) = M_y\{f(x + iy)e^{-i\lambda_k y}\}.$$

From (3.27) follows $A_k(x)e^{-\lambda_k x} = \text{const} = A_k$. Thus we can write (3.29) as

$$(3.30) \qquad A_k(x) = A_k e^{\lambda_k x}.$$

Therefore, the theorem has been proven.

We observe that (3.25) can be also written as

$$(3.31) \qquad f(z) \sim \sum_{k=1}^{\infty} A_k e^{\lambda_k z}.$$

The series on the right side of relation (3.31) is called the *Dirichlet series* of the analytic almost periodic function $f(z)$.

The Fourier exponents λ_k are also called *Dirichlet exponents* and the coefficients A_k are called the *Dirichlet coefficients* of the function $f(z)$.

As in the case of almost periodic functions of a real variable the following uniqueness theorem holds:

Theorem 3.16. *If two analytic functions almost periodic in the same strip have identical Dirichlet series, then the functions themselves are identical.*

Proof. Let us choose in a strip an arbitrary straight line Re $z = x_0$ and let us note that on this line they have the same Fourier series (they are considered to be functions of y). Thus (by the uniqueness theorem of Chapter I) the functions coincide on this line. According to the principle of identity for analytic functions, they coincide on the whole strip.

Theorem 3.17. *Let $f(z)$ be an analytic almost periodic function in the strip (a, b), so that*

(3.32) $$f(z) \sim \sum_{k=1}^{\infty} A_k e^{\lambda_k z}.$$

Then, for any x, $a < x < b$, we have

(3.33) $$M_y\{|f(x + iy)|^2\} = \sum_{k=1}^{\infty} |A_k|^2 e^{2\lambda_k x}.$$

The proof follows from the fact that for any x the Dirichlet series becomes a Fourier series. Applying Parseval's equality (1.18) to the function $f(x + iy)$, considered as an almost periodic in y, we obtain (3.33).

Theorem 3.18. *If $f(z)$ is an analytic almost periodic function in the strip (a, b), then the sequence of Bochner–Fejér polynomials*

(3.34) $$\sigma_m(z) = \sum_{k=1}^{n} r_{k,m} A_k e^{\lambda_k z}, \qquad n = n(m),$$

converges uniformly in any strip $[a_1, b_1] \subset (a, b)$.

Proof. For Re $z = a_1$ and Re $z = b_1$ the series $\sigma_m(z)$ converges uniformly to the function $f(z)$, i.e., to $f(a_1 + iy)$, $f(b_1 + iy)$. By the corollary of Theorem 3.3 it follows that $\sigma_m(z)$ converges uniformly in the strip $[a_1, b_1]$.

Theorem 3.19. *Let $f(z)$ be an analytic function in the strip (a, b), continuous in the strip $[a, b]$. If for Re $z = a$, Re $z = b$, $f(z)$ reduces to two almost periodic functions $f_a(y) = f(a + iy)$ and $f_b(y) = f(b + iy)$ respectively, such that*

$$f_a(y) \sim \sum_{k=1}^{\infty} A_k e^{\lambda_k a} e^{i\lambda_k y},$$

$$f_b(y) \sim \sum_{k=1}^{\infty} A_k e^{\lambda_k b} e^{i\lambda_k y},$$

then $f(z)$ is almost periodic in $[a, b]$, and we have

(3.35) $$f(z) \sim \sum_{k=1}^{\infty} A_k e^{\lambda_k z}.$$

Proof. We construct Bochner–Fejér polynomials for the series $\sum_{k=1}^{\infty} A_k e^{\lambda_k z}$. For Re $z = a$ and Re $z = b$, these trigonometric polynomials coincide with the Bochner–Fejér polynomials for the almost periodic functions $f_a(y)$ and $f_b(y)$. Consequently, this sequence converges uniformly on the lines Re $z = a$ and Re $z = b$. As in the proof of the preceding theorem, the sequence of Bochner–Fejér polynomials converges uniformly in the strip $[a, b]$. Since it converges to $f(x)$ for Re $z = a$ and Re $z = b$, by a simple argument it follows that this sequence will converge uniformly to $f(z)$ in the strip $[a, b]$. Obviously, the relation (3.35) holds and $f(z)$ is almost periodic.

Remark. We have seen in Theorem 3.11 that an analytic function in a strip, which reduces to an almost periodic function on a straight line situated in this strip, is almost periodic in any strip whatsoever contained in the given strip, if the function is bounded.

Theorem 3.19 gives us an almost periodicity criterion for an analytic function, without requiring boundedness. On the other hand, the function must be almost periodic on two straight lines.

Theorem 3.19 shows also that a Dirichlet series, which on the lines Re $z = a$ and Re $z = b$ reduces to the Fourier series of two almost periodic functions, can be considered as a Dirichlet series associated with an analytic almost periodic function. In the same context we include the following theorem:

Theorem 3.20. *Consider the Dirichlet series*

(3.36)
$$\sum_{k=1}^{\infty} A_k e^{\lambda_k z}$$

for which $\lambda_k < 0$, $k = 1, 2, \ldots$. If for Re $z = a$ the series (3.36) becomes the Fourier series of an almost periodic function $f_a(y)$, then there exists an analytic almost periodic function in the half-plane Re $z > a$ which for Re $z = a$ becomes $f_a(y)$, and has the series (3.36) as its Dirichlet series. This function tends to zero as Re $z \to +\infty$, uniformly with respect to y.

Proof. We form the Bochner–Fejér sums for $f_a(y)$ and for the series (3.36). Let $\sigma_m(y)$ and $\sigma_m(z)$ be the respective sums. We know that $\sigma_m(y) \to f_a(y)$ uniformly. Since the numbers λ_k are negative, according to the corollary of Theorem 3.4, it follows that $\sigma_m(z)$ converges uniformly in any strip $[a, b]$ to a function $f(z)$ which, like these polynomials, tends uniformly to zero as Re $z \to +\infty$. It is obvious that $f(z)$ is analytic in the half-plane Re $z > a$, almost periodic in the same half-plane, and that its Dirichlet series is the series (3.36) itself.

An immediate consequence of Theorem 3.20 is

Theorem 3.21. *If the function $f(z)$ is analytic and almost periodic in the strip (a, b) and*

$$f(z) \sim \sum_{k=1}^{\infty} A_k e^{\lambda_k z}$$

where $\lambda_k < 0$, then $f(z)$ can be extended to the half-plane $\mathrm{Re}\ z > a$, so that it becomes analytic almost periodic in this half-plane, tending to zero as $\mathrm{Re}\ z \to +\infty$, uniformly with respect to y.

Remark. Two similar theorems can be obtained for $\lambda_k > 0$, $k = 1, 2, \ldots$ substituting in this case the half-plane $\mathrm{Re}\ z > a$ by the half-plane $\mathrm{Re}\ z < a$.

Theorem 3.22. *Let $f(z)$ be an analytic almost periodic function defined in the strip (a, b), such that*

$$f(z) \sim \sum_{k=1}^{\infty} A_k e^{\lambda_k z}.$$

If there exist two numbers λ and Λ such that $\lambda < \lambda_k < \Lambda$, $k = 1, 2, \ldots$, then $f(z)$ is almost periodic in the strip $[a, b]$.

Proof. The function $f(z)e^{-\Lambda z}$ is almost periodic in the half-plane $\mathrm{Re}\ z > a$, and the function $f(z)e^{-\lambda z}$ is almost periodic in $\mathrm{Re}\ z < b$. Since $e^{-\Lambda z}$, $e^{-\lambda z} \neq 0$, our assertion follows immediately from the facts mentioned above and from $f(z) = f(z)e^{-\Lambda z} \cdot e^{\Lambda z} = f(z) \cdot e^{-\lambda z} \cdot e^{\lambda z}$.

Let us prove now a converse of Theorem 3.21:

Theorem 3.23. *Let $f(z)$ be an analytic almost periodic function defined and bounded in the half-plane $\mathrm{Re}\ z > a$. Then all Dirichlet exponents are $\leqslant 0$.*

The proof follows immediately from

$$\sup |f(z)| \geqslant |M_y\{f(x + iy)e^{-i\lambda_k y}\}| = |A_k|e^{\lambda_k x}.$$

This inequality is possible only if $\lambda_k = 0$ or $\lambda_k < 0$, since $x > a$ is arbitrary.

Theorem 3.24. *The class of analytic almost periodic functions in the half-plane $\mathrm{Re}\ z > a$ with Dirichlet exponents $\leqslant 0$ coincides with the class of analytic almost periodic functions bounded in the half-plane $\mathrm{Re}\ z > a$.*

Every function in this class tends uniformly with respect to y to a constant (i.e., to the constant term of its Dirichlet series) as $\mathrm{Re}\ z \to +\infty$.

Proof. If A_0 is the constant term of the Dirichlet series corresponding to the analytic almost periodic function $f(z)$, then

$$f(z) \sim \sum_{k=1}^{\infty} A_k e^{\lambda_k z} + A_0,$$

where $\lambda_k < 0$. The function $f(z) - A_0$ satisfies the conditions of Theorem 3.21. Therefore $f(z) \to A_0$ as $\operatorname{Re} z \to +\infty$, uniformly with respect to y. Hence $f(z)$ is bounded in the half-plane $\operatorname{Re} z > a$.

Conversely, if $f(z)$ is analytic, almost periodic and bounded in $\operatorname{Re} z > a$, then the conditions of Theorem 3.23 are satisfied. Hence $\lambda_k \leqslant 0$.

Before concluding this paragraph, let us look briefly at the convergence of the Dirichlet series.

It is clear that every convergence criterion for a Fourier series can lead us to a convergence criterion for the Dirichlet series associated with an analytic almost periodic function. However, we shall omit the criteria which can be obtained in this way. We shall give a criterion applying specifically to analytic functions.

Theorem 3.25. *Let $f(z)$ be an analytic function in the strip (a, b) and almost periodic in any strip $[a_1, b_1] \subset (a, b)$. If the series*

$$\sum_{k=1}^{\infty} e^{-|\lambda_k|\delta}$$

converges for an arbitrary $\delta > 0$, then

$$f(z) = \sum_{k=1}^{\infty} A_k e^{\lambda_k z},$$

the convergence being absolute at any point of (a, b).

Proof. Let x_0 be such that $a < x_0 < b$, and $\delta = \min\{x_0 - a, b - x_0\}$. The series

$$\sum_{k=1}^{\infty} A_k e^{\lambda_k(x_0 - \delta)} e^{i\lambda_k y}, \qquad \sum_{k=1}^{\infty} A_k e^{\lambda_k(x_0 + \delta)} e^{i\lambda_k y}$$

are the Fourier series corresponding to the almost periodic functions $f(x_0 - \delta + iy)$ and $f(x_0 + \delta + iy)$. Thus, there exists an $A > 0$, such that

$$|A_k e^{\lambda_k(x_0 - \delta)}|, \ |A_k e^{\lambda_k(x_0 + \delta)}| \leqslant A.$$

Consequently,

$$\sum_{k=1}^{\infty} |A_k e^{\lambda_k(x_0 + iy)}| \leqslant A \sum_{k=1}^{\infty} e^{-|\lambda_k|\delta},$$

which proves the theorem.

4. The behavior at infinity of analytic almost periodic functions in a half-plane

The analytic almost periodic functions in a half-plane can be easily classified by their behavior at infinity or by the nature of the set of the Dirichlet exponents. Let us prove first

Theorem 3.26. *If* $f(z)$ *is an analytic almost periodic function in the half-plane* Re $z > a$, *then only the following three cases can arise:*
(A) $f(z)$ *converges to a finite limit as* Re $z \to +\infty$;
(B) $f(z)$ *tends to infinity as* Re $z \to +\infty$;
(C) *In any half-plane* Re $z \geqslant a_1 > a$, $f(z)$ *takes values arbitrarily close to any complex number.*

Proof. We must show that if neither case (A) nor case (B) occurs, case (C) occurs necessarily. We give a proof by contradiction. If our assertion were not true, then we would find a complex number w_0 such that $|f(z) - w_0| \geqslant m > 0$ in the half-plane Re $z \geqslant a_1$. From this follows that the function

$$g(z) = \frac{1}{f(z) - w_0}$$

is almost periodic and bounded in the half-plane Re $z \geqslant a_1$. According to Theorem 3.24 the function $g(z)$ converges to a finite limit as Re $z \to +\infty$. But this is possible only if $f(z)$ approaches a finite limit or tends to infinity as Re $z \to +\infty$. Since this is impossible, it follows that this theorem is valid.

Remark. In case (C) we have a situation similar to the Weierstrass theorem concerning the behavior of a uniform analytic function in the neighborhood of an essential singular point.

We shall say in the sequel that a function, almost periodic and analytic in a half-plane Re $z > a$ belongs to class (A), if case (A) holds for this function. Analogously for cases (B) and (C). We observe that these classes are disjoint two by two.

We can turn now to the characterization of classes (A), (B), and (C) using the Dirichlet coefficients.

Theorem 3.27. *Let*

$$f(z) \sim \sum_{k=1}^{\infty} A_k e^{\lambda_k z}$$

be an analytic and almost periodic function in the half-plane Re $z \geqslant a$. *Then*

1) $f(z) \in$ (A) *if and only if* $\lambda_k \leqslant 0$;

2) $f(z) \in$ (B) *if and only if there exists a* $\lambda_k > 0$, *larger than all the other Dirichlet exponents*;

3) $f(z) \in$ (C) *if and only if among the positive Dirichlet exponents there occurs none which is larger than all the others.*

Proof. Theorem 3.24 expresses assertion 1).

We are to prove now assertion 2). Let λ be the largest among the positive Dirichlet exponents of the function $f(z)$. The function $g(z) = f(z)e^{-\lambda z}$ has all its Dirichlet exponents $\leqslant 0$, and the constant term of the Dirichlet series associated with it is nonzero. Thus, $g(z)$ converges to a finite limit (which is the constant term of the Dirichlet series) as Re $z \to +\infty$, uniformly with respect to y. Consequently $f(z) = g(z)e^{\lambda z}$ tends to ∞, uniformly with respect to y, as Re $z \to +\infty$.

Conversely, if $f(z)$ tends to ∞ as Re $z \to +\infty$ uniformly with respect to y, then among all the Dirichlet exponents there necessarily occur some which are positive (Theorem 3.24). It remains to show that among these exponents there exists one which is largest. If this were not true, then only the following situations would be possible: α) the Dirichlet coefficients have no upper bound; β) the Dirichlet coefficients do have an upper bound, but their upper bound is not a Dirichlet coefficient of the function $f(z)$. We shall show that, in the cases α) and β), $f(z)$ belongs to class (C).

Indeed, if case α) occurs and $C_1 > 0$ is arbitrary, one can determine a real number x_1 so that $|f(z)| > C_1$, if Re $z \geqslant x_1$. Consider now $x_2 > x_1$ and $C_2 > C_1$, so that we have $|f(z)| < C_2$ for Re $z = x_2$. Let us determine now the positive numbers c and k from the relations

$$C_1 = ce^{kx_1}, \qquad C_2 = ce^{kx_2}.$$

We shall have

$$|f(z)| > ce^{kx_1} \quad \text{for} \quad \text{Re } z \geqslant x_1,$$

$$|f(z)| < ce^{kx_2} \quad \text{for} \quad \text{Re } z = x_2.$$

Applying Theorem 3.2, we obtain

(3.37) $$|f(z)| < ce^{kx}, \quad \text{for} \quad x = \text{Re } z \geqslant x_2.$$

If n is such that $\lambda_n > k$, then

$$M(x) \geqslant |M_y\{f(x + iy)e^{-i\lambda_n y}\}| = |A_n| e^{\lambda_n x} > ce^{kx},$$

if x is sufficiently large. This however contradicts relation (3.37). This contradiction proves that in case α) the function $f(z)$ does not belong to class (B). Thus only the last possibility remains, i.e., $f(z)$ belongs to class (C).

Let us assume now that case β) occurs. As before, we shall have $|f(z)| > C_1$ for Re $z \geqslant x_1$. We choose c such that the equality $C_1 = ce^{\Lambda x_1}$ holds, where Λ is the upper bound of the Dirichlet coefficients. It follows that

(3.38) $|f(z)| > ce^{\Lambda x_1}$ for   Re $z \geqslant x_1$.

Now the function $g(z) = f(z)e^{-\Lambda z}$ has all the Dirichlet exponents negative, and by Theorem 3.21 tends to zero as Re $z \to +\infty$, uniformly with respect to y. Hence, there exists a number $x_2 > x_1$, such that for Re $z = x_2$

(3.39) $|f(z)| < \frac{1}{2}ce^{\Lambda x_2}$.

Given $\Lambda' < \Lambda$ such that $\frac{1}{2}e^{\Lambda x_2} < e^{\Lambda' x_1}$, it follows from (3.38) and (3.39) that

$$|f(z)| > ce^{\Lambda' x_1}   \text{for}   \text{Re } z \geqslant x_1,$$

$$|f(z)| < ce^{\Lambda' x_2}   \text{for}   \text{Re } z = x_2.$$

Applying again Theorem 3.2 we find

(3.40) $|f(z)| \leqslant ce^{\Lambda' x}$ for   Re $z \geqslant x_2$.

Since n may be determined so that $\Lambda' < \lambda_n$, this means that we have

$$M(x) \geqslant |A_n|e^{\lambda_n x} > ce^{\Lambda' x},$$

if x is sufficiently large. This contradicts (3.40). Thus, as in case β), the function $f(z)$ belongs to class (C).

With this assertion, 2) is completely established. 3) is obviously a consequence of 1) and 2). Thus Theorem 3.27 has been proved, and we have obtained a classification of analytic almost periodic functions in a half-plane; this classification makes use of the Dirichlet coefficients.

For functions belonging to class (C) we have a theorem similar to Picard's theorem for the exceptional values of analytic integral functions. We omit the proof of this theorem since it involves special considerations rarely used in the books on theory of analytic functions. The statement of the theorem is the following:

Theorem 3.28. *If $f(z)$ is an analytic almost periodic function in the half-plane* Re $z \geqslant a$ *and if $f(z) \in (C)$, then in any half-plane* Re $z \geqslant a_1 > a$, $f(z)$ *takes every complex value except at most one value.*

This theorem represents an improvement of the last assertion of Theorem 3.26.

It is an interesting fact that for Dirichlet exponents bounded above and whose upper bound is not a Dirichlet exponent, the function has no exceptional values. Before proving this assertion, let us establish

Lemma 3.2. *If the Dirichlet coefficients of an analytic almost periodic function $f(z)$ are all negative, and if among them there occurs a largest one, then there exists a number x_1 such that $f(z) \neq 0$ for Re $z > x_1$.*

If among the Dirichlet exponents there occurs no largest one, then $f(z)$ has zeros in any half-plane Re $z > x_1$.

Proof. Let us denote by Λ the upper bound of the Dirichlet exponents. In the first case the constant term of the Dirichlet series associated with the function $g(z) = f(z)e^{-\Lambda z}$ is nonzero. If we denote this term by A, then by Theorem 3.24 we shall have

$$\lim_{\text{Re } z \to +\infty} |f(z)e^{-\Lambda z}| = |A|$$

uniformly with respect to y. Thus we can determine x_1 such that

$$|f(z)e^{-\Lambda z}| > \tfrac{1}{2}|A|$$

for Re $z > x_1$, which proves that $f(z) \neq 0$ in the half-plane Re $z > x_1$. In the second case, the function $g(z)$ has only negative Dirichlet coefficients and therefore,

$$\lim_{\text{Re } z \to +\infty} g(z) = 0,$$

uniformly with respect to y. If in the half-plane Re $z > x_1$ the function $g(z)$ has no zeros, then the function $1/g(z)$ would be an analytic almost periodic function in any strip (x_1, x_2), $x_1 < x_2$. From Theorem 3.21 follows that $f(z)$ is almost periodic in the half-plane Re $z > x_1$. It belongs to class (B). Thus its Dirichlet series has a largest positive exponent, say M. If the corresponding term of the Dirichlet series is Be^{Mz}, the function $1/g(z)e^{-Mz}$ tends to B if Re $z \to +\infty$, uniformly with respect to y. We shall have

(3.41) $$\left| \frac{1}{g(z)} e^{-Mz} \right| > \frac{1}{2}|B|,$$

if Re z is sufficiently large, i.e.

(3.42) $$|f(z)| < \frac{2}{|B|} e^{(\Lambda - M)x},$$

if Re $z > a_1$. If n is such that $\lambda_n > \Lambda - M$, we obtain as in the proof of Theorem 3.27,

$$M(x) \geqslant |A_n|e^{\lambda_n x} > \frac{2}{|B|} e^{(\Lambda - M)x},$$

for a sufficiently large x. This inequality contradicts (3.42). Thus, in the second case $f(z)$ has zeros in any half-plane Re $z > x_1$.

Theorem 3.29. *If $f(z) \in (C)$ and its Dirichlet exponents are bounded above, $f(z)$ takes every complex value in any half-plane Re $z > x_1 > a$.*

Proof. If Λ is the upper bound of the Dirichlet exponents and w_0 is an arbitrary complex number, then the function

$$g(z) = [f(z) - w_0]e^{-\Lambda z}$$

satisfies the conditions of Lemma 3.2; we refer in particular to the conditions required in the last part. Thus, $g(z)$ vanishes in any half-plane Re $z > x_1 > a$.

Since $e^{-\Lambda z} \neq 0$, we obtain the assertion of the theorem.

Bibliographical notes

The analytic almost periodic functions were introduced by H. Bohr (106), who established their principal properties. Important results were obtained by A. S. Besicovitch (44, 46) and B. Jessen (364). One should especially mention B. Jessen and H. Tornehave's paper (372) which contains results regarding the distribution of values of analytic almost periodic functions. It is an interesting fact that H. Bohr (96) arrived at the concept of almost periodicity in connection with the study of properties of the Dirichlet series of analytic functions.

IV

Almost periodic solutions of ordinary differential equations

The problem of almost periodicity of solutions of ordinary differential equations has been investigated by many authors, and several interesting results can be found in monographs dedicated to the theory of differential equations. This chapter is only an introduction to this wide subject.

The results which we shall give below will be divided into two categories: some of them are almost periodicity criteria for bounded solutions (the existence of which is assumed) and others ensure the existence of almost periodic solutions under corresponding conditions.

In order to make the reading of this chapter independent of other expositions, we include some special results from algebra and from analysis, and also their proofs.

1. The primitive of an almost periodic function

Before passing to the study of almost periodic solutions of differential equations, let us find *the conditions under which the primitive of an almost periodic function is almost periodic.*

As we mentioned in Chapter I, §1, the primitive of an almost periodic function can be unbounded on the real line. In such a case it is clear that the primitive cannot be almost periodic. The complete answer to the above question can be stated as follows:

Theorem 4.1. *The primitive of an almost periodic function is almost periodic if and only if it is bounded on the real line.*

Proof. First, we note that we can confine ourselves to the real-valued functions. Indeed, if $f(x) = f_1(x) + if_2(x)$ is almost periodic, then $f_1(x)$ and $f_2(x)$ are almost periodic and conversely. Setting $F(x) = \int_0^x f(t)dt$, $F_1(x) = \int_0^x f_1(t)dt$, and $F_2(x) = \int_0^x f_2(t)dt$, we have $F(x) = F_1(x) + iF_2(x)$. Consequently, $F(x)$ is bounded if and only if $F_1(x)$ and $F_2(x)$ are bounded (on the real line).

Thus, let $f(x)$ be an almost periodic real-valued function such that

$$(4.1) \qquad m \leqslant F(x) \leqslant M, \qquad -\infty < x < +\infty,$$

where $F(x)$ has the same meaning as above. If $\varepsilon > 0$, then there exist two real numbers x_1 and x_2 such that

(4.2) $$F(x_1) < m + \frac{\varepsilon}{6}, \qquad F(x_2) > M - \frac{\varepsilon}{6}.$$

Set $d = |x_1 - x_2|$ and let $l_1 > 0$ be a number with the property that any interval of the real line of length l_1 contains an $(\varepsilon/6d)$-translation number of the function $f(x)$. If we set $l = l_1 + d$, then any $(\varepsilon/2l)$-translation number of $f(x)$ is an ε-translation number of the function $F(x)$.

First, let us show that any interval of length l of the real line contains two points z_1 and z_2, for which

(4.3) $$F(z_1) < m + \frac{\varepsilon}{2}, \qquad F(z_2) > M - \frac{\varepsilon}{2}.$$

Indeed, if $\xi = \min\{x_1, x_2\}$ and τ is an $(\varepsilon/6d)$-translation number of $f(x)$ such that $\xi + \tau \in (\alpha, \alpha + l_1)$, then the numbers $z_1 = x_1 + \tau$ and $z_2 = x_2 + \tau$ belong to the interval $(\alpha, \alpha + l)$.

We have

$$F(z_2) - F(z_1) = F(x_2) - F(x_1) + \int_{z_1}^{z_2} f(t)\, dt - \int_{x_1}^{x_2} f(t)\, dt$$

$$= F(x_2) - F(x_1) + \int_{x_1}^{x_2} \{f(t + \tau) - f(t)\}\, dt$$

$$\geqslant F(x_2) - F(x_1) - d\,\frac{\varepsilon}{6d} > M - m - \frac{2\varepsilon}{6} - \frac{\varepsilon}{6} = M - m - \frac{\varepsilon}{2}.$$

This inequality holds only if inequalities (4.3) hold.

Consider now an $(\varepsilon/2l)$-translation number of the function $f(x)$, say η. Let us show that

(4.4) $$|F(x + \eta) - F(x)| < \varepsilon, \qquad -\infty < x < +\infty.$$

Indeed, since x is a fixed real number, we choose a number z_1 in the interval $(x, x + l)$ with the property (4.3). We obtain

$$F(x + \eta) - F(x) = F(z_1 + \eta) - F(z_1) + \int_{x}^{x+\eta} f(t)\, dt - \int_{z_1}^{z_1+\eta} f(t)\, dt$$

$$= F(z_1 + \eta) - F(z_1) + \int_{x}^{z_1} f(t)\, dt - \int_{x+\eta}^{z_1+\eta} f(t)\, dt$$

$$\geqslant m - \left(m + \frac{\varepsilon}{2}\right) - \left|\int_{x}^{z_1} \{f(t + \eta) - f(t)\}\, dt\right| > -\frac{\varepsilon}{2} - l \cdot \frac{\varepsilon}{2l}$$

$$= -\varepsilon.$$

Analogously we also obtain the inequality

$$F(x + \eta) - F(x) < \varepsilon,$$

from which follows (4.4), where x is arbitrary.

Thus, Theorem 4.1 has been proved.

Remark. An interesting boundedness criterion for the primitive of an almost periodic function can be obtained making use of the Fourier series.

One can show that the primitive of an almost periodic function is bounded, if there exists a number $\lambda > 0$ such that $|\lambda_k| > \lambda$, where λ_k are the Fourier exponents of the given function.

2. Linear systems with constant coefficients

In this paragraph we shall consider systems of the form

(4.5) $$\frac{dy_i}{dx} = \sum_{j=1}^{n} a_{ij} y_j + f_i(x), \qquad i = 1, 2, \ldots, n,$$

where a_{ij} are complex numbers and $f_i(x)$ are almost periodic (complex-valued) functions.

Arguments similar to those that follow can be made in the case of systems of the same form, where a_{ij} and $f_i(x)$ are real.

A solution of system (4.5) will be called almost periodic, if all its components $y_1, y_2, \ldots, y_n$ are almost periodic functions of x.

Of course, to find almost periodic solutions of systems such as (4.5) we must look among their bounded solutions. Since the systems of the form (4.5) have in general unbounded solutions, one cannot expect all the solutions to be almost periodic. We shall show however that the bounded solutions, if such solutions exist, are almost periodic.

We need

Lemma 4.1. *Given an arbitrary square matrix* $A = (a_{ij})$, *there exists a matrix* $\alpha = (\alpha_{ij})$ *of the same order, such that*

1) $\det \alpha \neq 0$;

2) $\alpha^{-1} A \alpha$ *is triangular, i.e.*

(4.6) $$\alpha^{-1} A \alpha = \begin{pmatrix} \lambda_1 & b_{12} & \cdots & b_{1n} \\ 0 & \lambda_2 & \cdots & b_{2n} \\ \cdot & \cdot & \cdots & \cdot \\ 0 & 0 & \cdots & \lambda_n \end{pmatrix},$$

where $\lambda_1, \ldots, \lambda_n$ *are the eigenvalues of* A.

This can be proved by induction on the order of the matrices.

First, let us prove the lemma for square matrices of second order. Equality (4.6) can in this case be written as

(4.7)
$$\begin{pmatrix} a_{11} & a_{12} \\ a_{21} & a_{22} \end{pmatrix} \cdot \begin{pmatrix} \alpha_{11} & \alpha_{12} \\ \alpha_{21} & \alpha_{22} \end{pmatrix} = \begin{pmatrix} \alpha_{11} & \alpha_{12} \\ \alpha_{21} & \alpha_{22} \end{pmatrix} \cdot \begin{pmatrix} \lambda_1 & b_{12} \\ 0 & \lambda_2 \end{pmatrix}.$$

The elements a_{ij} in (4.7) are given and all the other quantities must be determined. From (4.7) we obtain

$$a_{11}\alpha_{11} + a_{12}\alpha_{21} = \lambda_1\alpha_{11},$$

$$a_{21}\alpha_{11} + a_{22}\alpha_{21} = \lambda_1\alpha_{21},$$

or

$$(a_{11} - \lambda_1)\alpha_{11} + a_{12}\alpha_{21} = 0,$$

$$a_{21}\alpha_{11} + (a_{22} - \lambda_1)\alpha_{21} = 0.$$

These equations show that λ_1 must be an eigenvalue of the matrix A, and the vector (α_{i1}) must be an eigenvector of the matrix A corresponding to λ_1. We then choose the vector (α_{i2}) such that $\det \alpha \neq 0$. Consequently,

$$A = \begin{pmatrix} \lambda_1 & b_{12} \\ 0 & b_{22} \end{pmatrix}.$$

Since A and $\alpha^{-1}A\alpha$ have the same eigenvalues, b_{22} must be equal to λ_2, the second eigenvalue of A (which may coincide with λ_1).

Let us assume now that the lemma holds for square matrices of order n, and let us prove that it remains valid for square matrices of order $n + 1$. Let A be a square matrix of order $n + 1$, and let λ_1 be a nonzero eigenvalue of this matrix (if A does not have nonzero eigenvalues, then A is the zero matrix and therefore it is triangular). We shall denote by (α_{i1}) the eigenvector of the matrix A which corresponds to λ_1. Consider a nonsingular matrix α_1 of order $n + 1$, the first column of which consists of the vector (α_{i1}) itself. We shall have

(4.8)
$$\alpha_1^{-1}A\alpha_1 = \begin{pmatrix} \lambda_1 & b_{12} & \cdots & b_{1,n+1} \\ 0 & & & \\ \vdots & & B_n & \\ \vdots & & & \\ 0 & & & \end{pmatrix},$$

B_n being a square matrix of order n. We see that the characteristic equation of the matrix on the right side of formula (4.8) is

(4.9)
$$(\lambda_1 - \lambda)\det(B_n - \lambda E) = 0,$$

where E represents the identity matrix. Equation (4.9) must have the same roots as the characteristic equation of the matrix A, and thus, the eigenvalues of B_n will be $\lambda_2, \ldots, \lambda_{n+1}$. By the induction hypothesis we can assert that there exists a matrix α_n of order n, det $\alpha_n \neq 0$, such that

$$\alpha_n{}^{-1}B_n\alpha_n = \begin{pmatrix} \lambda_2 & c_{12} & \cdots & c_{1n} \\ 0 & \lambda_3 & \cdots & c_{2n} \\ \cdots\cdots\cdots\cdots\cdots\cdots \\ 0 & 0 & \cdots & \lambda_{n+1} \end{pmatrix}.$$

Let us consider the nonsingular matrix

$$\alpha_{n+1} = \begin{pmatrix} 1 & 0 \\ 0 & \alpha_n \end{pmatrix} = \begin{pmatrix} 1 & 0 & \cdots & 0 \\ 0 & & & \\ \vdots & & \alpha_n & \\ 0 & & & \end{pmatrix}.$$

If we set $\alpha = \alpha_1\alpha_{n+1}$, we obtain

$$\alpha^{-1}A\alpha = \alpha_{n+1}^{-1}(\alpha_1{}^{-1}A\alpha_1)\alpha_{n+1} = \begin{pmatrix} \lambda_1 & b_{12} & \cdots & b_{1,n+1} \\ 0 & \lambda_2 & \cdots & b_{2,n+1} \\ \cdots\cdots\cdots\cdots\cdots\cdots \\ 0 & 0 & \cdots & \lambda_{n+1} \end{pmatrix}.$$

which proves Lemma 4.1.

We can now prove the result mentioned at the beginning of this paragraph.

Theorem 4.2. *If the functions $f_i(x)$, $i = 1, 2, \ldots, n$, are almost periodic, then any bounded solution (on the whole real line) of the system (4.5) is almost periodic.*

Proof. Let us first observe that one can assume without loss of generality that A is a triangular matrix. Indeed, if A were not triangular, then we would introduce some unknown functions by the change of variables

(4.10) $$y_i = \sum_{j=1}^{n} \alpha_{ij} z_j,$$

the matrix $\alpha = (\alpha_{ij})$ being chosen so that $\alpha^{-1}A\alpha$ is triangular. Differentiating both sides of (4.10), substituting in (4.5), and then solving with respect to dz_i/dx, we obtain a system of the form

$$\frac{dz_1}{dx} = \lambda_1 z_1 + b_{12} z_2 + \cdots + b_{1n} z_n + f_1^*(x),$$

$$\frac{dz_2}{dx} = \qquad \lambda_2 z_2 + \cdots + b_{2n} z_n + f_2^*(x),$$

(4.11)

$$. \quad . \quad . \quad . \quad . \quad . \quad . \quad . \quad . \quad . \quad . \quad . \quad . \quad . \quad .$$

$$\frac{dz_n}{dx} = \qquad \lambda_n z_n + f_n^*(x),$$

where the functions $f_i^*(x)$ are uniquely determined by the relations

(4.12) $$\sum_{j=1}^{n} \alpha_{ij} f_j^*(x) = f_i(x), \qquad i = 1, 2, \ldots, n.$$

Furthermore, if $y_1, \ldots, y_n$ is a bounded solution of system (4.5), then the corresponding solution $z_1, \ldots, z_n$ of system (4.11) is also bounded and conversely. This follows immediately from (4.10) and from the fact that $\alpha = (\alpha_{ij})$ has an inverse [so that we can apply Cramer's rule to solve system (4.10) with respect to z]. Finally, (4.12) shows that $f_i^*(x)$ are almost periodic, if and only if $f_i(x)$ are almost periodic.

Let us consider the equation

(4.13) $$\frac{dz}{dx} = \lambda z + f(x),$$

where λ is a complex number and $f(x)$ is an almost periodic function; let us show that any bounded solution of such an equation is almost periodic. Consider $\lambda = \mu + iv$. We distinguish the following cases: 1) $\mu > 0$; 2) $\mu < 0$; 3) $\mu = 0$.

The general solution of equation (4.13) is

(4.14) $$z(x) = e^{\lambda x} \left[C + \int_0^x e^{-\lambda t} f(t)\, dt \right],$$

where C is an arbitrary constant.

In case 1), $|e^{\lambda x}| = e^{\mu x} \to +\infty$ as $x \to +\infty$. For $z(x)$ to be bounded on the real line (since only such solutions are of interest for us), one must have $C + \int_0^x e^{-\lambda t} f(t)\, dt \to 0$ as $x \to +\infty$. This means that we must take in (4.14)

(4.15) $$C = -\int_0^\infty e^{-\lambda t} f(t)\, dt.$$

We note that the improper integral in (4.15) is convergent, since

$$|e^{-\lambda t}f(t)| \leqslant \sup|f(t)|\,e^{-\mu t}, \qquad t \geqslant 0.$$

Thus, the unique bounded solution of equation (4.13) could be

(4.16)
$$z_0(x) = -\int_x^\infty e^{\lambda(x-t)}f(t)\,dt.$$

We have however

(4.17)
$$|z_0(x)| \leqslant Me^{\mu x}\int_x^\infty e^{-\mu t}\,dt = \frac{M}{\mu},$$

where $M = \sup|f(x)|$, $-\infty < x < +\infty$. From (4.17) it follows that $z_0(x)$ is bounded. We then have

$$|z_0(x+\xi) - z_0(x)| \leqslant \frac{1}{\mu}\sup_x|f(x+\xi) - f(x)|, \qquad -\infty < x < +\infty.$$

This inequality proves that any $\varepsilon\mu$-translation number of $f(x)$ is an ε-translation number of $f(x)$.

In case 2 one proceeds analogously. One sees that

(4.18)
$$z_1(x) = \int_{-\infty}^x e^{\lambda(x-t)}f(t)\,dt$$

is the unique bounded solution of equation (4.13). We have

(4.19)
$$|z_1(x)| \leqslant \frac{M}{|\mu|}$$

and as in the previous case, any $\varepsilon|\mu|$-translation number of $f(x)$ is an ε-translation number of $z_1(x)$.

Finally, let us consider case 3. We have

(4.20)
$$z(x) = e^{ivx}\left[C + \int_0^x e^{-ivt}f(t)\,dt\right].$$

From (4.20) it follows that $z(x)$ is bounded if and only if

$$\int_0^x e^{-ivt}f(t)\,dt$$

is bounded on the real line. Since the function under the integral sign is almost periodic, Theorem 4.1 implies that this indefinite integral is almost periodic, if it is bounded. Thus, as a consequence of (4.20), any bounded solution (and therefore, any solution) is also almost periodic.

The assertion of the theorem is proved for equations of the form (4.13).

Let us return now to system (4.11) and assume that $z_1, \ldots, z_n$ is a bounded solution of this system [i.e., every function $z_i(x)$ is bounded]. From the last equation of a system of the form (4.13) it follows that $z_n(x)$ is almost periodic. We substitute now $z_n(x)$ obtained from the last equation in the equation preceding the last, which becomes therefore of the form (4.13). From the above considerations it follows that $z_{n-1}(x)$ is almost periodic and so on. Thus, applying n times successively the assertion proved above for equations of the form (4.13), it will follow that the bounded solution considered is almost periodic.

Theorem 4.2 is now completely proved.

We must note however that Theorem 4.2 asserts the almost periodicity of any bounded solution under the hypothesis that such a solution exists. It indicates no condition which will ensure the existence of bounded solutions. We shall prove a theorem which will specify when a system of the form (4.5) possesses bounded solutions.

Theorem 4.3. *If the functions* $f_i(x)$, $i = 1, 2, \ldots, n$, *are almost periodic and the matrix* $A = (a_{ij})$ *has no eigenvalues with real part zero, then the system* (4.5) *admits a unique almost periodic solution.*

If $y_1(x), \ldots, y_n(x)$ *is the almost periodic solution, and*

$$M = \max_i \{\sup |f_i(x)|\}, \quad -\infty < x < +\infty,$$

then

(4.21) $|y_i(x)| \leqslant KM, \quad -\infty < x < +\infty, \quad i = 1, 2, \ldots, n,$

where K *is a positive constant depending only on the matrix* A.

Proof. An argument used in the proof of Theorem 4.2 allows us to confine our attention, regarding the uniqueness and the existence of an almost periodic solution only to systems of the form (4.11). It will be sufficient to show that a system of the form (4.11), where Re $\lambda_i \neq 0$, $i = 1, 2, \ldots, n$, has a unique bounded solution on the whole real line. Indeed, the last equation of this system shows [see cases 1 and 2 in the proof of Theorem 4.2] that for any $z_n(x)$ there exists only one possibility, namely

(4.22) $z_n(x) = -\displaystyle\int_x^{\infty} e^{\lambda_n(x-t)} f_n^*(t)\, dt,$

if Re $\lambda_n > 0$ and

(4.23) $z_n(x) = \displaystyle\int_{-\infty}^{x} e^{\lambda_n(x-t)} f_n^*(t)\, dt,$

if Re $\lambda_n < 0$. Substituting the value for $z_n(x)$ obtained in this way in the equation second to the last, we get for $z_{n-1}(x)$ an equation of the form (4.13). Since Re $\lambda_{n-1} \neq 0$, it follows that for $z_{n-1}(x)$ we have a unique possibility and so on. We have proved therefore the first assertion of Theorem 4.3.

Let us prove now that a bounded solution satisfies an inequality such as (4.21). One should note that in this case, too, it suffices to consider only systems of the form (4.11). If for the systems of the form (4.11) an inequality of the type (4.21) holds, from the substitution (4.10) it follows that for the system (4.5) such an estimation is valid.

Let $\mu > 0$ so that

(4.24)
$$|\text{Re } \lambda_i| \geqslant \mu, \qquad i = 1, 2, \ldots, n.$$

From (4.22) or from (4.23) it follows that

(4.25)
$$|z_n(x)| \leqslant \frac{M}{\mu} = K_n M.$$

Let us consider the equation which gives us $z_{n-1}(x)$:

$$\frac{dz_{n-1}}{dx} = \lambda_{n-1} z_{n-1} + b_{n-1,n} z_n + f_{n-1}^*.$$

Expressing $z_{n-1}(x)$ by a formula of the form (4.22) or (4.23) and estimating the integral again, we obtain

(4.26)
$$|z_{n-1}(x)| \leqslant \frac{1}{\mu} \left(|b_{n-1,n}| \frac{M}{\mu} + M \right) = K_{n-1} M.$$

Proceeding further in the same manner one obtains

(4.27)
$$|z_i(x)| \leqslant K_i M, \qquad i = n - 2, n - 3, \ldots, 1.$$

If we set $K = \max\{K_i\}$, $i = 1, 2, \ldots, n$, then from (4.25), (4.26) and (4.27) will follow for every real x

(4.28)
$$|z_i(x)| \leqslant KM, \qquad i = 1, 2, \ldots, n.$$

Inequality (4.28) is nothing else but the inequality which appears in the statement in the case of systems of the form (4.11). Thus Theorem 4.3 is proved.

Remark 1. It follows from the proof that the constant K might depend on the transformation (4.10): i.e., on the matrix $\alpha = (\alpha_{ij})$ for which $\alpha^{-1} A \alpha$ has a triangular form. Once we fix the matrix α, K depends only on A. It is important that it no longer depends on $f_i(x)$.

Remark 2. Theorem 4.3 is not valid if the condition Re $\lambda_i \neq 0$ is not satisfied. More precisely, let us show by an example that Theorem 4.3 is not true, if there exists at least one eigenvalue λ_k for which Re $\lambda_k = 0$. Without loss of generality one can consider that $\lambda_k = \lambda_n = iv$, where v is real. The last equation of system (4.11) can then be written

$$(4.29) \qquad \frac{dz_n}{dx} = ivz_n + f_n^*(x).$$

Since $f_n^*(x)$ is an arbitrary almost periodic function, we shall take

$$f_n^*(x) = e^{ivx}.$$

One sees that the general integral of equation (4.29) is

$$z_n(x) = (x + C)e^{ivx}.$$

Consequently the equation in z_n has no bounded solution on the whole real line.

Before concluding this paragraph let us consider an example.

In the theory of oscillations one frequently encounters equations such as

$$(4.30) \qquad \frac{d^2y}{dx^2} + 2A\frac{dy}{dx} + By = f(x),$$

where $f(x)$ is an almost periodic function. Most frequently $f(x)$ is a sine function or a finite sum of such functions. If the ratios of the periods are rational numbers, then $f(x)$ is periodic; in the contrary case $f(x)$ is almost periodic.

The linear system which is equivalent to equation (4.30) is

$$\frac{dy}{dx} = z,$$

$$\frac{dz}{dx} = -By - 2Az + f(x).$$

The characteristic roots are $-A \pm \sqrt{A^2 - B}$. If we assume A and B real, $AB \neq 0$, then the real parts of the characteristic roots are distinct from zero. Indeed, the real part is $-A$ for both roots if $A^2 - B \leqslant 0$; if $A^2 - B > 0$, then the roots are real, distinct and none of them is zero. Thus, by Theorem 4.3, equation (4.30) has a unique almost periodic solution. In the case $A = 0$, $B \neq 0$ any bounded solution of equation (4.30) will be almost periodic as follows from Theorem 4.2. This last case includes equations of the form

$$\frac{d^2y}{dx^2} + \omega^2 y = f(x).$$

The above considerations are entirely valid if $f(x)$ is real-valued. The mentioned solution is also real.

3. *Quasilinear systems*

In this paragraph we shall apply the method of successive approximations in order to prove the existence and the uniqueness of the almost
periodic solutions for systems of the form

(4.31) $$\frac{dy_i}{dx} = \sum_{j=1}^{n} a_{ij} y_i + f_i(x) + \mu g_i(x, y_1, \ldots, y_n),$$

where μ is a small parameter, and the functions $f_i(x)$ and $g_i(x, y_1, \ldots, y_n)$,
$i = 1, 2, \ldots, n$, satisfy certain conditions which will be specified further.
Let us first make a few remarks.

To begin with, we could have considered the functions g_i of the form
$g_i(x, y_1, \ldots, y_n, \mu)$. The considerations which follow are also valid for such
functions. We shall consider only the functions g_i which are independent
of μ in order to simplify the statements and the proofs.

Moreover, we shall consider the case in which the linear system corresponding to $\mu = 0$ (called the *generating system*) satisfies the conditions
of Theorem 4.3. As we have already seen, the conditions of Theorem 4.3
ensure the existence of a unique almost periodic solution for the generating
system. The case when the generating system has a family of almost
periodic solutions is more complicated and will be omitted in our exposition.

Let us state the principal result of this paragraph:

Theorem 4.4. *We assume that system* (4.31) *satisfies the following conditions:*

1) *the matrix $A = (a_{ij})$ has no eigenvalues with the real part zero;*
2) *the functions $f_i(x)$, $i = 1, \ldots, n$, are almost periodic;*
3) *if we denote by $y_i^{(0)}(x)$, $i = 1, 2, \ldots, n$, the bounded solution of the
generating system, then the functions $g_i(x, y_1, \ldots, y_n)$, $i = 1, 2, \ldots, n$, are
continuous on the set*

(4.32) $(\bar{\Delta})$: $|y_i - y_i^{(0)}(x)| \leqslant a$, $-\infty < x < +\infty$,

and almost periodic in x, uniformly with respect to $y_1, y_2, \ldots, y_n$ satisfying
(4.32), *and so that*

(4.33) $|g_i(x, y_1, \ldots, y_n) - g_i(x, z_1, \ldots, z_n)|$

$$\leqslant L \sum_{j=1}^{n} |y_j - z_j|, \, i = 1, 2, \ldots, n,$$

in $\bar{\Delta}$; L is a positive constant.

Then there exists a unique almost periodic solution of system (4.31) *whose graph belongs to* $\bar{\Delta}$ *if the parameter* μ *is sufficiently small. This solution approaches the almost periodic solution of the generating system, uniformly on the real line, as* $\mu \to 0$.

Proof. First, we note that from Theorem 2.2 it follows that the functions $g_i(x, y_1, y_2, \ldots, y_n)$ are bounded on $\bar{\Delta}$:

$$|g_i(x, y_1, \ldots, y_n)| \leqslant M, \qquad i = 1, 2, \ldots, n,$$

$$(x, y_1, \ldots, y_n) \in \bar{\Delta}.$$

Let us now construct the sequence of successive approximations, starting from the solution of the generating system $y_i^{(0)}(x)$, and taking as $y_i^{(k)}$ the bounded solution of the system

(4.34) $$\frac{dy_i^{(k)}}{dx} = \sum_{j=1}^{n} a_{ij} y_j^{(k)} + f_i(x) + \mu g_i(x, y_1^{(k-1)}, \ldots, y_n^{(k-1)}),$$

$$i = 1, 2, \ldots, n, \qquad k = 1, 2, 3, \ldots$$

The superscript k indicates the rank in the process of successive approximations. As one can see, in each step of successive approximations we must integrate a system of the form (4.5), since in g_i appear components of the approximation of the immediately lower order. Let us show first that this process makes sense, i.e., the approximations defined by (4.34) belong to Δ if μ is sufficiently small. As we said, the first approximation $y_i^{(0)}(x)$ is the solution of the generating system and belongs by hypothesis to $\bar{\Delta}$. Let us assume that the approximation of the order $k - 1$ has its graph in $\bar{\Delta}$. It follows from (4.34) that

$$\frac{d[y_i - y_i^{(0)}]}{dx} = \sum_{j=1}^{n} a_{ij} [y_j^{(k)} - y_j^{(0)}] + \mu g_i(x, y_1^{(k-1)}, \ldots, y_n^{(k-1)}).$$

Let K be the constant whose existence was shown in Theorem 4.3. According to (4.21) we have

(4.35) $$|y_i^{(k)}(x) - y_i^{(0)}(x)| \leqslant |\mu| \cdot KM,$$

where M has the same meaning as above. Therefore, $y_i^{(k)}$ will belong to $\bar{\Delta}$ if

(4.36) $$|\mu| \leqslant \frac{a}{KM}.$$

This is a first restriction for μ.

Let us also note that all the approximations are almost periodic. Indeed, $y_i^{(0)}(x)$ is almost periodic by hypothesis. According to Theorem 2.8, $g_i(x, y_1^{(0)}(x), \ldots, y_n^{(0)}(x))$ are almost periodic. Applying Theorem 4.3 we obtain as a result that $y_i^{(1)}(x)$ is almost periodic. By induction one sees easily that all the approximations $y_i^{(k)}(x)$ are almost periodic.

It remains only to show that the sequences $\{y_i^{(k)}(x)\}$, $i = 1, 2, \ldots, n$, are uniformly convergent on the real line, if μ is sufficiently small. From (4.34) it follows that the differences $y_i^{(k+1)}(x) - y_i^{(k)}(x)$ satisfy the following system of equations:

$$\frac{d[y_i^{(k+1)} - y_i^{(k)}]}{dx} = \sum_{j=1}^{n} a_{ij}[y_j^{(k+1)} - y_j^{(k)}]$$

$$+ \mu[g_i(x, y_1^{(k)}, \ldots, y_n^{(k)}) - g_i(x, y_1^{(k-1)}, \ldots, y_n^{(k-1)})].$$

Applying now inequality (4.21), considering that g_i satisfies (4.33), we obtain

$$|y_i^{(k+1)}(x) - y_i^{(k)}(x)| \leq |\mu|KL \sum_{j=1}^{n} \sup_x |y_j^{(k)}(x) - y_j^{(k-1)}(x)|,$$

$$-\infty < x < +\infty, \qquad i = 1, 2, \ldots, n, \qquad k = 1, 2, \ldots$$

Let us denote by $c_k = \max_i \{\sup_x |y_j^{(k+1)}(x) - y_j^{(k)}(x)|\}$, $-\infty < x < +\infty$. We obtain

(4.37) $$c_k \leq |\mu| \, nKLc_{k-1}, \qquad k = 1, 2, \ldots$$

We have seen above [inequality (4.35) for $k = 1$] that

(4.38) $$c_0 \leq |\mu| \, nKM.$$

From (4.37) and (4.38) it follows that

(4.39) $$c_k \leq \frac{M}{L}(|\mu| \, nKL)^{k+1}, \ k = 0, 1, 2, \ldots$$

Let us assume now that $|\mu| \, nKL < 1$, i.e.

(4.40) $$|\mu| < \frac{1}{nKL}.$$

According to (4.36) and to (4.40) it follows that we must take

(4.41) $$|\mu| < \min\left\{\frac{a}{KM}, \frac{1}{nKL}\right\}.$$

In this case the process of successive approximations makes sense and is convergent, since the series

(4.42) $$\sum_{k=0}^{\infty} c_k,$$

which is a majorizing series for each of the function series

$$\sum_{k=0}^{\infty} |y_i^{(k+1)}(x) - y_i^{(k)}(x)|, \qquad i = 1, 2, \ldots, n,$$

is convergent. Thus setting

$$y_i(x) = \lim_{k \to +\infty} y_i^{(k)}(x), \qquad i = 1, 2, \ldots, n,$$

from (4.34) it follows that $y_1(x), \ldots, y_n(x)$ is a solution of system (4.31). It obviously belongs to $\bar{\Delta}$.

We must show now the uniqueness of this almost periodic solution. If this solution were not unique, then one could find another almost periodic solution $z_1(x), \ldots, z_n(x)$ of system (4.31). Writing down that $z_1(x), \ldots, z_n(x)$ satisfies system (4.31) and using (4.34), we obtain

$$\frac{d[z_i - y_i^{(k)}]}{dx} = \sum_{j=1}^{n} a_{ij}[z_j - y_j^{(k)}]$$
$$+ \mu[g_i(x, z_1, \ldots, z_n) - g_i(x, y_1^{(k-1)}, \ldots, y_n^{(k-1)})].$$

Applying (4.21) one obtains

$$|z_i(x) - y_i^{(k)}(x)| \leqslant |\mu| \, KL \sum_{j=1}^{n} \sup_{x} |z_j(x) - y_j^{(k-1)}(x)|,$$

$$-\infty < x < +\infty, \quad i = 1, 2, \ldots, n, \quad k = 1, 2, \ldots$$

If we set $d_k = \max_{i}\{\sup_{x} |z_i(x) - y_i^{(k)}(x)|\}$, $-\infty < x < +\infty$, we obtain

(4.43) $$d_k \leqslant |\mu| \, nKL \, d_{k-1}, \qquad k = 1, 2, \ldots$$

It follows from (4.43) that

$$d_k \leqslant (|\mu| \, nKL)^k \, d_0, \qquad k = 1, 2, \ldots$$

and consequently $d_k \to 0$ for $k \to +\infty$. This means that

$$\lim_{k \to +\infty} y_i^{(k)}(x) = z_i(x), \qquad i = 1, 2, \ldots n,$$

which proves the uniqueness of the solution.

Finally, let us prove that the solution, the existence and the uniqueness of which have been proved, converges to the solution $y_i^{(0)}(x)$ of the generating system, uniformly on the real line. For this purpose we shall use inequalities (4.35). Since we proved the convergence of the sequence of successive approximations, we can pass to the limit in these inequalities. One obtains

$$(4.44) \qquad |y_i(x) - y_i^{(0)}(x)| \leqslant |\mu| \, KM, \qquad i = 1, 2, \ldots, n.$$

Inequalities (4.44) show that the solution of system (4.31) can differ as little as desired (on the whole real line) from the solution of the generating system, with the only condition that $|\mu|$ be sufficiently small. Thus we complete the proof of Theorem 4.4.

Returning to the example of the preceding paragraph, we can assert that the equation

$$\frac{d^2 y}{dx^2} + 2A \frac{dy}{dx} + By = f(x) + \mu g(x, y),$$

where A and B are real, and $AB \neq 0$, and $f(x)$, $g(x, y)$ are almost periodic functions in x, the last of which satisfies a Lipschitz condition with respect to y, has a unique almost periodic solution, for sufficiently small $|\mu|$.

4. An Amerio criterion for almost periodicity of bounded solutions

In this paragraph we shall consider nonlinear differential systems of the form

$$\frac{dy_i}{dx} = f_i(x, y_1, \ldots, y_m), \qquad i = 1, 2, \ldots, m,$$

or in vector notation

$$(4.45) \qquad \frac{dY}{dx} = f(x, Y).$$

In all the following statements we shall assume that the function $f(x, Y)$ satisfies the following conditions:

1) It is continuous on the open set $B \subset R^{m+1}$, where $B = R \times A$, $A \subset R^m$ being an open set, with values in R^m. As usual, R is the real line and R^m is the m-dimensional Euclidean space.

2) It is almost periodic in x for any $Y \in A$, uniformly with respect to $Y \in C$, for any compact set (bounded and closed) $C \subset A$.

If the compact set $C \subset A$ is given and $\{h_n\}$ is a sequence of real numbers, then from the sequence $\{f(x + h_n, Y)\}$ can be extracted a subsequence

$\{f(x + h_{1n}, Y)\}$ uniformly convergent on $R \times C$ to a function $g(x, Y)$ which is almost periodic in x, uniformly with respect to $Y \in C$.

Let us consider now a sequence of compact sets $\{C_n\}$, $C_n \subset A$, such that

$$C_n \subset C_{n+1}, \quad \bigcup_{n=1}^{\infty} C_n = A.$$

If $\{h_n\}$ is an arbitrary sequence of real numbers, then one can determine a subsequence $\{h_{1n}\}$ such that $\{f(x + h_{1n}, Y)\}$ is uniformly convergent on $R \times C_1$. One can extract a subsequence $\{h_{2n}\}$ from $\{h_{1n}\}$ such that $\{f(x + h_{2n}, Y)\}$ is uniformly convergent on $R \times C_2$ and so on. If one considers the diagonal sequence $\{h_{nn}\}$, then one sees without difficulty that $\{f(x + h_{nn}, Y)\}$ converges on B to a function $g(x, Y)$, uniformly on any set $R \times C$, where $C \subset A$ is compact. Consequently, $g(x, Y)$ is almost periodic in x, uniformly with respect to $Y \in C$, for any compact set $C \subset A$.

Below we shall denote by $g(x, Y)$ any function obtained by the above described procedure, i.e., such that one can determine a sequence of real numbers $\{k_n\}$ for which

(4.46) $$\lim_{n \to \infty} f(x + k_n, Y) = g(x, Y),$$

the convergence being uniform on any set $R \times C$, $C \subset A$ being compact.

Briefly, $g(x, Y)$ is a function of the closure of the family $\{f(x + h, Y)\}$, $h \in R$, with respect to the uniform convergence on any set of the form $R \times C$, where $C \subset A$ is compact.

We shall denote by $\mathscr{F}$ this family of vector functions. For any $g(x, Y) \in \mathscr{F}$ there exists a real sequence $\{k_n\}$ for which (4.46) is true. One sees easily that $\mathscr{F}$ is generated by each of its elements in the same way in which it is generated by $f(x, Y)$.

We consider equations (4.45) and equations

(4.47) $$\frac{dY}{dx} = g(x, Y), \qquad g \in \mathscr{F}.$$

As we shall see, there is a strong relation between (4.45) and each of equations (4.47).

Let us now define the notion of separated solution of a differential system of the form (4.45) in a set $D = R \times C$, where $C \subset A$ is compact. We shall consider solutions defined on the whole real line.

Let $Z(x)$ be a solution of system (4.45) such that $Z(x) \in C$, $x \in R$. This occurs if and only if the graph of this solution in R^{m+1} belongs to the set D. We shall say that $Z(x)$ is *separated* in D if one of the following cases occur:

α) $Z(x)$ is the unique solution of system (4.45), whose graph belongs to D.

β) If $Y(x)$ is another solution of system (4.45), whose graph belongs to D, one can determine a number $\rho > 0$, depending on $Y(x)$, such that

$$(4.48) \qquad\qquad |Z(x) - Y(x)| \geqslant \rho, \qquad x \in R.$$

Here $|Z(x) - Y(x)|$ represents the length of the vector between the bars.

Lemma 4.2. *If all the solutions of system* (4.45) *with graph in the set* D *are separated in this set, then their number is finite.*

Proof. Indeed, the set $\{Y(x)\}$ of solutions of the system (4.45) with their graphs in D is a set of uniformly bounded [since $Y(x) \in C$, $x \in R$] and equi-continuous functions. The equi-continuity is a consequence of the fact that $\{Y'(x)\}$ is also uniformly bounded since $Y'(x) = f(x, Y(x))$ and

$$Y(x') - Y(x) = \int_x^{x'} Y'(t)\, dt.$$

If there were infinitely many solutions with graphs in D, then one could extract from this set a sequence which converges uniformly on any finite interval to a vector function $Z(x)$. It is obvious that $Z(x)$ is also a solution of system (4.45) with graph in D, but nonseparated.

The contradiction which has been reached shows that the number of separated solutions cannot be infinite.

Lemma 4.3. *If system* (4.45) *has a solution* $Y(x)$, *such that* $Y(x) \in C$ *for* $x \in R$, *then each system* (4.47) *has a solution with graph in* D.

Proof. Suppose that equation (4.47) corresponds to the vector function $g(x, Y)$ for which (4.46) holds.

Consider the vector functions $Y_n(x) = Y(x + k_n)$. We note that they are the solutions of the equations $Y'_n = f(x + k_n, Y_n)$, and their graphs are in D. As above, one sees that $\{Y'_n(x)\}$ is a family of uniformly bounded and equi-continuous vector functions on R, and this permits us to assert that there is a subsequence $\{k_{1n}\}$ of $\{k_n\}$ such that $\{Y_{1n}(x)\}$ is uniformly convergent on any finite interval of the real line. Let $Z(x) = \lim_{n \to \infty} Y_{1n}(x)$. Consequently, $Z'(x) = g(x, Z(x))$, $Z(x) \in C$ for $x \in R$, which proves Lemma 4.3.

Theorem 4.5. *If system* (4.45) *has a solution* $Y(x)$, *such that* $Y(x) \in C$ *for* $x \geqslant x_0$, *then each system* (4.47) *has a solution defined for* $x \in R$, *with its graph in* D.

Proof. We shall consider the sequence $Y_n(x) = Y(x + n)$, $n = 1, 2, \ldots.$
Since $Y'_n = f(x + n, Y_n)$, it follows that $Y_n(x)$ is a solution of this equation
on the interval $[x_0 - n, +\infty)$. These solutions are uniformly bounded and
equi-continuous on any interval $[X, +\infty)$; of course, only those functions
$Y_n(x)$ for which $X_0 - n \leqslant X$ are defined on $[X, +\infty)$. There exists a sequence of natural numbers $\{n_r\}$ such that the sequences $\{Y_{n_r}(x)\}$ and
$\{f(x + n_r, Y)\}$ converge on R and on B, respectively, to the vector functions
$\overline{Y}(x)$ and $\bar{f}(x, Y)$ the convergence being uniform on any finite interval of
the real line, and on any $D \subset B$ respectively. Since $\overline{Y}'(x) = \bar{f}\{x, \overline{Y}(x)\}$, and
$\overline{Y}(x) \in C$ for $x \in R$, this means that there exists an element in $\mathscr{F}$ for which
the conditions of the preceding lemma are satisfied. Thus any equation
(4.47) has a solution defined on R with graph in D.

*Lemma 4.4. If for every system (4.47) the solutions with graph in D are
separated in D, then there exists a number $\sigma > 0$, independent of $g \in \mathscr{F}$, such
that for any pair of solutions $Y_1(x)$, $Y_2(x)$ of system (4.47) we have*

$$(4.49) \qquad |Y_1(x) - Y_2(x)| \geqslant \sigma, \qquad x \in R.$$

Proof. Let us consider the system (4.47) corresponding to a certain
$g(x, Y) \in \mathscr{F}$. There exists a largest number $\sigma > 0$, such that two arbitrary
solutions of (4.45) separated in D are situated at a distance $\geqslant \sigma$. It suffices
to show that any two separated solutions of system (4.47) are also situated
at a distance $\geqslant \sigma$ from each other.

As seen in Lemma 4.3, if $Z(x)$ is a solution of system (4.45) with graph
in D, then we can extract from the sequence $\{k_n\}$ appearing in the formula
(4.46) a subsequence $\{k_{1n}\}$ such that $\lim_{n \to \infty} Z(x + k_{1n}) = Y(x)$, $x \in R$,
where $Y(x)$ is a solution of system (4.47) with graph in D. If we shall
consider two distinct solutions $Z_1(x)$ and $Z_2(x)$ of system (4.45) separated
in D, then we shall have

$$\inf_{x \in R} |Z_1(x + k_n) - Z_2(x + k_n)| = \inf_{x \in R} |Z_1(x) - Z_2(x)| \geqslant \sigma.$$

Since from the sequence $\{k_n\}$ can be extracted a subsequence $\{k_{1n}\}$ such
that $\{Z_1(x + k_{1n})\}$ and $\{Z_2(x + k_{1n})\}$ are convergent to the solutions $Y_1(x)$
and $Y_2(x)$ of system (4.47), for these solutions we shall have

$$(4.50) \qquad \inf_{x \in R} |Y_1(x) - Y_2(x)| \geqslant \sigma.$$

Therefore $Y_1(x)$ and $Y_2(x)$ are distinct.

Consequently, to distinct solutions of the system (4.45) correspond
distinct solutions of system (4.47). This means that the number of solutions

separated in D of (4.47) is greater or equal to the number of similar separated solutions in D of (4.45). We noticed that the family $\mathscr{F}$ is generated by each of its elements in the same manner in which it is generated by $f(x, Y)$. Hence, we can say that the number of solutions of system (4.45) separated in D is greater or equal to the number of similar solutions of system (4.47). In other words, each system (4.47) has the same number of separated solutions in D; this permits us to assert that the above method leads to the determination of all the solutions of system (4.47) separated in D, starting from the system (4.45). From (4.50) we obtain Lemma 4.4.

Theorem 4.6. If each of the systems (4.47) *has its solutions separated in* D, *then all these solutions are almost periodic.*

Proof. Let $Y(x)$ be a solution of system (4.45), separated in D. It suffices to show that from any sequence $\{Y(x + h_n)\}$ can be extracted a subsequence which is uniformly convergent on the real line. As we remarked in Chapter II, §1 this ensures the almost periodicity of the vector function $Y(x)$.

According to Lemma 4.3 we can assert without loss of generality that $\{Y(x + h_n)\}$ converges to a function $Z(x)$, uniformly on any finite interval. Moreover, we shall have

$$(4.51) \qquad \lim_{n \to \infty} f(x + h_n, Y) = h(x, Y),$$

uniformly for $(x, Y) \in D$ and

$$(4.52) \qquad \frac{dZ}{dx} = h(x, Z).$$

We are to prove now that the sequence $\{Y(x + h_n)\}$ is uniformly convergent on the real line to $Z(x)$.

According to Lemma 4.4, there exists a number $\rho > 0$ such that given two solutions of system (4.47) $Y_1(x)$ and $Y_2(x)$ contained in D, we shall have $|Z_1(x) - Z_2(x)| \geqslant 2\rho$, $x \in R$.

Let us assume now that $\{Y(x + h_n)\}$ does not converge uniformly to $Z(x)$ on R. Let us denote

$$(4.53) \qquad \varphi_{n,p}(x) = |Y(x + h_n) - Y(x + h_p)|, \qquad n < p,$$

$$(4.54) \qquad I_{n,p} = \{x; x \in R, \varphi_{pn}, (x) \leqslant \rho\}.$$

An immediate consequence is that $\varphi_{n,p}(x)$ is continuous and $I_{n,p}$ is closed (possibly empty). We note that by the definition of convergence we have

$$\varphi_{n,p}(0) = |Y(h_n) - Y(h_p)| \leq \rho, \qquad N \leq n < p.$$

Hence, $I_{n,p}$ contains the point $x = 0$ if $N \leq n < p$. If we omit the first $N - 1$ terms of the sequence $\{Y(x + h_n)\}$, we can admit that the sets $I_{n,p}$ are nonvoid. Let

$$(4.55) \qquad \delta_{n,p} = \sup \varphi_{n,p}(x), \qquad x \in I_{n,p}.$$

It follows that $\delta_{n,p} \leq \rho$. We now claim that

$$(4.56) \qquad \lim_{(n,p) \to \infty} \delta_{n,p} = 0$$

cannot hold. Indeed, if (4.56) were true, then for $\varepsilon > 0$, $0 < \varepsilon < \rho$ there would exist an N_ε such that $N_\varepsilon \leq n < p$ implies $\delta_{n,p} < \varepsilon$. It would follow then that $\varphi_{n,p}(x) < \varepsilon$ for $x \in I_{n,p}$. By definition we have $\varphi_{n,p}(x) > \rho$ in the complementary set of $I_{n,p}$. Since $\varphi_{n,p}(x)$ is continuous on R, it would follow that $I_{n,p} = R$, i.e., $\{Y(x + h_n)\}$ converges uniformly on R. This contradicts the hypothesis and consequently,

$$(4.57) \qquad \limsup_{(n,p) \to \infty} \delta_{n,p} = 2\alpha > 0.$$

The relation (4.57) shows that there exist two sequences $\{n_r\}$ and $\{p_r\}$ such that $\delta_{n_r,p_r} \geq 3\alpha/2$. From (4.53), (4.54), and (4.57) it follows that there exists a sequence of real numbers $\{x_r\}$, such that

$$\varphi_{n_r, p_r}(x_r) \geq \alpha,$$

which permits us to write

$$(4.58) \qquad \alpha \leq |Y(x_r + h_{n_r}) - Y(x_r + h_{p_r})| \leq \rho.$$

Let us extract from the sequences $\{h_{n_r}\}$ and $\{h_{p_r}\}$ two subsequences $\{\lambda_s\}$ and $\{\mu_s\}$ corresponding to the values r_s of r, such that, letting $x'_s = x_{r_s}$, we obtain

$$(4.59) \qquad \lim_{s \to \infty} Y(x'_s + \lambda_s) = U, \qquad \lim_{s \to \infty} Y(x'_s + \mu_s) = V;$$

where U and V are two constant vectors, such that

$$(4.60) \qquad \alpha \leq |U - V| \leq \rho.$$

Let us consider now the sequences $\{Y(x + x'_s + \lambda_s)\}$ and $\{Y(x + x'_s + \mu_s)\}$. From these sequences one can extract two subsequences $\{Y(x + x'_{1s} + \lambda_{1s})\}$ and $\{Y(x + x'_{1s} + \mu_{1s})\}$, which converge on R (uniformly on any finite interval) to the solutions $Z_1(x)$ and $Z_2(x)$ of some equations from the class (4.47):

(4.61) $$\frac{dZ_1}{dx} = h_1(x, Z_1), \quad \frac{dZ_2}{dx} = h_2(x, Z_2),$$

where

(4.62) $$\begin{cases} \lim_{s \to \infty} f(x + x'_{1s} + \lambda_{1s}, Y) = h_1(x, Y), \\ \lim_{s \to \infty} f(x + x'_{1s} + \mu_{1s}, Y) = h_2(x, Y). \end{cases}$$

According to (4.59) and (4.60) we have

(4.63) $$\alpha \leqslant |Z_1(0) - Z_2(0)| \leqslant \rho.$$

Let us show that

(4.64) $$h_1(x, Y) = h_2(x, Y), \quad (x, Y) \in B.$$

Indeed, from (4.51) we obtain

$$\lim_{s \to \infty} f(x + \lambda_s, Y) = \lim_{s \to \infty} f(x + \mu_s, Y) = h(x, Y),$$

since $\{\lambda_s\}$ and $\{\mu_s\}$ are subsequences of $\{h_n\}$. The convergence is uniform on D. Consider $\varepsilon > 0$. For $s \geqslant S_1(\varepsilon)$ we shall have

(4.65) $$|f(x + \lambda_{1s}, Y) - f(x + \mu_{1s}, Y)| < \varepsilon, \quad (x, Y) \in D.$$

Then, for $s \geqslant S(\varepsilon) \geqslant S_1(\varepsilon)$ it will follow that

$$|h_1(x, Y) - h_2(x, Y)| \leqslant |h_1(x, Y) - f(x + x'_{1s} + \lambda_{1s}, Y)|$$
$$+ |f(x + x'_{1s} + \lambda_{1s}, Y) - f(x + x'_{1s} + \mu_{1s}, Y)|$$
$$+ |f(x + x'_{1s} + \mu_{1s}, Y) - h_2(x, Y)| < 3\varepsilon,$$

for any $(x, Y) \in D$, using (4.62) and (4.65). Relation (4.64) is thus established since $D \subset B$ is chosen arbitrarily. It follows that $Z_1(x)$ and $Z_2(x)$ are solutions of the same equation (4.47) and consequently,

$$|Z_1(x) - Z_2(x)| \geqslant 2\rho, \quad x \in R,$$

which is incompatible with (4.63).

This contradiction shows us that $\{Y(x + h_n)\}$ converges uniformly to $Z(x)$ on R, which means that $Y(x)$ is almost periodic.

Corollary. Consider the system

$$(4.66) \qquad \frac{dy_i}{dx} = \sum_{j=1}^{m} a_{ij}(x)y_j + f_i(x),$$

where $a_{ij}(x)$ and $f_i(x)$, $i, j = 1, 2, \ldots, m$, are almost periodic real-valued functions.

Let us consider all the sequences of real numbers $\{h_n\}$ for which $\{a_{ij}(x + h_n)\}$ converge uniformly on the real line. We shall denote by $b_{ij}(x)$, $i, j = 1, 2, \ldots, m$ the limit functions of sequences of the form $\{a_{ij}(x + h_n)\}$.

If each system of the form

$$(4.67) \qquad \frac{dy_i}{dx} = \sum_{j=1}^{m} b_{ij}(x)y_j$$

does not possess any other bounded solutions, except the trivial one, then any bounded solution of (4.66) *is almost periodic.*

We notice first that (4.66) cannot have more than one bounded solution. Indeed, the existence of two bounded solutions would imply the existence of a nontrivial solution of (4.67). According to Lemma 4.3, each of the systems

$$\frac{dy_i}{dx} = \sum_{j=1}^{m} b_{ij}(x)y_j + g_i(x),$$

where $g_i(x)$ are the limits of the sequences $\{f_i(x + h_n)\}$, $i = 1, 2, \ldots, m$, has a bounded solution. From the assumption made, it follows that each such solution is unique, i.e., it is separated (on a set of the form $D = R \times S$, S being a sufficiently large sphere in R^m).

Thus, the requirements of Theorem 4.6 are satisfied and this proves the corollary.

This result was established for the first time by J. Favard.

5.   The equation of nonlinear oscillations

We shall study in this paragraph the equation

$$(4.68) \qquad \frac{d^2x}{dt^2} + f(x)\frac{dx}{dt} + g(x) = ke(t),$$

where $e(t)$ is an almost periodic real function, satisfying the condition

$$(4.69) \qquad |e(t)| \leqslant 1,$$

and $k > 0$ is a real parameter. Let us assume for the time being that $f(x)$ and $g(x)$ are continuous for $x \in R$.

Let us set

$$F(x) = \int_0^x f(v)\, dv,$$

We shall consider the differential systems

(4.70) $$\frac{dx}{dt} = y - F(x), \qquad \frac{dy}{dt} = -g(x) + kp(t),$$

where $p(t)$ is any function from the closure of the set $\{e(t + h)\}$, $h \in R$, with respect to the uniform convergence on R.

If $p(t) = e(t)$, then system (4.70) is equivalent to equation (4.68).

We shall assume throughout this paragraph the following:

1) there exist some positive numbers a, b, c, d such that $a > b$, $c < d$, $g(c) = k$, $g(-b) = -k$;

2) $k < \min\{[F(d) - F(c)]f(x) + g(-a), [F(-b) - F(-a)]f(x) - g(d)\}$, where $-a \leqslant x \leqslant d$;

3) $g(x)$ has a continuous derivative, $g(0) = 0$, $0 < g'(x) \leqslant \beta$ for $x \neq 0$;

4) $f(x) \geqslant \alpha > 0$ for $x \in R$;

5) $\beta < \alpha^2$.

Let us consider a domain Ω in the plane (x, y) bounded by the following arcs of curves:

$$\Gamma_1 : y = F(x) + \beta_1, \qquad -a \leqslant x \leqslant c;$$
$$\Gamma_2 : y = F(d), \qquad c \leqslant x \leqslant d;$$
$$\Gamma_3 : x = d, \qquad F(d) - \beta_2 \leqslant y \leqslant F(d);$$
$$\Gamma_4 : y = F(x) - \beta_2, \qquad -b \leqslant x \leqslant d;$$
$$\Gamma_5 : y = F(-a). \qquad -a \leqslant x \leqslant -b;$$
$$\Gamma_5 : y = -a, \qquad F(-a) \leqslant y \leqslant F(-a) + \beta_1.$$

In the formulas above $\beta_1 = F(d) - F(c)$, $\beta_2 = F(-b) - F(-a)$, which permits us to assert that the six arcs form a simple Jordan curve.

Before we establish the existence and uniqueness of the almost periodic solution of equation (4.68) with graph in Ω we must prove

Lemma 4.5. *Let $m(t)$ be a continuous function on the real line, such that $0 < \alpha \leqslant m(t) \leqslant M$. Then the equation*

(4.71) $$\frac{dz}{dt} = z[m(t) - z]$$

has a solution which is definite on the real axis such that $\alpha \leqslant z(t) \leqslant M$.

Proof. Let $\{t_n\}$ be a sequence of numbers which converges to $-\infty$, and let $\{z_n\}$ be a sequence of numbers such that $\alpha \leqslant z_n \leqslant M$. Since equation (4.71) is of Bernoulli type, one can see without any difficulty that the functions

$$(4.72) \qquad z_n(t) = \frac{z_n \exp\left(\int_{t_n}^{t} m(\tau)\, dt\right)}{1 + z_n \int_{t_n}^{t} \exp\left(\int_{0}^{\tau} m(\theta)\, d\theta\right) d\tau}, \qquad n = 1, 2, \ldots,$$

are solutions of the considered equation, satisfying the inequalties

$$(4.73) \qquad \alpha \leqslant z_n(t) \leqslant M, \qquad t \geqslant t_n, \quad n = 1, 2, \ldots$$

It follows that the sequence of solutions $\{z_n(t)\}$ is uniformly bounded and equi-continuous on any half-line $[t_0, +\infty)$. Therefore, one can extract a subsequence which converges (uniformly on any finite interval) to a solution $z(t)$ of equation (4.71).

As a consequence of (4.73) we have $\alpha \leqslant z(t) \leqslant M$, which is the statement of Lemma 4.5.

Theorem 4.7. *If conditions* 1 *through* 5 *are satisfied, then equation* (4.68) *has a unique almost periodic solution, whose derivative is also almost periodic, so that* $\{x(t), x'(t)\} \in \Omega, t \in R$.

Proof. As will follow from the following considerations, each of the systems (4.70) possesses a unique almost periodic solution with graph in Ω.

Let us consider system (4.70) and show that any arbitrary solution of this system, which passes through a point of Ω at the time t_0, remains in Ω for all $t > t_0$. In other words, a solution which passes through a point of Ω cannot leave Ω (through a boundary point).

At an arbitrary point on Γ_1, the trajectory of system (4.70) has the slope

$$\frac{dy}{dx} = \frac{-g(x) + kp(t)}{\beta_1} \leqslant \frac{-g(-a) + k}{\beta_1}.$$

If we consider condition 2), it follows that

$$\frac{dy}{dx} < f(x) = F'(x).$$

Thus, the slope of this trajectory is less than the slope of Γ_1 at the same point. Since $dx/dt = \beta_1 > 0$ on Γ_1, the considered trajectory enters Ω, when t increases.

If we are situated at a point on Γ_2, such that $c < x < d$, then we have $dx/dt = -g(x) + kp(t) < -g(c) + k = 0$. Consequently, the trajectories of system (4.70) which cross this segment enter Ω if t increases. A similar argument can be applied to the points of Γ_3 for which $F(d) - \beta_2 \leqslant y < F(d)$ since at these points $dx/dt = y - F(x) < 0$ (i.e., x decreases).

At the point $(d, F(d))$ we have $dx/dt = 0$, $dy/dt < 0$. Hence, the trajectory which passes through this point is tangent to the straight line $x = d$. Since $y < F(x)$ under the curve $y = F(x)$, this means that at the points of Ω situated under the above-mentioned curve, $dx/dt = y - F(x) < 0$. Thus the trajectory will enter Ω when t increases.

At the point $(c, F(d))$, $dy/dt = -g(c) + kp(t) \leqslant 0$ as $dx/dt > 0$. As a result, the trajectory which passes through this point enters Ω or is tangent to the straight line $y = F(d)$. We note that around the point $(c, F(d))$ of Ω and to the right of $x = c$ [since $(dx/dt) > 0$ indicates that x increases together with t], dy/dt is negative. Therefore the trajectory will enter Ω in this case as well.

The other parts of the boundary of Ω can be discussed in the same manner. Since no new situations occur except those already investigated, we shall omit the other arguments.

Applying Theorem 4.5 it follows that each system (4.70) has a solution defined for $t \in R$, so that the graph belongs to Ω.

Let us show that each system (4.70) has only one solution with this property.

Indeed, let $(x(t), y(t))$ and $(\bar{x}(t), \bar{y}(t))$, $t \in R$, be two distinct solutions of system (4.70), with graphs in Ω. From the uniqueness theorem we obtain $(x(t), y(t)) \neq (\bar{x}(t), \bar{y}(t))$, for any $t \in R$.

If we set

$$u(t) = x(t) - \bar{x}(t), \qquad v(t) = y(t) - \bar{y}(t),$$

we obtain

(4.74) $$\frac{du}{dt} = v - m(\bar{x}, u)u, \qquad \frac{dv}{dt} = -h(\bar{x}, u)u,$$

where

(4.75) $$m(\bar{x}, u) = \begin{cases} \dfrac{F(\bar{x} + u) - F(\bar{x})}{u}, & u \neq 0, \\ f(\bar{x}), & u = 0, \end{cases}$$

(4.76) $$h(\bar{x}, u) = \begin{cases} \dfrac{g(\bar{x} + u) - g(\bar{x})}{u}, & u \neq 0, \\ g'(\bar{x}), & u = 0. \end{cases}$$

It follows immediately that $\alpha \leqslant m(\bar{x}, u) \leqslant M$, $t \in R$, M being a number for which $f(\bar{x}(t)) \leqslant M$, $t \in R$.

For $m(t) = m(\bar{x}, u)$, equation (4.71) has a solution $z(t)$ defined on the real line, such that $\alpha \leqslant z(t) \leqslant M$ (Lemma 4.5).

Consider

$$(4.77) \qquad D(t) = \{v^2 + (v - zu)^2\}^{1/2},$$

where $z(t)$ is the above specified solution. Since $(u(t), v(t)) \neq (0, 0)$ for $t \in R$, we have $D > 0$ for $t \in R$.

Considering that $z(t)$ satisfies equation (4.71) with $m(t) = m(\bar{x}, u)$, and u and v satisfy (4.74), we obtain

$$(4.78) \qquad DD' = -pu^2 + 2quv - zv^2,$$

where

$$(4.79) \quad p = z(z^2 - h), \qquad q = z^2 - h, \qquad q^2 - pz = -h(z^2 - h).$$

Relation (4.78) can also be written as

$$(4.80) \qquad DD' = -z\left[\left(v - \frac{q}{z}u\right)^2 + \left(\frac{pz - q^2}{z^2}\right)u^2\right].$$

If we observe that $pz - q^2 = h(z^2 - h) \geqslant 0$, it follows that the equality sign holds only if $h = 0$, i.e., $u = \bar{x} = 0$. This means that DD' is negative for $(u, v) \neq (0, 0)$, and can vanish only if $(u, v) = (0, 0)$.

A consequence of the above considerations is that $D' < 0$ for $(u, v) \neq (0, 0)$. Thus D^2 is a strictly decreasing function of t, and consequently $D^2 \to D_0{}^2 =, 0 \leqslant D_0 < +\infty$, for $t \to -\infty$ (D^2 is bounded on the real line because u, v, and z are bounded). But $D^2 \to D_0{}^2 \neq 0$ as $t \to -\infty$ implies the existence of a sequence diverging to $-\infty$, such that $DD' \to 0$ on this sequence. In the contrary case, we would have $DD' \leqslant -\lambda < 0$, for $t \leqslant t_1$, which implies $D^2 \to +\infty$ for $t \to -\infty$.

It follows from (4.79) that on the above sequence one should have $v - (q/z)u \to 0$, $(pz - q^2)u \to 0$, since $z \geqslant \alpha > 0$. This can happen only if $u \to 0$ and $v \to 0$ on the considered sequence or $(pz - q^2) = h(z^2 - h) \to 0$ and $v - (q/z)u \to 0$ on the same sequence.

If $u \to 0$ and $v \to 0$, this means that $D \to 0$ on that sequence, which is incompatible with $D^2 \to D_0{}^2 \neq 0$.

If $h(z^2 - h) \to 0$ and $v - (q/z)u \to 0$, it follows that $h \to 0$ (since $z^2 - h \geqslant \alpha^2 - \beta > 0$) and $v - (q/z)u \to 0$ on the above specified sequence. If the limit of u is distinct from zero on this sequence, then there exists a subsequence on which $u \to u_0 \neq 0$. We can assume without loss of generality that on this subsequence $\bar{x} \to \bar{x}_0$. Because $h(\bar{x}, u)$ is continuous, $h \to h(\bar{x}_0, u_0)$ on this subsequence. Since $h \neq 0$ for $u_0 \neq 0$ we have reached a contradiction. This means that $u \to 0$ on the considered sequence, and since $DD' \to 0$ on the same sequence, according to (4.78) we have $v \to 0$, i.e., $D \to 0$. As we have already seen, this contradicts the fact that $D^2 \to D_0{}^2 \neq 0$ as $t \to -\infty$.

By Theorem 4.6 each system (4.70) has only one almost periodic solution with graph in Ω. In particular, equation (4.68) has a unique almost periodic solution, together with its derivative, so that $(x(t), x'(t)) \in \Omega$ for $t \in R$.

Remark 1. One can assert that the second derivative is almost periodic too. This fact follows immediately from equation (4.68). Consequently, we can say that equation (4.68) describes a unique almost periodic motion.

Remark 2. It can be shown without difficulty that the almost periodic solution is asymptotically stable. Indeed, if we consider the "distance" D between any arbitrary solutions, with graph in Ω for $t \geqslant t_0$, defined by (4.77), it follows that $DD' < 0$, and thus $D^2 \to D_1{}^2 \geqslant 0$ as $t \to +\infty$. A consequence of a simple argument is the fact that the hypothesis $D_1{}^2 > 0$ is not acceptable.

Remark 3. The conditions of the theorem are satisfied if

$$\lim_{|x| \to \infty} \inf |g(x)| > k$$

and if conditions 3, 4, and 5 are satisfied.

Bibliographical notes

The results presented in this chapter are contributions of I. G. Malkin (496, 497), L. Amerio (5), and of C. E. Langenhop and G. Seifert (437).

The almost periodicity of the primitive of a bounded almost periodic function was established by H. Bohr, who observed that P. Bohl's proof (95) of the corresponding theorem for quasiperiodic functions remains valid without essential changes. There exist other proofs of this theorem based on the normality of almost periodic functions [see, for instance, J. Favard (247)].

The first theorem of the same type as Theorem 4.2 was established by E. Esclangon (233) for quasiperiodic solutions of linear equations of higher order. It was extended by H. Bohr and O. Neugebauer (141) to almost periodic solutions. Our exposition is close to that of I. G. Malkin (497). This paper contains some results concerning the existence of almost periodic solutions in the case when the characteristic equation has purely imaginary roots.

Important results in this field were obtained by I. Bărbălat (32), B. P. Demidovich (207, 208), A. Halanay (305–308), Y. Sibuya (588, 589), by

Ş. Şandor (571, 572), J. J. Schäffer (573), I. Z. Štokalo (613–615), and others.

Recently, Z. Opial (532) showed that any bounded solution of the equation $x' = f(t, x)$, where $f(t, x)$ is almost periodic in t, uniformly with respect to x, $|x| \leq M$ and monotone with respect to x, is almost periodic. This result generalizes the theorem regarding the almost periodicity of the primitive of an almost periodic function. If $f(t, x)$ is strictly monotone with respect to x, then the almost periodic solution is unique.

The equation $x' = a(t)x + b(t)$, where $a(t)$ and $b(t)$ are almost periodic functions, was studied by R. H. Cameron (182), who proved that there exists a unique almost periodic solution for any $b(t)$, if the mean of the function $a(t)$ is not zero. J. L. Massera (505) extended this result to linear and quasi-linear systems.

The homogeneous and nonhomogeneous linear systems were widely studied by S. Bochner, R. H. Cameron, and J. Favard (see List of References).

Remarkable results regarding the existence of almost periodic solutions may be found in N. N. Bogoliubov's and Yu. A. Mitropolskiĭ's monograph (93).

The research in this field continues with intensity, and numerous results have been included in recently published monographs on the theory of differential equations. We mention those of J. K. Hale (316), A. Halanay (314), T. Yoshizawa (689), V. I. Zubov (704), and J. L. Massera and J. J. Schäffer (507).

The almost periodicity of the solutions of certain differential systems of special form encountered in the theory of automatic control was established by V. A. Jakubovič (362, 363).

We recall the works of M. A. Krasnoselskiĭ and A. I. Perov (426) concerning the existence of almost periodic solutions of systems of nonlinear differential equations.

Results similar to those from §5 have been obtained by Z. Opial (532, 533) and M. L. Cartwright (184).

Numerous works were dedicated to the almost periodicity of the solutions of equations with retarded argument or of certain functional equations related to the latter. We mention the works of J. K. Hale (317), A. Halanay (310–313), and S. N. Šimanov (592, 593).

The works of R. K. Miller and G. Seifert (see References) contain numerous interesting results in connection with the almost periodic character of the solutions of differential systems. We also mention the works of D. Wexler (657–659), who studies linear systems of the form $\dot{x} = A(t)x + f(t)$, where $f(t)$ is a distribution. He also establishes results concerning the almost periodicity of the solutions of these systems.

Contributions to the problem of almost periodicity of solutions of ordinary differential equations have been numerous during the last 20 years. The bibliography by A. M. Fink (730) contains a good many references to this kind of work.

For almost periodicity criteria concerning ordinary differential equations, see C. Corduneanu (719), B. P. Demidovich (724), A. M. Fink (727), El. Hanebaly (732), V. H. Harasahal (734), Russel Johnson (740), (741), Russel Johnson and J. Moser (742), M. A. Krasnoselskii et al. (744), G. Seifert (761), C. Simirad (764), Janusz Traple (766) who considers the concept of weak almost periodicity as defined by W. F. Eberlein (225) and T. Yoshizawa (771).

The problem of the almost periodicity of solutions of integral or integro-differential equations has been also investigated by many authors. We mention in this direction the papers by C. Corduneanu (714), (717), R. K. Miller (748), and O. Staffans (765).

Almost periodicity in dynamical systems has been also considered by many authors. We mention the contribution of C. M. Dafermos (722), in which further references can be found.

Of course, results concerning the almost periodicity of ordinary differential equations are obtainable from similar results on equations in Banach spaces. See in this direction the books by A. Haraux (735), B. M. Levitan and V. V. Zhikov (745), and S. Zaidman (773).

Using his new characterization of almost periodic functions, A. Haraux (736) provides almost periodicity results for periodic systems. See also A. Haraux (737) for evolution equations.

V

Almost periodic solutions of certain partial differential equations

Although the number of papers concerning the almost periodicity of solutions of ordinary differential equations is quite considerable, the number of papers dedicated to the problem of almost periodicity of solutions of partial differential equations is relatively small.

J. Favard developed the theory of harmonic almost periodic functions. This theory is very similar to the theory of analytic almost periodic functions. Miron Nicolescu, in his studies of polyharmonic functions, established a series of results with regard to the almost periodicity of this type of function. In the case of harmonic functions, as well as in the case of polyharmonic functions, the functions defined in a strip emerge naturally. The almost periodicity is required with respect to one of the variables, uniformly with respect to the other.

For hyperbolic equations it was necessary to introduce the notion of almost periodicity in the mean. C. Muckenhoupt approached the problem of the almost periodicity of solutions of hyperbolic equations in two variables. His results were later extended by Bochner. In 1945 S. L. Sobolev returned to this problem by using the methods of functional analysis. Only recently, the problem of almost periodicity of solutions of partial differential equations was considered in a systematic manner, especially by L. Amerio and his school. These results assume a special knowledge of functional analysis and therefore will be omitted.

1. Certain properties of harmonic and polyharmonic functions

A function $u(x, y)$ is said to be *harmonic* in the domain D, if everywhere in this domain it satisfies the equation

$$(5.1) \qquad \Delta u = \frac{\partial^2 u}{\partial x^2} + \frac{\partial^2 u}{\partial y^2} = 0.$$

This definition implies that the product of a harmonic function with a constant, or the sum of two harmonic functions, is also a harmonic function.

We shall recall a few of the properties of harmonic functions, omitting the proofs.*

* I. G. Petrovsky: *Lectures on Partial Differential Equations*, Interscience, New York 1954.

Theorem 5.1. *Let $u(x, y)$ be a harmonic function in the domain D and let $(x_0, y_0) \in D$. If the disk of radius r with center at (x_0, y_0) and its boundary are contained in D, then Poisson's formula*

(5.2) $u(x_0 + \rho \cos \varphi, y_0 + \rho \sin \varphi)$

$$= \frac{1}{2\pi} \int_{-\pi}^{\pi} u(x_0 + r \cos t, y_0 + r \sin t) \frac{r^2 - \rho^2}{r^2 - 2r\rho \cos(t - \varphi) + \rho^2} \, dt$$

is true for any $\rho < r$.

Corollary. Under the same hypotheses as in Theorem 5.1 the following formulas are true:

(5.3)
$$\frac{\partial u}{\partial x_0} = \frac{1}{\pi r} \int_{-\pi}^{\pi} u(x_0 + r \cos t, y_0 + r \sin t) \cos t \, dt,$$

$$\frac{\partial u}{\partial y_0} = \frac{1}{\pi r} \int_{-\pi}^{\pi} u(x_0 + r \cos t, y_0 + r \sin t) \sin t \, dt.$$

In order to obtain these equalities we differentiate (5.2) with respect to ρ and then we set $\rho = 0$. One makes use of the fact that φ is arbitrary. $\partial u/\partial x_0$ and $\partial u/\partial y_0$ are the derivatives at the point (x_0, y_0).

Theorem 5.2. *The limit of a uniformly convergent sequence of harmonic functions is also harmonic.*

Theorem 5.3. *A harmonic function possesses partial derivatives of any order and these derivatives are harmonic functions.*

The existence of the derivatives can be deduced in a relatively simple manner from (5.2). The fact that they are harmonic functions follows from

$$\Delta \frac{\partial^{m+n} u}{\partial x^m \partial y^n} = \frac{\partial^{m+n}}{\partial x^m \partial y^n} \Delta u = 0.$$

Another useful theorem is

Theorem 5.4. *If $u(x, y)$ is a harmonic function in the simply-connected domain D, then there exists a harmonic function $w(x, y)$ in D, such that*

(5.4)
$$\frac{\partial w}{\partial x} = u(x, y).$$

Proof. Let $v(x, y)$ be a harmonic function conjugate to $u(x, y)$, i.e., such that

$$\frac{\partial v}{\partial x} = -\frac{\partial u}{\partial y}, \qquad \frac{\partial v}{\partial y} = \frac{\partial u}{\partial x}.$$

If (x_0, y_0) is a fixed point in the domain D, then we can take

$$v(x, y) = \int_{(x_0, y_0)}^{(x, y)} -\frac{\partial u}{\partial y} \, dx + \frac{\partial u}{\partial x} \, dy,$$

the path of integration being arbitrary. The integral does not depend on the path because u satisfies the equation $\Delta u = 0$. Let us consider now the function of a complex variable $f(z) = u(x, y) + iv(x, y)$. This function is holomorphic in D and we can set $F(z) = \int_{z_0}^{z} f(z) \, dz \, (z_0 = x_0 + iy_0, z = x + iy)$. The function $F(z)$ is holomorphic in D and we have $F'(z) = f(z)$. Set $w(x, y) = \operatorname{Re} F(z)$. Then $w(x, y)$ satisfies (5.4), which proves the theorem.

We are now to establish two more theorems concerning functions harmonic in a strip.

Theorem 5.5. *Let $u(x, y)$ be a bounded, harmonic function in the strip $a \leqslant x \leqslant b$. If on the straight lines $x = a$, $x = b$ we have*

(5.5) $$m \leqslant u(x, y) \leqslant M,$$

then these inequalities hold in the entire strip.

Proof. We shall use Theorem 3.3. If $v(x, y)$ is a harmonic function conjugate to $u(x, y)$, then we consider the complex function

(5.6) $$\varphi(z) = e^{u(x,y) + iv(x,y)}.$$

This function is holomorphic in the considered strip. Since $|\varphi(z)| = e^{u(x,y)}$, $1/|\varphi(z)| = e^{-u(x,y)}$, i.e., the functions $\varphi(z)$ and $1/\varphi(z)$ are bounded in the strip $[a, b]$. Since

$$|\varphi(z)| \leqslant e^{M}, \qquad \frac{1}{|\varphi(z)|} \leqslant e^{-m}$$

on the sides of the strip, we have

$$e^{m} \leqslant |\varphi(z)| = e^{u(x,y)} \leqslant e^{M}$$

for $a \leqslant x \leqslant b$. From these equalities, (5.5) follows on the whole strip $a \leqslant x \leqslant b$.

Theorem 5.6. *Let $u(x, y)$ be a harmonic function in the strip $a < x < b$, bounded in any strip $[a_1, b_1] \subset (a, b)$. Then its derivatives are also bounded in any strip $[a_1, b_1] \subset (a, b)$.*

Proof. Let us consider a positive number r, such that

$$r < \min\{a_1 - a, \quad b - b_1\}.$$

By hypothesis, there exists a number $K > 0$, such that

$$(5.7) \qquad |u(x, y)| \leqslant K$$

in the strip $[a_1 - r, b_1 + r] \subset (a, b)$. For any point of $[a_1, b_1]$ the disk of radius r with the center at this point is fully contained in the strip $[a_1 - r, b_1 + r]$. According to formulas (5.3) it follows that

$$(5.8) \qquad \left| \frac{\partial u}{\partial x} \right| \leqslant \frac{2K}{r}, \qquad \left| \frac{\partial u}{\partial y} \right| \leqslant \frac{2K}{r}$$

at any point of $[a_1, b_1]$. Therefore, by (5.8) the derivatives of first order are bounded in any strip $[a_1, b_1] \subset (a, b)$. According to Theorem 5.3 these derivatives are harmonic in (a, b) so that we can apply to them the above argument. Thus the derivatives of second order are bounded in any strip $[a_1, b_1] \subset (a, b)$ and so on.

Corollary. Under the conditions of Theorem 5.6 the function $u(x, y)$, together with its derivatives, is uniformly continuous in any strip $[a_1, b_1] \subset (a, b)$.

For the function $u(x, y)$ this is a consequence of the fact that the first-order derivatives are bounded in any strip $[a_1, b_1] \subset (a, b)$. One proceeds in the same manner for other derivatives, considering the preceding theorem.

Let us consider now the polyharmonic functions. First, let us define the iterations of the Laplace operator:

$$\Delta^n u = \Delta(\Delta^{n-1} u); \qquad \Delta^1 u = \Delta u.$$

A function $u(x, y)$ is called *polyharmonic* of order n in a domain D, if everywhere in this domain it satisfies the equation

$$(5.9) \qquad \Delta^n u = 0.$$

In particular, the polyharmonic functions of first order are nothing else but harmonic functions, whereas the polyharmonic functions of second order are called *biharmonic*.

One can show that the sum of two polyharmonic functions of orders m and n, $m \leqslant n$ is also a polyharmonic function of order n, $n > 1$.

The structure of polyharmonic functions in a simply connected domain D was made clear by E. Almansi. A theorem of interest for us is the structure theorem established by Miron Nicolescu.

Theorem 5.7. *If $u(x, y)$ is a polyharmonic function of order n in a simply-connected domain D, then one can determine n harmonic functions in D, let us denote them by $u_1(x, y)$, $u_2(x, y)$, ..., $u_n(x, y)$, such that*

$$(5.10) \qquad u(x, y) = u_1(x, y) + xu_2(x, y) + \cdots + x^{n-1}u_n(x, y),$$

for any $(x, y) \in D$.

Proof. We shall apply the principle of induction. The statement is obviously true for polyharmonic functions of first order. Let us assume that it is true for polyharmonic functions of order $<n$, and then let us show that it remains valid for the polyharmonic functions of order n. Since

$$\Delta^n u = \Delta^{n-1}(\Delta u) = 0,$$

Δu is a polyharmonic function of order $n - 1$ if u is a polyharmonic function of order n. Therefore, according to the induction hypothesis we can write

$$(5.11) \qquad \Delta u = v_1(x, y) + xv_2(x, y) + \cdots + x^{n-2}v_{n-1}(x, y),$$

where $v_1(x, y)$, $v_2(x, y)$, ..., $v_{n-1}(x, y)$ are harmonic functions in the domain D. It remains to show that one can choose the functions $u_1(x, y)$, $u_2(x, y)$, ..., $u_n(x, y)$ in (5.10), so that (5.11) holds. Applying the operator Δ to both sides of equality (5.10), we obtain

$$\Delta u = 2\left(\frac{\partial u_2}{\partial x} + u_3\right) + x\left(4\frac{\partial u_3}{\partial x} + 6u_4\right) + \cdots + x^{n-2}\cdot 2(n - 1)\frac{\partial u_n}{\partial x}.$$

Equality (5.11) for Δu leads us to set

$$2\frac{\partial u_2}{\partial x} + u_3 = v_1,$$

$$(5.12) \qquad 4\frac{\partial u_3}{\partial x} + 6u_4 = v_2,$$

$$\cdots\cdots\cdots\cdots\cdots$$

$$2(n - 1)\frac{\partial u_n}{\partial x} = v_{n-1}.$$

If we consider the last equation in (5.12) and apply Theorem 5.4, it will follow that there exists a harmonic function $u_n(x, y)$ for which this equation is satisfied. Then we substitute $u_n(x, y)$ determined in this way in the equation second to the last in (5.12). For the harmonic function $u_{n-1}(x, y)$ we shall obtain an equation of type (5.4), from which this function can then be determined. Thus we can determine all the functions $u_k(x, y)$, $k = 2, \ldots, n$. It remains to determine the function $u_1(x, y)$ from the equality

$$u_1(x, y) = u(x, y) - xu_2(x, y) - \cdots - x^{n-1}u_n(x, y).$$

From the previous considerations we get $\Delta u_1 = 0$ and with this the theorem has been proved.

Remark. Expansion (5.10) is not the unique expansion with the property that $u_1(x, y)$, $u_2(x, y)$, $\ldots$, $u_n(x, y)$ are harmonic functions. If we consider that xy is a harmonic function and set $u'_1(x, y) = u_1(x, y) - xy$, and $u'_2(x, y) = u_2(x, y) + y$, it follows that

$$u(x, y) = u'_1(x, y) + xu'_2(x, y) + \cdots + x^{n-1}u_n(x, y),$$

where $u'_1(x, y)$ and $u'_2(x, y)$ are harmonic functions as well.

2.   Harmonic and polyharmonic almost periodic functions

In this paragraph we shall present certain results regarding the harmonic and polyharmonic functions. For the harmonic almost periodic functions we can develop a theory analogous to that of analytic almost periodic functions. For instance, we can associate Fourier series with these functions (by taking the real parts of the terms of the Dirichlet series corresponding to certain analytic functions); the uniqueness and the approximation theorems remain valid, etc. However, we shall consider only those properties of harmonic almost periodic functions that follow immediately. On the other hand, we shall establish two theorems concerning certain boundary value problems.

A function $u(x, y)$ harmonic in the strip $[a, b]$ is said to be almost periodic in this strip, if it is almost periodic in y, uniformly with respect to x, $a \leqslant x \leqslant b$.

An analogous definition can be stated for the functions harmonic in a strip (a, b).

The following properties of the almost harmonic periodic functions are consequences of the theorems established for the almost periodic functions depending on parameters, and of the statements of the preceding paragraph.

Theorem 5.8. *A harmonic almost periodic function in the strip $[a, b]$ is bounded and uniformly continuous in this strip.*

Corollary. If a function $u(x, y)$ is harmonic and almost periodic in the strip (a, b), then it is bounded and uniformly continuous in any strip $[a_1, b_1] \subset (a, b)$.

Theorem 5.9. *The sum of two harmonic almost periodic functions in the strip $[a, b]$ is a harmonic almost periodic function in the same strip.*

Corollary. The sum of two harmonic almost periodic functions in the strip (a, b) is a harmonic function in the same strip, almost periodic in any strip $[a_1, b_1] \subset (a, b)$.

Theorem 5.10. *The limit of a uniformly convergent sequence of harmonic almost periodic functions in a strip (a, b) is a function harmonic in the same strip, almost periodic in any strip $[a_1, b_1] \subset (a, b)$.*

Theorem 5.11. *If $u(x, y)$ is a function harmonic in the strip (a, b), almost periodic in any strip $[a_1, b_1] \subset (a, b)$, then its derivatives are almost periodic in any strip $[a_1, b_1] \subset (a, b)$.*

Proof. For the derivative $\partial u/\partial y$ the statement is a consequence of Theorems 2.5 and 5.6 (Corollary). Regarding the derivative $\partial u/\partial x$, we consider the functions

$$\frac{u(x + n^{-1}, y) - u(x, y)}{n^{-1}} = \frac{\partial u(x + \theta_n n^{-1}, y)}{\partial x}, \qquad 0 < \theta_n < 1,$$

which are almost periodic in y, uniformly with respect to x, in any strip $[a_1, b_1] \subset (a, b)$, if n is sufficiently large. One applies Theorem 5.10 since

$$\lim_{n \to +\infty} \frac{\partial u(x + \theta_n n^{-1}, y)}{\partial x} = \frac{\partial u(x, y)}{\partial x},$$

uniformly in any strip $[a_1, b_1] \subset (a, b)$, $\partial u/\partial x$ being uniformly continuous in $[a_1, b_1]$.

Theorem 5.12. *Let $u(x, y)$ be a harmonic function in the strip (a, b), bounded in any strip $[a_1, b_1] \subset (a, b)$. If the function $\partial u/\partial x$ is almost periodic in any strip $[a_1, b_1] \subset (a, b)$, then $u(x, y)$ is also almost periodic in any strip $[a_1, b_1] \subset (a, b)$.*

Proof. Let us set

$$h(x, y) = \frac{\partial u}{\partial x}.$$

According to the preceding theorem it follows that

(5.13)
$$\frac{\partial h(x, y)}{\partial y} = \frac{\partial^2 u}{\partial x^2}$$

is almost periodic in any strip $[a_1, b_1] \subset (a, b)$. Let us fix a number x_0, $a < x_0 < b$. We have

(5.14)
$$u(x, y) = \int_{x_0}^x h(\xi, y) \, d\xi + u(x_0, y).$$

Consequently,

$$\frac{\partial^2 u}{\partial y^2} = \int_{x_0}^x \frac{\partial^2 h(\xi, y)}{\partial y^2} \, d\xi + u''(x_0, y).$$

Since $h(x, y)$ is a harmonic function, we find that

$$\frac{\partial^2 u}{\partial y^2} = - \int_{x_0}^x \frac{\partial^2 h(\xi, y)}{\partial \xi^2} \, d\xi + u''(x_0, y),$$

whence

(5.15)
$$\frac{\partial^2 u}{\partial y^2} = - \frac{\partial h(x, y)}{\partial x} + \frac{\partial h(x, y)}{\partial x} \bigg|_{x = x_0} + u''(x_0, y).$$

Considering (5.13) and (5.15) and the fact that $u(x, y)$ is harmonic, we obtain

$$u''(x_0, y) + \frac{\partial h(x, y)}{\partial x} \bigg|_{x = x_0} = 0,$$

i.e.

(5.16)
$$u''(x_0, y) = - \frac{\partial h(x, y)}{\partial x} \bigg|_{x = x_0}$$

This means that $u''(x_0, y)$ is an almost periodic function of y. By Theorem 5.6, $u'(x_0, y)$ is bounded. Thus it is almost periodic in y. From (5.14) it follows that $u(x, y)$ is almost periodic in y, uniformly with respect to x in any strip $[a_1, b_1] \subset (a, b)$, since $\int_{x_0}^x h(\xi, y) d\xi$ has this property and $u(x_0, y)$ for a fixed x_0 can be considered almost periodic in y, uniformly with respect to x in (a, b).

Theorem 5.13. *Let $u(x, y)$ be a function harmonic and bounded in the strip $[a, b]$. If on the straight lines $x = a$ and $x = b$ it reduces to two almost periodic functions, then it is almost periodic in the strip $[a, b]$.*

Proof. According to Theorem 2.6 concerning the almost periodic functions which depend uniformly on parameters, it suffices to show that the family $\{u(x, y + t)\}$, t being an arbitrary real number, is normal in the strip $[a, b]$. According to Theorem 5.5 it is sufficient to prove that from any sequence of real numbers $\{t_n\}$ a subsequence $\{t'_n\}$ can be extracted so that the sequences $\{u(a, y + t'_n)\}$ and $\{u(b, y + t'_n)\}$ are uniformly convergent for $-\infty < y < +\infty$. This fact is obvious since the functions $u(a, y)$ and $u(b, y)$ are almost periodic. This concludes the proof of the theorem.

Remark. The question which arises naturally is whether the almost periodicity of $u(x, y)$ for $x = x_0$ ensures the almost periodicity in any closed strip contained in the given strip.

J. Favard considered the function

$$u(x, y) = \text{Im} \frac{1}{(z - 1)(z + 1)}$$

which is harmonic and vanishes for $x = 0$. One sees without difficulty that it is not almost periodic for any of the other values of x, $-1 < x < 1$ (i.e., on the strip in which it is harmonic).

In the following two theorems we shall show the existence of solutions of two boundary problems for the harmonic almost periodic functions.

Theorem 5.14. *Given two almost periodic functions $f_1(y)$ and $f_2(y)$, there exists a unique harmonic function $u(x, y)$, bounded in the strip $[a, b]$, such that*

(5.17) $$u(a, y) = f_1(y), \qquad u(b, y) = f_2(y).$$

This harmonic function is also almost periodic.

Proof. One sees easily that the function

$$\zeta = \text{tg}\left(\frac{\pi}{4} \cdot \frac{2z - (b + a)}{b - a}\right)$$

defines a conformal mapping between the strip (a, b) and the circle $|\zeta| < 1$. The straight line $x = a$ is taken into the open semicircle $\zeta = e^{i\alpha}$, $\pi/2 < \alpha < 3\pi/2$, and the straight line $x = b$ into the (open) semicircle $\zeta = e^{i\alpha}$, $-\pi/2 < \alpha < \pi/2$. Therefore, the problem of determination of a harmonic function in the considered strip reduces to the problem of determination of a function harmonic inside the unit circle, if the values of these functions on the two open semicircles mentioned above are known. This is a problem of the Dirichlet type, in which the data on the circumference form a bounded function, piecewise continuous (in the present case there are two points of discontinuity at $\zeta = \pm i$). We know that this problem admits a unique bounded solution. The uniqueness follows from Theorem 5.5. The almost periodicity of the harmonic function, the existence of which was established above, is a result of Theorem 5.13.

Theorem 5.15. *Let $f(y)$ be an almost periodic function. There exists a harmonic function $u(x, y)$, bounded in the half-plane $x > 0$, continuous in the half-plane $x \geq 0$, such that*

$$(5.18) \qquad\qquad u(0, y) = f(y).$$

This function is almost periodic in the half-plane $x > 0$.

Proof. Let us consider the function

$$(5.19) \qquad\qquad u(x, y) = \frac{1}{\pi} \int_{-\infty}^{\infty} f(\eta) \frac{x \, d\eta}{x^2 + (y - \eta)^2}$$

for $x > 0$, and let us show that it satisfies the requirements of the statement. First, let us show that the integral converges uniformly with respect to (x, y), if this point describes a domain in the half-plane $x > 0$ whose boundary does not cross the straight line $x = 0$. This fact follows immediately if we notice that $f(y)$ is bounded on the real line, and

$$(5.20) \qquad\qquad \frac{1}{\pi} \int_{-\infty}^{\infty} \frac{x \, d\eta}{x^2 + (y - \eta)^2} = 1,$$

for any (x, y) of the half-plane $x > 0$. Analogously, one can show that it is permissible to differentiate the integrand with respect to x and to y. Since

$$u_0(x, y) = \frac{x}{x^2 + (y - \eta)^2}$$

is a harmonic function in the half-plane $x > 0$, one establishes that $u(x, y)$ is also harmonic. If $|f(y)| \leq M$, $-\infty < y < +\infty$, then

$$|u(x, y)| \leq M \frac{1}{\pi} \int_{-\infty}^{\infty} \frac{x \, d\eta}{x^2 + (y - \eta)^2} = M,$$

which proves the boundedness of $u(x, y)$ in the half-plane $x > 0$.

Let us prove now that boundary condition (5.18) is satisfied. According to (5.20) we can write

$$u(x, y) - f(y) = \frac{1}{\pi} \int_{-\infty}^{\infty} [f(\eta) - f(y)] \frac{x \, d\eta}{x^2 + (y - \eta)^2}.$$

Substituting $y - \eta = -tx$ we obtain

(5.21) $$u(x, y) - f(y) = \frac{1}{\pi} \int_{-\infty}^{\infty} [f(tx + y) - f(y)] \frac{dt}{1 + t^2}.$$

Consider $\varepsilon > 0$ and $\delta > 0$ such that $|f(y_1) - f(y_2)| < \varepsilon/3$ if $|y_1 - y_2| < \delta$. Let η_1 be a positive number such that

$$\frac{2M}{\pi \eta_1} < \frac{\varepsilon}{3}.$$

We assume that x is sufficiently small so that $x\eta_1 < \delta$. We can now write (5.20) as

(5.22) $$u(x, y) - f(y) = \frac{1}{\pi} \left\{ \int_{-\infty}^{-\eta_1} + \int_{-\eta_1}^{\eta_1} + \int_{\eta_1}^{\infty} \right\} = I_1 + I_2 + I_3.$$

Since $|tx| < \delta$ in the interval $[-\eta_1, +\eta_1]$, this means that

(5.23) $$|I_2| < \frac{\varepsilon}{3}.$$

Let us now evaluate I_1:

(5.24) $$|I_1| \leqslant \frac{2M}{\pi} \int_{-\infty}^{-\eta_1} \frac{dt}{1 + t^2} < \frac{\varepsilon}{3}, \qquad \text{if} \quad x < h,$$

since η_1 can be taken arbitrarily large with the condition that x be sufficiently small (so that we have $x\eta_1 < \delta$). Analogously

(5.25) $$|I_3| < \frac{\varepsilon}{3}, \qquad \text{if} \quad x < h.$$

From the last three inequalities and from (5.22) follows that

$$|u(x, y) - f(y)| < \varepsilon, \qquad \text{if} \quad x < h.$$

This shows that

$$\lim_{x \to 0} u(x, y) = f(y),$$

uniformly with respect to y on the whole axis. This means (5.18) holds.

Finally, let us show that $u(x, y)$ is almost periodic in the half-plane $x > 0$. If η is an ε-translation number of the function $f(y)$, then it follows that

$$|u(x, y + \eta) - u(x, y)| \leqslant \frac{1}{\pi} \int_{-\infty}^{\infty} |f(y + \eta) - f(y)| \frac{x \, du}{x^2 + (y - u)^2} < \varepsilon.$$

This means that any ε-translation number of the function $f(y)$ is an ε-translation number of the function $u(x, y)$, uniformly with respect to x.

The theorem is completely proved.

Remark. The proof has an artificial character since at the beginning we define a function whose existence must be proved, and then we verify that it fulfills the conditions required by the theorem. The reader who is familiar with the theory of partial differential equations will note that formula (5.19) represents the solution of the Dirichlet problem for a half-plane. For the sake of brevity we did not specify the method for obtaining (5.19).

Let us consider now the *polyharmonic almost periodic functions.* Their definition is analogous to that stated at the beginning of the paragraph, with the difference that the word harmonic must be replaced by polyharmonic. Because they are almost periodic functions which depend uniformly on a parameter, we can immediately obtain a number of their properties. We shall direct our attention to the problem of almost periodicity of harmonic functions contained in the representation (5.10) under the hypothesis that the polyharmonic function u is almost periodic.

Let us first note that if the polyharmonic function $u(x, y)$ has an expansion of the form (5.10), where $u_1(x, y), \ldots, u_n(x, y)$ are harmonic almost periodic functions, then it is almost periodic. Conversely, if $u(x, y)$ is a polyharmonic almost periodic function in a certain strip, then it is possible that the functions $u_1(x, y), \ldots, u_n(x, y)$ are unbounded (see the example of the preceding paragraph, from which follows that for any polyharmonic function there exist representations in which $u_1(x, y)$ is not bounded). It is useless to ask whether $u_1(x, y), \ldots, u_n(x, y)$ are almost periodic, unless they are bounded.

Theorem 5.16. *Let* $u(x, y)$ *be a polyharmonic function of order* n *in the strip* (a, b), *almost periodic in any strip* $[a_1, b_1] \subset (a, b)$. *If the functions* $u_1(x, y), \ldots, u_n(x, y)$ *in an arbitrary representation of the form* (5.10) *are bounded in any strip* $[a_1, b_1] \subset (a, b)$, *then they are harmonic almost periodic functions in any strip* $[a_1, b_1] \subset (a, b)$.

Proof. We shall prove the theorem by induction on n. The theorem is obvious for harmonic functions. Let us assume that is is true for polyharmonic functions of order $< n$, and then we shall show that it remains valid for polyharmonic functions of order n. Let $u_1(x, y), \ldots, u_n(x, y)$ be bounded and harmonic in any strip $[a_1, b_1] \subset (a, b)$. We have

$$\Delta u = v_1(x, y) + x v_2(x, y) + \cdots + x^{n-2} v_{n-1}(x, y),$$

where $v_1(x, y), \ldots, v_{n-1}(x, y)$ are given by (5.12). Since $u_1(x, y), \ldots, u_n(x, y)$

are bounded with their derivatives in any strip $[a_1, b_1] \subset (a, b)$, it follows from (5.12) that $v_1(x, y)$, $v_2(x, y)$, ..., $v_{n-1}(x, y)$ are harmonic functions, bounded in any strip $[a_1, b_1] \subset (a, b)$. Δu is a polyharmonic function of order $n - 1$, and by the induction hypothesis it follows that $v_1(x, y)$, ..., $v_{n-1}(x, y)$ are almost periodic in any strip $[a_1, b_1] \subset (a, b)$. As a consequence of the last equation of (5.12), $\partial u_n / \partial x$ is almost periodic in any strip $[a_1, b_1] \subset (a, b)$. Applying now Theorem 5.12 it follows that $u_n(x, y)$ is almost periodic in any strip $[a_1, b_1] \subset (a, b)$. Analogously we obtain that $u_{n-1}(x, y)$, ..., $u_2(x, y)$ are almost periodic in any strip $[a_1, b_1] \subset (a, b)$. Finally, for $u_1(x, y)$ this fact follows from its definition.

3. Almost periodic solutions of hyperbolic equations

Let us consider now the hyperbolic equation

$$(5.26) \qquad \frac{\partial^2 u}{\partial t^2} = \frac{\partial}{\partial x}\left[p(x)\frac{\partial u}{\partial x}\right] - q(x)u,$$

with the boundary conditions

$$(5.27) \qquad u(0, t) = 0, \qquad u(l, t) = 0$$

and initial conditions

$$(5.28) \qquad u(x, 0) = f(x), \qquad \left.\frac{\partial u(x, t)}{\partial t}\right|_{t=0} = g(x).$$

The method of separation of variables leads us to the Sturm–Liouville system

$$(5.29) \qquad \frac{d}{dx}\left[p(x)\frac{dy}{dx}\right] + [\lambda - q(x)]y = 0,$$

$$(5.30) \qquad y(0) = 0, \qquad y(l) = 0,$$

with the eigenvalues λ_k and the eigenfunctions $\varphi_k(x)$, $k = 1, 2, \ldots$, which permit us to determine the solution of equation (5.26) with boundary value conditions (5.27), and initial conditions (5.28).

If we use certain regularity conditions (which we do not find necessary to include here)* then the solution of our problem can be represented as a sum of a uniformly convergent series

$$(5.31) \qquad u(x, t) = \sum_{k=1}^{\infty} (A_k \cos \sqrt{\lambda_k}\, t + B_k \sin \sqrt{\lambda_k}\, t)\varphi_k(x),$$

* See, for instance, V. I. Smirnov, *A course of higher mathematics*, Vol. IV, Pergamon Press, Oxford, 1964.

where A_k and B_k are given by the formulas

$$(5.32) \qquad A_k = \int_0^l f(x)\varphi_k(x)\, dx, \qquad B_k = \frac{1}{\sqrt{\lambda_k}} \int_0^l g(x)\varphi_k(x)\, dx,$$

assuming that the sequence $\varphi_k(x)$ is normalized.

Since series (5.31) is uniformly convergent with respect to t and x, $t \in R$, $x \in [0, l]$, it follows that the solution $u(x, t)$ is almost periodic in t, uniformly with respect to $x \in [0, l]$. This is a consequence of the fact that each of the functions

$$(5.33) \qquad u_n(x, t) = \sum_{k=1}^{n} (A_k \cos \sqrt{\lambda_k}\, t + B_k \sin \sqrt{\lambda_k}\, t)\varphi_k(x), \qquad n = 1, 2, \ldots$$

is almost periodic in t, uniformly with respect to $x \in [0, l]$, and of Theorem 2.4.

The conditions that ensure the uniform convergence of series (5.31), as well as the series obtained by differentiation twice with respect to t or x, are very restrictive and are not satisfied in certain problems of mathematical physics.

Therefore, it was necessary to extend the concept of solution; mathematical physics operates nowadays with generalized solutions, whose existence is ensured by less restrictive conditions than those required for the existence of a solution in the classical sense.

One of the most interesting applications of the notion of function almost periodic in the mean is connected with the generalized solutions of hyperbolic equations.

Let us give now the definition of the concept of a generalized solution of equation (5.26) with conditions (5.27) and (5.28), following S. L. Sobolev.

Let $\{u_n(x, t)\}$ be a sequence of solutions (in the classical sense) of equation (5.26), satisfying boundary value conditions (5.27). We shall say that $u(x, t)$ is a *generalized solution* of equation (5.26), corresponding to boundary conditions (5.27) and initial conditions (5.28), if

$$(5.34) \qquad \lim_{n \to \infty} \int_0^l [u(x, t) - u_n(x, t)]^2 \, dx = 0,$$

$$(5.35) \qquad \lim_{n \to \infty} \int_0^l [f(x) - u_n(x, 0)]^2 \, dx = 0,$$

$$(5.36) \qquad \lim_{n \to \infty} \int_0^l \left[g(x) - \frac{\partial u_n(x, t)}{\partial t}\bigg|_{t=0} \right]^2 dx = 0.$$

We assume that (5.34) holds uniformly with respect to t.

Obviously, a generalized solution might be even discontinuous. Anyway, according to well-known theorems in the theory of functions of a real variable, $u(x, t)$ is a square-summable function on $[0, l]$ for any t.

It is possible to prove a uniqueness theorem for generalized solutions, but we shall not do this.

Let us now specify the conditions assuring the existence of a generalized solution for equation (5.26) with conditions (5.27) and (5.28).

We suppose that the following are satisfied:

1) $p(x) > 0$ and admits a continuous derivative in $[0, l]$;
2) $q(x) \geqslant 0$ and is continuous in $[0, l]$;
3) $f(x)$ has a square-summable derivative in $[0, l]$; $f(0) = f(l) = 0$.
4) $g(x)$ is a square-summable function in $[0, l]$.

We note first that the Sturm–Liouville system (5.29), (5.30) admits a sequence of eigenvalues λ_k, such that

$$(9.37) \qquad \lambda_k > 0, \qquad \lim_{k \to \infty} \lambda_k = +\infty,$$

and the corresponding eigenvectors $\varphi_k(x)$ form an orthonormal system on the interval $[0, l]$, closed in $L_2[0, l]$.

Theorem 5.17. *If conditions* 1, 2, 3, *and* 4 *are satisfied, then the series*

$$\sum_{k=1}^{\infty} (A_k \cos \sqrt{\lambda_k}\, t + B_k \sin \sqrt{\lambda_k}\, t) \varphi_k(x),$$

where A_k and B_k are defined by (5.32), converges in the mean on the interval $[0, l]$, uniformly with respect to $t \in R$. The sum of the series is a generalized solution of equation (5.26), corresponding to conditions (5.27) and (5.28), and is almost periodic in the mean.

Proof. From (5.32) it follows that

$$(5.38) \qquad \sum_{k=1}^{\infty} A_k^2 < +\infty, \qquad \sum_{k=1}^{\infty} \lambda_k B_k^2 < +\infty.$$

According to (5.37) it follows that

$$(5.39) \qquad \sum_{k=1}^{\infty} B_k^2 < +\infty.$$

In order to prove the convergence in the mean of the series in the above statement, it suffices to show that for any $\varepsilon > 0$, there exists an $N(\varepsilon) > 0$, such that

(5.40) $$\int_0^l \left\{ \sum_{k=m}^n (A_k \cos \sqrt{\lambda_k}\, t + B_k \sin \sqrt{\lambda_k}\, t) \varphi_k(x) \right\}^2 dx < \varepsilon,$$

if $n, m \geqslant N(\varepsilon)$, $t \in R$. Since the sequence $\{\varphi_k(x)\}$ is orthonormal, (5.40) can be also expressed as

(5.41) $$\sum_{k=m}^n (A_k \cos \sqrt{\lambda_k}\, t + B_k \sin \sqrt{\lambda_k}\, t)^2 < \varepsilon.$$

Since

$$(A_k \cos \sqrt{\lambda_k}\, t + B_k \sin \sqrt{\lambda_k}\, t)^2 \leqslant A_k{}^2 + B_k{}^2,$$

(5.41) will follow from

(5.42) $$\sum_{k=m}^n (A_k{}^2 + B_k{}^2) < \varepsilon;$$

the above inequality is possible for n, m sufficiently large, since we have

$$\sum_{k=1}^\infty (A_k{}^2 + B_k{}^2) < +\infty.$$

Conditions (5.35) and (5.36) are fulfilled since the Fourier series of a square-summable function, with respect to an orthonormal closed system, converges in the mean to that function.

The almost periodicity in the mean of the generalized solution is an immediate consequence of Theorem 2.12 and of the fact that every term of the considered series is almost periodic in t, uniformly with respect to $x \in [0, l]$.

Remark. Since equation (5.26) describes a motion—for instance, the motion of a string—the almost periodic character of the motion is of interest. In other words, we are interested to know whether the velocity and the acceleration are also almost periodic functions.

This occurs if the conditions stated above are fulfilled.

If in the integral

$$I(y) = \int_0^l [p(x)y'^2 + q(x)y^2]\, dx \geqslant 0$$

we set

$$y = f(x) - \sum_{k=1}^n A_k \varphi_k(x),$$

then after elementary calculations we obtain

$$\sum_{k=1}^{n} \lambda_k A_k^2 \leqslant \int_0^l [p(x)f'^2(x) + q(x)f^2(x)]\, dx.$$

This shows that

(5.43) $$\sum_{k=1}^{\infty} \lambda_k A_k^2 < +\infty.$$

From (5.38) and (5.43) it follows by an argument analogous to that used in the proof of Theorem 5.17 that our above statements are true.

For the same purpose we may use Theorem 2.13, but we shall not give the details.

4. Almost periodic solutions of certain parabolic equations

In this paragraph we shall consider the nonlinear parabolic equation

(5.46) $$\frac{\partial^2 u}{\partial x^2} = \frac{\partial u}{\partial t} + f(x, t, u),$$

in the set

(Δ) $$-\infty < x < +\infty, \qquad 0 \leqslant t \leqslant T,$$

where $T > 0$. We assume that the function $f(x, t, u)$ is continuous in the set

(Δ₁) $$-\infty < x < +\infty, \qquad 0 \leqslant t \leqslant T, \qquad |u| \leqslant A,$$

where $A > 0$, and that it satisfies the Lipschitz condition

(5.47) $$|f(x, t, u) - f(x, t, v)| \leqslant L|u - v|.$$

Let $u(x, t)$ be a solution of equation (5.46), defined in (Δ). It follows that $|u(x, t)| \leqslant A$. Let us now put

(5.48) $$u(x, 0) = \varphi(x), \qquad -\infty < x < +\infty.$$

Before we present a method for approximating the solutions of equation (5.46) by the solutions of certain systems of ordinary differential equations, let us note that the equation

(5.49) $$\frac{d^2 y}{dx^2} - \alpha^2 y = f(x),$$

where α is a positive constant, and $f(x)$ is a bounded continuous function on the real line, admits a unique solution, bounded on the real line, given by

$$(5.50) \qquad y_0(x) = -\frac{1}{2\alpha}\left\{e^{\alpha x}\int_x^\infty e^{-\alpha t}f(t)\,dt + e^{-\alpha x}\int_{-\infty}^x e^{\alpha t}f(t)\,dt\right\}.$$

We may get this formula by considerations similar to those from Chapter IV, §3. One may easily verify that (5.50) is a bounded solution of equation (5.46), if $f(x)$ is bounded on the real line.

If we denote by $\|y\| = \sup|y(x)|$, $-\infty < x < +\infty$, then it follows from (5.50) that

$$(5.51) \qquad \|y_0\| \leqslant \frac{1}{\alpha^2}\|f\|.$$

Let us now consider the recurrent system of differential equations

$$(5.52) \qquad \frac{d^2u_k}{dx^2} = h^{-1}[u_k - u_{k-1}] + f(x, t_k, u_{k-1}), \qquad k = 1, 2, \ldots, n.$$

taking $u_0(x) = \varphi(x)$. The number h is determined by $nh = T$, and $t_k = kh$. One may expect that the unique bounded solution of this system, which exists according to the above considerations referring to equation (5.50), if $\varphi(x)$ and $f(x, t, u)$ are bounded, will approximate the bounded solution of equation (5.46) with condition (5.48).

Let us introduce the notations

$$(5.53) \qquad \varepsilon_k(x) = u(x, t_k) - u_k(x), \qquad k = 1, 2, \ldots, n.$$

Taking into account (5.48) and $u_0(x) = \varphi(x)$, it is natural to set $\varepsilon_0(x) = 0$.

Lemma 5.1. *We assume that the function $f(x, t, u)$ is continuous, bounded in the set (Δ_1) and satisfies the Lipschitz condition (5.47). Let $u(x, t)$ be a solution of equation (5.46) defined on (Δ), uniformly continuous together with the derivative $\partial u/\partial t$.*

Then the method of approximation described above is convergent. More precisely, there exists a positive constant M such that

$$(5.54) \qquad \|\varepsilon_k\| \leqslant M\omega(h), \qquad k = 1, 2, \ldots, n,$$

where $\omega(h)$ is the modulus of continuity (uniformly with respect to x and t) of the functions $u(x, t)$ and $\partial u/\partial t$.

Proof. If in equation (5.46) we set $t = t_k$, and then subtract from it equation (5.52), we obtain

$$(5.55) \qquad \frac{d^2\varepsilon_k}{dx^2} - h^{-1}\varepsilon_k = -h^{-1}\varepsilon_{k-1} + r_k(x), \qquad k = 1, 2, \ldots, n,$$

where

$$(5.56) \quad r_k(x) = \frac{\partial u(x, t_k)}{\partial t} - h^{-1}[u(x, t_k) - u(x, t_{k-1})] + f(x, t_k, u(x, t_k))$$

$$- f(x, t_k, u_{k-1}(x)), \qquad k = 1, 2, \ldots, n.$$

It follows from (5.55) that every $\varepsilon_k(x)$, $k = 1, 2, \ldots, n$, satisfies a differential equation of the type (5.49). Since every $\varepsilon_k(x)$ is a bounded function on the real line, this means that we may apply formula (5.51). We obtain thus

$$(5.57) \quad \|\varepsilon_k\| \leqslant \|\varepsilon_{k-1}\| + h \|r_k\|, \qquad k = 1, 2, \ldots, n.$$

Considering the fact that $f(x, t, u)$ satisfies the Lipschitz condition, and $|u(x, t) - u(x, \tau)| \leqslant \omega(h)$, $|\partial u(x, t)/\partial t - u(x, \tau)/\partial t| \leqslant \omega(h)$ for (x, t), $(x, \tau) \in \Delta$, $|t - \tau| \leqslant h$, we obtain

$$(5.58) \quad \|\varepsilon_k\| \leqslant (1 + hL)\|\varepsilon_{k-1}\| + (1 + L)h\omega(h), \qquad k = 1, 2, \ldots, n.$$

By recursion from (5.58) we get

$$(5.59) \quad \|\varepsilon_k\| \leqslant \omega(h)(1 + L)L^{-1}[(1 + hL)^n - 1], \qquad k = 1, 2, \ldots, n.$$

Since $nh = T$, from (5.59) we obtain (5.54) with $M = (1 + L)L^{-1}[e^{LT} - 1]$.

Now we can establish an almost periodicity criterion for the parabolic equation (5.46), assuming obvious corresponding conditions for the functions $f(x, t, u)$ and $\varphi(x)$.

Theorem 5.18. *Let $u(x, t)$ be a bounded solution of (5.46), defined on (Δ), satisfying the conditions of Lemma 5.1. If $f(x, t, u)$ is almost periodic in x, uniformly with respect to (t, u) and $\varphi(x)$ is almost periodic, then $u(x, t)$ is almost periodic in x, uniformly with respect to t, $0 \leqslant t \leqslant T$.*

Proof. Let $\varepsilon > 0$. We must show that there exists an $l(\varepsilon) > 0$, such that any interval of length $l(\varepsilon)$ of the x-axis will contain at least one point ξ with the property

$$(5.60) \quad |u(x + \xi, t) - u(x, t)| < \varepsilon, \qquad (x, t) \in \Delta.$$

Consider that n is sufficiently large so that $nh = T$ implies

$$(5.61) \quad 4(M + 1)\omega(h) < \varepsilon.$$

Consider also system (5.52) for such an n. Since all the functions $u_k(x)$ are almost periodic, then as a consequence of the admitted hypotheses and from Theorem 4.2, there is an $l(\varepsilon)$ with the property that any interval on the

x-axis of length $l(\varepsilon)$ contains at least one ξ which is a common $(\varepsilon/2)$-translation number of the functions $u_k(x)$, $k = 0, 1, \ldots, n$ (see Chapter VI, §2).

For any $t \in [0, T]$ there exists a t_k such that $|t - t_k| < h$. If ξ is a common $(\varepsilon/2)$-translation number of the functions $u_k(x)$, $k = 0, 1, \ldots, n$, then we shall have

$$|u(x + \xi, t) - u(x, t)| \leqslant |u(x + \xi, t) - u(x + \xi, t_k)|$$

$$+ |u(x + \xi, t_k) - u_k(x + \xi)| + |u_k(x + \xi) - u_k(x)|$$

$$+ |u_k(x) - u(x, t_k)| + |u(x, t_k) - u(x, t)|$$

$$\leqslant 2(M + 1)\omega(h) + |u_k(x + \xi) - u_k(x)|.$$

From these inequalities follows (5.60), if we consider (5.61) and also the manner in which ξ was chosen. This proves the theorem.

Remark. In physics equation (5.46) may be interpreted as describing the phenomenon of heat propagation. The meaning of the preceding theorem is the following: If the temperature distribution in an infinite rod at the initial moment is almost periodic and the heat sources inside the rod have also an almost periodic distribution, then at any time t the temperature distribution is given by an almost periodic function.

Let us now consider equation (5.46) in the set

(D) $0 \leqslant x \leqslant \pi, -\infty < t < +\infty,$

the only conditions to be imposed on the solution being the boundary-value conditions

(5.62) $u(0, t) = 0, u(\pi, t) = 0, -\infty < t < +\infty.$

If we want a solution in (D), it is obvious that we cannot impose an initial condition. In fact, we want to determine a solution of the equation (5.46) when its values are assigned on the boundary of the domain in which the solution is sought. This is a Dirichlet-type problem, usually encountered in the case of elliptic equations. The function $f(x, t, u)$ is assumed continuous in the set

(D₁) $0 \leqslant x \leqslant \pi, -\infty < t < +\infty, -\infty < u < +\infty,$

together with its partial derivative with respect to u, $\partial f/\partial u$. It is also assumed that $f(x, t, u)$ is almost periodic in the mean in t, uniformly with respect to u, in a sense to be made precise later.

Of course, we expect some almost periodicity properties of a solution of the equation (5.46), provided such solution is "bounded" in the sense required by the type of almost periodicity involved in our problem. In other words, it is natural to assume that the solution defined in (D), is bounded in the mean:

$$(5.63) \qquad \int_0^\pi u^2(x, t)\, dx \leqslant A < +\infty, \qquad \text{for any real } t.$$

Before we prove that any solution of (5.46) is almost periodic in the mean if it satisfies (5.63), we need to establish a simple result regarding a differential inequality on the whole real axis. This result is of a qualitative nature, because we will impose a certain qualitative property on the solution (its boundedness on R).

Lemma 5.2. *Let $v(t)$ be a real-valued differentiable function defined on the whole real axis, such that $v(t) \geqslant 0$, and*

$$(5.64) \qquad v'(t) + av(t) \leqslant b\, \sqrt{v(t)}, \quad -\infty < t < +\infty,$$

where a and b are positive constants. Then, the boundedness of $v(t)$ on the real axis implies the estimate

$$(5.65) \qquad \sup v(t) \leqslant (b/a)^2.$$

Proof. If the supremum is attained at a certain t_0, then $v'(t_0) = 0$, and (5.64) implies (5.65). If $v(t) \to \sup v(t)$ as $t \to +\infty$ or $t \to -\infty$, then $v'(t) \to 0$, and taking the limit in (5.64), one again obtains (5.65). If none of the above-mentioned situations occurs, then there exists a sequence of relative maxima of $v(t)$, attained at some values t_n, such that $t_n \uparrow +\infty$ or $t_n \downarrow -\infty$. At such points we have $v'(t_n) = 0$, and from (5.64) we easily obtain (5.65), keeping in mind that $v(t_n) \to \sup v(t)$.

Lemma 5.2 is thereby proved.

Let us consider now an arbitrary real number τ, and let $v(x, t) = u(x, t + \tau) - u(x, t)$, where $u(x, t)$ is a $C^{(2)}$-solution of (5.46), satisfying (5.63). From the equation (5.46) and the similar equation in which t is replaced by $t + \tau$, we obtain by subtraction the equation

$$(5.66) \qquad \frac{\partial^2 v}{\partial x^2} = \frac{\partial v}{\partial t} + f(x, t + \tau, u(x, t + \tau)) - f(x, t, u(x, t)).$$

If we multiply both sides of (5.66) by $v(x, t)$, and integrate the resulting equation with respect to x from 0 to π, we obtain the following relation:

$$(5.67) \quad \int_0^\pi v \frac{\partial^2 v}{\partial x^2} \, dx = \int_0^\pi v \frac{\partial v}{\partial t} \, dx$$

$$+ \int_0^\pi v\{f(x, t + \tau, u(x, t + \tau)) - f(x, t, u(x, t))\} \, dx.$$

If we integrate by parts in the integral in the left-hand side of (5.67) and take into account the conditions (5.62), then (5.67) leads to

$$(5.68) \quad - \int_0^\pi \left(\frac{\partial v}{\partial x}\right)^2 dx = \frac{1}{2} \frac{d}{dt} \int_0^\pi v^2 \, dx$$

$$+ \int_0^\pi v\{f(x, t + \tau, u(x, t + \tau)) - f(x, t, u(x, t))\} \, dx.$$

At this point we need to rely on the so-called Poincaré's inequality, which in our case can be written as

$$(5.69) \quad \int_0^\pi \left(\frac{\partial v}{\partial x}\right)^2 dx \geqslant \int_0^\pi v^2 \, dx,$$

which holds true under assumptions (5.62). For the reader who is not familiar with the variational theory of eigenvalues and eigenfunctions, we suggest a quick check of Poincaré's inequality by using Fourier series (in x) and Parseval's formula. If we take (5.69) into account, then (5.68) leads immediately to the inequality

$$(5.70) \quad \frac{1}{2} \frac{d}{dt} \int_0^\pi v^2 \, dx + \int_0^\pi v^2 \, dx$$

$$\leqslant \int_0^\pi v \{f(x, t + \tau, u(x, t + \tau)) - f(x, t + \tau, u(x, t))\} \, dx$$

$$+ \int_0^\pi v\{f(x, t + \tau, u(x, t)) - f(x, t, u(x, t))\} \, dx.$$

We shall now make another assumption on the function $f(x, t, u)$, namely

$$(5.71) \quad \frac{\partial f}{\partial u}(x, t, u) \leqslant \mu < 1, \, (x, t, u) \in D_1.$$

Applying the mean-value theorem to the difference in the integrand of the first integral in the right-hand side of (5.70) and the usual Cauchy inequality to the second integral, we obtain the inequality

$$(5.72) \quad \frac{1}{2} \frac{d}{dt} \int_0^\pi v^2 \, dx + (1 - \mu) \int_0^\pi v^2 \, dx \leqslant \left\{ \int_0^\pi v^2 \, dx \right\}^{1/2}$$

$$\cdot \left\{ \int_0^\pi |f((x, t + \tau, u(x, t)) - f(x, t, u(x, t))|^2 \, dx \right\}$$

Since $f(x, t, u)$ is almost periodic in the mean, given $\epsilon > 0$ there exists $l(\epsilon) > 0$, such that any interval of the real axis of length l contains a number τ with the property

$$(5.73) \qquad \int_0^\pi |f(x, t + \tau, u) - f(x, t, u)|^2 \, dx \leqslant \epsilon^2,$$

for all real $u = u(x) \, \epsilon \, L^2(0, \pi)$ satisfying (5.63). For such a value of τ one obtains from (5.72) the following inequality for $v(x, t) = u(x, t + \tau) - u(x, t)$:

$$(5.74) \qquad \int_0^\pi |u(x, t + \tau) - u(x, t)|^2 \, dx \leqslant \epsilon^2 (1 - \mu)^{-2}.$$

Indeed, taking into account (5.73), the inequality (5.72) yields an inequality of the form (5.64), and Lemma 5.2 leads to (5.74).

From (5.74) one sees that the solution $u(x, t)$ is almost periodic in the mean.

Summing up the above discussion, we can state the following result.

Theorem 5.19. *Let* $u(x, t)$ *be a* $C^{(2)}$-*solution of equation (5.46) in (D), satisfying the boundary value conditions (5.62) and the boundedness condition (5.63). If the nonlinearity* $f(x, t, u)$ *is almost periodic in the mean, uniformly with respect to* $u = u(x)$ *satisfying (5.63) and its derivative satisfies the inequality (5.71), then* $u(x, t)$ *is almost periodic in the mean (as a function of t).*

It is useful to note that condition (5.71) is satisfied by functions that are strongly nonlinear in u, such as $f(x, t, u) = \mu u - u^3 - e^u + f(x, t)$, where $f(x, t)$ is an almost periodic function in the mean in (D), and μ satisfies $\mu < 1$.

Bibliographical notes

The results of the first two paragraphs are due to J. Favard (243, 244) and M. Nicolescu (525, 526). In §3 we cited a particular case of a theorem by O. A. Ladyzhenskaya (434).

The almost periodicity of solutions of hyperbolic equations emerges in connection with the study of oscillations, and was approached for the first time by C. F. Muckenhoupt (518). A. S. Avakian (28) has made a study similar to that of C. F. Muckenhoupt, considering the two-dimensional wave equation.

S. Bochner (71) and S. Bochner and J. von Neumann (83) have used methods of functional analysis in order to approach this difficult problem. S. L. Sobolev (595) succeeded in showing that sufficiently regular solutions of the homogeneous wave equation in any number of dimensions are almost periodic in the mean. S. Zaidman (690) has obtained some results regarding the nonhomogeneous wave equation by methods of the theory of semigroups of operators.

In a long series of works, L. Amerio (4, 5, 7, 8, 13, 19) (see also References) has brought remarkable contributions to the problem of almost periodicity of the solutions of hyperbolic equations, systematically using the methods of functional analysis. The articles (9–12, 14, 15, 19) of the same author consider functional equations which contain as a special case the hyperbolic or parabolic equations. In particular, he considers the equation $\dot{x} = Ax + f(t)$, where x and f are elements of a Hilbert space, and A is a linear operator in this space.

The almost periodicity of the solutions of hyperbolic equations was also studied by H. Günzler (298, 304).

For certain elliptic equations, the almost periodicity of the solutions has been studied by S. Zaidman (701).

We mention here the works by C. Foiaş and S. Zaidman (258), G. Prouse (550), and S. Zaidman (693), regarding the almost periodicity of the solutions of parabolic equations.

Lately, some works have appeared regarding the almost periodicity of the solutions of nonlinear partial differential equations. For example, for the Navier–Stokes equations some results were obtained by C. Foias (257) and G. Prouse (553). The latter has considered (554, 555) the hyperbolic equations containing a dissipation term.

A major source on the problem of almost periodicity of partial differential equations (including nonlinear ones) is the book by L. Amerio and G. Prouse (705). See also H. D. Fattorini (726), A. Haraux (735), B. M. Levitan and V. V. Zhikov (745), A. A. Pankov (754), and S. Zaidman (773), (775).

The number of journal papers is considerable. We mention here a few papers in a series dealing with almost periodicity in the case of more than one variable: Y. Sibuya (763), G. R. Sell (762), and D. Petrovanu (755).

Usually, the almost periodicity with respect to the time-variable is investigated for parabolic and hyperbolic equations. In the papers quoted above one looks at the case when the property of almost periodicity is with respect to some space variables.

See also C. Corduneanu (715), (720), and M. Yamaguchi and K. Nishihara (770) for more results in the case of elliptic and hyperbolic equations. In (718), C. Corduneanu uses qualitative inequalities for obtaining almost periodicity criteria. The connection between the almost periods of the forcing term and those of the solution is obtained in a rather simple fashion.

VI

Almost periodic functions with values in Banach spaces

As seen in Chapter I, where almost periodic complex-valued functions of a real variable have been defined by the property that they can be approximated uniformly and with any desired accuracy on the whole line by trigonometric polynomials, there exist two characteristic properties of almost periodic functions: property A and property B. If we want to define almost periodic functions with values in a Banach space, then three possibilities are at hand, since the three properties mentioned earlier can be formulated in terms of Banach spaces. It is unimportant in principle which of these properties is chosen as definition, since, as will be shown, these properties are equivalent in Banach spaces, too. It would thus be possible to follow the method given in Chapter I. However, we think that it is useful for the reader to be aware of another presentation of the theory of almost periodic functions. Therefore, we choose here as definition property B from Chapter I. This in fact is the definition given by H. Bohr in the case of numerical functions.

1. *Definitions and general properties*

In this chapter we shall denote by X a complex Banach space with the norm topology; $x, y, \ldots$ are elements from X of norm $\|x\|, \|y\|, \ldots$; R is the set of real numbers; f, or $f : R \to X$, or $t \to f(t)$, or $x = f(t)$, where $t \in R$, is a function defined on the set of real numbers R with values in the Banach space X. If f is a function defined on R with values in X, and h is a fixed real number, then we shall call the h-translate of f the function $f_h : R \to X$ defined by

$$f_h(t) = f(t + h), \qquad t \in R.$$

A function $T : R \to X$, whose values are given by

$$(6.1) \qquad T(t) = \sum_{k=1}^{n} c_k e^{i\lambda_k t}, \qquad t \in R,$$

where λ_k are real numbers, c_k are elements from X, and i is the imaginary unit, is called *trigonometric polynomial* with values in X.

A continuous function on R, $f : R \to X$ is called *almost periodic*, if for any number $\varepsilon > 0$, one can find a number $l(\varepsilon) > 0$ such that any interval on

153

the real line of length $l(\varepsilon)$ contains at least one point of abscissa τ with the property that

(6.2) $\|f(t + \tau) - f(t)\| < \varepsilon, \qquad t \in R.$

Remark. A number τ for which inequality (6.2) holds true is called ε-translation number of the function f. The above property says that for every $\varepsilon > 0$, the function f has a set of ε-translation numbers which is relatively dense in R.

A continuous function $f : R \rightarrow X$ is called *normal* if any set of translates of f has a subsequence, uniformly convergent on R in the sense of the norm.

A function $f : R \rightarrow X$ is called a function with the *approximation property*, if for any number $\varepsilon > 0$, one can determine a trigonometric polynomial T_ε with values in X, such that

(6.3) $\|f(t) - T_\varepsilon(t)\| < \varepsilon, \qquad t \in R.$

Remark. A function with the approximation property is obviously continuous. The approximation property is equivalent to the fact that there exists at least one sequence of trigonometric polynomials T_n with values in X, uniformly convergent on R to f.

We shall now establish some general properties of almost periodic functions, and the fact that the almost periodic functions, the normal functions and the functions with approximation property coincide.

Theorem 6.1. *An almost periodic function with values in X is bounded in X (i.e., bounded in the norm).*

Proof. Let f be an almost periodic function, and $l = l(1)$ the number corresponding to $\varepsilon = 1$ from the definition of an almost periodic function. On the interval $[0, l]$ the real function $t \rightarrow \|f(t)\|$ is continuous and thus bounded. Let $M > 0$ be such that

(6.4) $\|f(t)\| \leqslant M, \qquad t \in [0, l].$

If t is arbitrary, then the interval $[-t, -t + l]$ contains at least one translation number τ, corresponding to $\varepsilon = 1$. Hence

$$\|f(t)\| \leqslant \|f(t + \tau) - f(t)\| + \|f(t + \tau)\| < 1 + M,$$

since $t + \tau \in [0, l]$, which shows that the function is bounded.

Theorem 6.2. *An almost periodic function is uniformly continuous on R.*

Proof. Let $l = l(\varepsilon/3) > 0$ which corresponds to $\varepsilon/3$, $\varepsilon > 0$ in the definition of an almost periodic function. On the interval $[-1, 1 + l]$ the function f

is continuous and hence it is also uniformly continuous. Let $\delta = \delta(\varepsilon/3) > 0$, $\delta < 1$, such that for t', $t'' \in [-1, \ 1 + l]$ with $|t' - t''| < \delta$ we have that $\|f(t') - f(t'')\| < \varepsilon/3$.

Finally, let t_1, t_2 be such that $|t_1 - t_2| < \delta$, and let τ be an $(\varepsilon/3)$-translation number of f contained in the interval $[-t_1, \ -t_1 + l]$. Since $|t_2 - t_1| < \delta$ and $0 \leqslant t_1 + \tau \leqslant l$, it follows that $t_2 + \tau$ is situated on the interval $[-1, 1 + l]$. Thus

$$\|f(t_2) - f(t_1)\| \leqslant \|f(t_2) - f(t_2 + \tau)\| + \|f(t_2 + \tau) - f(t_1 + \tau)\|$$

$$+ \|f(t_1 + \tau) - f(t_1)\| < \frac{\varepsilon}{3} + \frac{\varepsilon}{3} + \frac{\varepsilon}{3} = \varepsilon,$$

which concludes the proof.

Theorem 6.3. *If f is almost periodic with values in X, then λf, where λ is a complex number and any translate f_h are almost periodic functions. The numerical function $t \to \|f(t)\|$ is also almost periodic.*

Proof. The first two properties are obvious. The last follows from the inequality $|\|f(t + \tau)\| - \|f(t)\|| \leqslant \|f(t + \tau) - f(t)\|$.

Theorem 6.4. *If f_n is a sequence of almost periodic functions with values in X, and if*

$$\lim_{n \to \infty} f_n(t) = f(t),$$

uniformly on R in the sense of convergence in the norm, then f is almost periodic.

Proof. The continuity of f is evident. If $\varepsilon > 0$, then there exists a natural number $N(\varepsilon)$ with the property

(6.5) $\qquad \|f_n(t) - f(t)\| < \frac{\varepsilon}{3}, \qquad t \in R, \qquad n \geqslant N(\varepsilon).$

We fix n_0 for which (6.5) is true, and consider $l(\varepsilon/3)$ determined from the almost periodicity of f_{n_0} and τ an $(\varepsilon/3)$-translation number of f_{n_0}. For any $t \in R$ we shall have

$$\|f(t + \tau) - f(t)\| \leqslant \|f(t + \tau) - f_{n_0}(t + \tau)\| + \|f_{n_0}(t + \tau) - f_{n_0}(t)\|$$

$$+ \|f_{n_0}(t) - f(t)\| < \frac{\varepsilon}{3} + \frac{\varepsilon}{3} + \frac{\varepsilon}{3} = \varepsilon,$$

which proves the almost periodicity of f.

Theorem 6.5. *The set of values of an almost periodic function with values in X is relatively compact in X.*

Proof. Since in the Banach spaces the relatively compact sets coincide with the precompact sets, it is sufficient to show that for any $\varepsilon > 0$, the set of values of the function can be imbedded in a finite number of spheres of radius ε. Let $l = l(\varepsilon/2) > 0$ be the number which corresponds to $\varepsilon/2$ from the definition of an almost periodic function f. Since f is continuous on $[0, l]$, the set of its values on this interval is compact.

Let us denote by $x_1, x_2, \ldots, x_p$ the centers of the spheres of radius $\varepsilon/2$ which cover the set $\{f(t); t \in [0, l]\}$.

For an arbitrary t let us take an $(\varepsilon/2)$-translation number τ in the interval $[-t, -t + l]$, say $t + \tau \in [0, l]$, and let x_i be the center of the sphere of radius $\varepsilon/2$ which contains $f(t + \tau)$. We have

$$\|f(t) - x_i\| \leqslant \|f(t + \tau) - f(t)\| + \|f(t + \tau) - x_i\| < \frac{\varepsilon}{2} + \frac{\varepsilon}{2} = \varepsilon,$$

which shows that the set of values of the function is covered for any $\varepsilon > 0$ by a finite number of spheres of radius ε.

Remark. If f is almost periodic with values in X, then any sequence of values $\{f(t_n)\}$ has convergent subsequences. This observation follows from the fact that in Banach spaces the necessary and sufficient condition for a set to be relatively compact is that any sequence of elements of the set have convergent subsequences.

Let us now show that the set of almost periodic functions coincides with the set of normal functions.

Theorem 6.6. *The necessary and sufficient condition for a continuous function $f: R \to X$ to be almost periodic is that it be normal.*

Proof. The sufficiency of the condition can be established in the same way as in Theorem 1.10 with the difference that the modulus must be replaced by norm when one deals with the values of f.

The necessary condition can be proven by the following well-known method of diagonal extraction. Let

(6.6) $f_{h_1}, f_{h_2}, \ldots, f_{h_n}, \ldots$

be a sequence of translates of f, and let $S = \{s_n\}$ be a dense sequence in R.

From the sequence $\{f_{h_n}(s_1)\} = \{f(s_1 + h_n)\}$, which is a sequence of values of f, let us choose a convergent subsequence (Remark from Theorem 6.5).

Let us denote by

(6.7) $f_{h_{11}}, f_{h_{12}}, \ldots, f_{h_{1n}}, \ldots$

the subsequence of $\{f_{h_n}\}$ convergent in s_1. Applying the previous argument to the sequence (6.7), let us choose a subsequence $\{f_{h_{2n}}\}$ converging in s_2. Continuing in the same manner, we then form the diagonal sequence

(6.8) $f_{h_{11}}, f_{h_{22}}, \ldots, f_{h_{nn}}, \ldots$

which converges pointwise in S. Let us denote this sequence by $\{f_{k_n}\}$. Let us show that this sequence converges uniformly on R in the sense of the norm.

Taking a number $\varepsilon > 0$, let $l = l(\varepsilon/5) > 0$ from the definition of an almost periodic function, and let $\delta = \delta(\varepsilon/5)$ be determined from the uniform continuity of f; we cover the interval $[0, l]$ with a finite number of intervals of length less than δ, and in each of these intervals we choose one point of the set S. We obtain thus the set $S_0 = \{r_1, r_2, \ldots, r_p\}$. Since the set S_0 is finite, the sequence $\{f_{k_n}\}$ is uniformly convergent in S. Therefore, there exists a natural number $N(\varepsilon/5)$, such that for $m, n \geqslant N(\varepsilon/5)$ we have

(6.9) $\| f_{k_n}(r_i) - f_{k_m}(r_i) \| < \dfrac{\varepsilon}{5}, \qquad i = 1, 2, \ldots, p.$

For any t, let τ be an $(\varepsilon/5)$-translation number from the interval $[-t, -t + l]$, i.e., $t + \tau \in [0, l]$, and let r_i be a point from S_0 for which $|t + \tau - r_i| < \delta$. For $n, m \geqslant N(\varepsilon/5)$, it follows that

$$\| f_{k_n}(t) - f_{k_m}(t) \| = \| f(t + k_n) - f(t + k_m) \|$$

$$\leqslant \| f(t + k_n) - f(t + k_n + \tau) \| + \| f(t + k_n + \tau) - f(r_i + k_n) \|$$

$$+ \| f(r_i + k_n) - f(r_i + k_m) \| + \| f(r_i + k_m) - f(t + k_m + \tau) \|$$

$$+ \| f(t + k_m + \tau) - f(t + k_m) \| < 5 \frac{\varepsilon}{5} = \varepsilon,$$

which shows that the sequence $\{f_{k_n}\}$, which is a subsequence of $\{f_{h_n}\}$, satisfies the Cauchy uniform convergence condition on R. The theorem is completely proved.

Theorem 6.7. *The sum of two almost periodic functions with values in X is an almost periodic function.*

Proof. Let f and g be almost periodic functions, and let $\{h_n\}$ be an arbitrary sequence of real numbers. From the sequence of translates $\{f_{h_n}\}$, according to Theorem 6.6 we choose a uniformly convergent subsequence on R, say $\{f_{k_n}\}$. From the sequence of translates $\{g_{k_n}\}$ we choose a subsequence uniformly convergent on R, say $\{g_{l_n}\}$. Then the sequence $\{f_{l_n} + g_{l_n}\}$, which is a subsequence of the sequence $\{f_{h_n} + g_{h_n}\}$, is uniformly convergent on R and the theorem is proved.

Theorem 6.8. *A function $f : R \to X$ with approximation property is almost periodic.*

Proof. According to Theorems 6.4 and 6.7 it is sufficient to prove that any function φ of the form

(6.10) $$\varphi(t) = ce^{i\lambda t}, \qquad t \in R,$$

is almost periodic, where λ is real and c is an element from X. But $\varphi(t)$ is obviously a periodic function.

2.   The Banach space of almost periodic functions

Let us denote by $AP(X)$ the set of almost periodic functions on R with values in the Banach space X. According to Theorems 6.3 and 6.7, $AP(X)$ is a vector space over the field of complex numbers with respect to the usual operations of addition of functions and multiplication by scalars. Since an almost periodic function is bounded in the norm, the mapping defined by

(6.11) $$f \to |f|, \qquad f \in AP(X),$$

where

(6.12) $$|f| = \sup_{t \in R} \| f(t) \|,$$

is a norm on $AP(X)$, a fact which can be easily verified.

One sees without difficulty that the topology of the norm of $AP(X)$ coincides with the topology of uniform convergence on R of sequences of almost periodic functions, which leads to the conclusion that $AP(X)$ is a Banach space (see Theorem 6.4).

The main goal pursued in this paragraph is to establish a compactness criterion for families of almost periodic functions in the topology of $AP(X)$. For this we need an additional statement of an intrinsic interest.

Theorem 6.9. *Let $f_1(t), f_2(t), \ldots, f_m(t)$ be almost periodic functions from R into X. For every $\varepsilon > 0$, there exist common ε-translation numbers for these functions.*

Proof. Let us remark that we can associate with this set of functions an almost periodic function from R into the Banach space X^m. We recall that the norm in X^m can be defined by

$$\|x\| = \sum_{i=1}^{m} \|x_i\|,$$

where $x = (x_1, x_2, \ldots, x_m) \in X^m$. In order to prove the almost periodicity of $f(t) = (f_1(t), f_2(t), \ldots, f_m(t))$, it suffices to prove its normality (see Theorem 6.6). This is a very simple exercise.

From the almost periodicity of $f(t)$ it follows that for every $\varepsilon > 0$, there exists $l(\varepsilon) > 0$ such that any interval of length l on the real axis contains at least one number l with the property

(6.13) $$\|f(t + \tau) - f(t)\| < \varepsilon, \qquad t \in R.$$

Taking in account the definition of the norm in X^m, from (6.13) it follows that

(6.14) $$\|f_i(t + \tau) - f_i(t)\| < \varepsilon, \qquad t \in R, \quad i = 1, 2, \ldots, m,$$

which proves the theorem.

Before stating the compactness criterion, we notice that the definitions for a family of equi-continuous and equi-almost periodic functions can be formulated in the same way as in the case of numerical functions (see Chapter I, §5).

Theorem 6.10. *The necessary and sufficient condition that a family $\mathscr{F}$ of functions from $AP(X)$ be relatively compact is that the following properties hold true:*

1) *$\mathscr{F}$ is equi-continuous;*
2) *$\mathscr{F}$ is equi-almost periodic;*
3) *for any $t \in R$, the set of values of functions from $\mathscr{F}$ be relatively compact in X.*

Necessity. We shall use the fact that a set from a complete metric space is relatively compact, if and only if for any $\varepsilon > 0$ there exists an ε-finite grid (Hausdorff's theorem). If the set $\mathscr{F} \subset AP(X)$ is relatively compact, and $\varepsilon > 0$, then one can derive from the set $\mathscr{F}$ a finite ε-grid $\{f_1, f_2, \ldots, f_m\}$. Since a finite family of almost periodic functions is equi-almost periodic (Theorem 6.9) and equi-continuous, it follows from the definition of an ε-grid that the family of functions $\mathscr{F}$ is equi-almost periodic and equi-continuous. This is a consequence of the fact that for any $f \in \mathscr{F}$ there exist f_p, $1 \leqslant p \leqslant m$, such that $|f - f_p| < \varepsilon$ and

$$(6.15) \quad \|f(t+\tau) - f(t)\| \leqslant \|f(t+\tau) - f_p(t+\tau)\| + \|f_p(t+\tau) - f_p(t)\|$$
$$+ \|f_p(t) - f(t)\|.$$

Since from any sequence $\{f_n\} \subset \mathscr{F}$ one may extract a convergent subsequence in $AP(X)$, i.e., uniformly convergent on R, it follows that the last condition of the theorem is also necessary.

Sufficiency. Since for every $t \in R$ the set of values assumed by the functions of the family is relatively compact, this means that from every sequence of functions of the family one may choose a subsequence $\{f_n\}$, which converges pointwise on a countable set M, dense in R. We show that the sequence $\{f_n\}$ converges uniformly on R.

Let $l = l(\varepsilon/5) > 0$ be determined from the condition of equi-almost periodicity of $\mathscr{F}$, and $\delta = \delta(\varepsilon/5)$ from the condition of equicontinuity of $\mathscr{F}$. We cover the interval $[0, l]$ with p intervals of length smaller than δ, and let $M_0 = \{s_1, s_2, \ldots, s_p\} \subset M$, such that every subinterval of the covering contains an s_i. The sequence $\{f_n\}$ is uniformly convergent on the set M_0, and let $N(\varepsilon/5)$ be determined by the uniform convergence of the sequence $\{f_n\}$ on M_0. For any $t \in R$, the interval $[-t, -t + l]$ contains at least one point of abscissa τ which is an $(\varepsilon/5)$-translation number of all the functions from $\mathscr{F}$. Since $t + \tau \in [0, l]$, let $s_i \in M_0$ be the point for which $|t + \tau - s_i| < \delta$. Finally, for any $n, m \geqslant N(\varepsilon/5)$, it follows that

$$(6.16) \quad \|f_n(t) - f_m(t)\| \leqslant \|f_n(t) - f_n(t+\tau)\| + \|f_n(t+\tau) - f_n(s_i)\|$$
$$+ \|f_n(s_i) - f_m(s_i)\| + \|f_m(s_i) - f_m(t+\tau)\|$$
$$+ \|f_m(t+\tau) - f_m(t)\| < 5 \cdot \frac{\varepsilon}{5} = \varepsilon.$$

The proof of the theorem is complete.

Remark. The space $AP(X)$ is a closed vector subspace in the Banach space $C(X)$ of function, continuous and bounded in R, with values in X, where the norm is given by $\|f\| = \sup_{t \in R} \|f(t)\|$. Therefore Theorem 6.6 can also be stated as follows: The necessary and sufficient condition that a continuous function $f: R \to X$ be almost periodic is that the set of its translates be relatively compact in $C(X)$.

3. The Fourier series associated with an almost periodic function with values in a Banach space. The approximation theorem

After deriving the previous general results regarding the almost periodic functions with values in a Banach space X, we shall now prove the existence of the mean, a fact which will permit us to attach to every function from $AP(X)$ a Fourier series.

Theorem 6.11. *For any almost periodic function f with values in the Banach space X, there exists the mean value*

$$M\{f\} = \lim_{T \to \infty} \frac{1}{T} \int_0^T f(t)\, dt \in X. \tag{6.17}$$

Proof. Let $\varepsilon > 0$ and let $l = l(\varepsilon/2)$ be the number associated with $\varepsilon/2$ in the definition of the almost periodic function f, and $A = \sup_{t \in R} \| f(t) \|$.

If α is a real arbitrary number and ξ is an $(\varepsilon/2)$-translation number from the interval $(\alpha, \alpha + l)$, then

$$\int_\alpha^{\alpha + T} f(t)\, dt = \int_\alpha^\xi f(t)\, dt + \int_\xi^{\xi + T} f(t)\, dt + \int_{\xi + T}^{\alpha + T} f(t)\, dt,$$

whence it follows that

$$\left\| \frac{1}{T} \int_0^T f(t)\, dt - \frac{1}{T} \int_\alpha^{\alpha + T} f(t)\, dt \right\| \leqslant \left\| \frac{1}{T} \int_0^T f(t)\, dt - \frac{1}{T} \int_\xi^{\xi + T} f(t)\, dt \right\|$$

$$+ \left\| \frac{1}{T} \int_\alpha^\xi f(t)\, dt \right\| + \left\| \frac{1}{T} \int_{\xi + T}^{\alpha + T} f(t)\, dt \right\| \leqslant \frac{1}{T} \int_0^T \| f(t) - f(t + \xi) \|\, dt$$

$$+ \frac{1}{T} \left| \int_\alpha^\xi \| f(t) \|\, dt \right| + \frac{1}{T} \left| \int_{\xi + T}^{\alpha + T} \| f(t) \|\, dt \right| < \frac{\varepsilon}{2} + \frac{2A}{T}\, l.$$

Let us take $\alpha = (k - 1)T$, $k = 1, 2, \ldots, n, \ldots$. Consequently,

$$\left\| \frac{1}{T} \int_0^T f(t)\, dt - \frac{1}{T} \int_{(k-1)T}^{kT} f(t)\, dt \right\| < \frac{\varepsilon}{2} + \frac{2Al}{T}, \quad k = 1, 2, \ldots. \tag{6.18}$$

But

$$\left\| \frac{1}{T} \int_0^T f(t)\, dt - \frac{1}{nT} \int_0^{nT} f(t)\, dt \right\|$$

$$= \left\| \frac{1}{T} \int_0^T f(t)\, dt - \frac{1}{nT} \sum_{k=1}^n \int_{(k-1)T}^{kT} f(t)\, dt \right\|$$

$$\leqslant \frac{1}{n} \sum_{k=1}^n \left\| \frac{1}{T} \int_0^T f(t)\, dt - \frac{1}{T} \int_{(k-1)T}^{kT} f(t)\, dt \right\| < \frac{\varepsilon}{2} + \frac{2Al}{T}, \tag{6.19}$$

if we consider (6.18).

Let now T_1 and T_2 be two positive numbers such that $m_1 T_1 = m_2 T_2$, where m_1 and m_2 are two natural numbers. Taking into account (6.19), we obtain

$$\left\| \frac{1}{T_1} \int_0^{T_1} f(t)\, dt - \frac{1}{T_2} \int_0^{T_2} f(t)\, dt \right\| < \varepsilon + 2A \left(\frac{1}{T_1} + \frac{1}{T_2} \right) l. \tag{6.20}$$

This last inequality remains valid for arbitrary T_1, T_2 due to the continuity of the left side with respect to these arguments. If we take $T_1, T_2 > 4Al/\varepsilon$, then it follows that

$$\left\| \frac{1}{T_1} \int_0^{T_1} f(t)\,dt - \frac{1}{T_2} \int_0^{T_2} f(t)\,dt \right\| < 2\varepsilon,$$

which proves the existence of the limit (6.17).

Remark. One may see that the following equality holds

$$\lim_{T \to \infty} \frac{1}{T} \int_a^{a+T} f(t)\,dt = M\{f\}$$

uniformly with respect to $a \in R$.

Indeed, one may easily check that for any translate f_a, $M\{f_a\} = M\{f\}$. It suffices to prove that for every $\varepsilon > 0$ there exists a $T_0(\varepsilon) > 0$, independent of a, such that

$$(6.21) \qquad \left\| \frac{1}{T} \int_0^T f(t+a)\,dt - M\{f_a\} \right\| < \varepsilon, \qquad T > T_0(\varepsilon).$$

If in (6.19) we make $n \to \infty$, we obtain

$$\left\| \frac{1}{T} \int_0^T f(t)\,dt - M\{f\} \right\| \leqslant \frac{\varepsilon}{2} + \frac{2Al}{T}.$$

It follows immediately that (6.21) is true because A and l are independent of a.

Taking $a = -T$, we obtain the following expression for the mean

$$(6.22) \qquad M\{f\} = \lim_{T \to \infty} \frac{1}{2T} \int_{-T}^T f(t)\,dt.$$

We shall not dwell upon elementary properties of the mean value (see Theorem 1.3), although we shall use some of these properties.

The immediate goal is to construct the Fourier series for almost periodic functions with values in a Banach space. We can easily convince ourselves that the method used in Chapter I, §3 cannot be applied in this case. The way which will be followed requires some auxiliary statements. The connection between the Fourier exponents and the almost periods will be made clear.

Lemma 6.1. *Let*

$$(6.23) \qquad \varphi(t) \sim \sum_{k=1}^\infty a_k e^{i\lambda_k t}$$

be a numerical almost periodic function. For any $\varepsilon > 0$, one may determine *a natural number n and a number* $\delta > 0$, $\delta < \pi$, *such that any real number* τ *which is a solution of the system of diophantine inequalities*

(6.24) $|\lambda_k \tau| < \delta \pmod{2\pi}$, $k = 1, 2, \ldots, n$,

is an ε-translation number of the function φ.

Proof. Let $\varepsilon > 0$ be arbitrary. There exists a trigonometric polynomial $S(t) = \sum_{k=1}^{n} b_k e^{i\lambda_k t}$ such that

(6.25) $|\varphi(t) - S(t)| < \dfrac{\varepsilon}{3}$, $t \in R$.

According to Theorem 1.24, we may admit that the exponents of $S(t)$ are among the Fourier exponents of the function φ.

From (6.25) it follows easily that any $(\varepsilon/3)$-translation number of $S(t)$ is an ε-translation number for $\varphi(t)$.

On the other hand, if τ is a solution of the system of inequalities (6.24), this implies that

$$|e^{i\lambda_k \tau} - 1| = \sqrt{(1 - \cos \lambda_k \tau)^2 + \sin^2 \lambda_k \tau} = 2 \left| \sin \frac{\lambda_k \tau}{2} \right| < \delta.$$

Therefore,

$$|S(t + \tau) - S(t)| \leqslant \sum_{k=1}^{n} |b_k| \cdot |e^{i\lambda_k \tau} - 1| < \delta \sum_{k=1}^{n} |b_k| < \frac{\varepsilon}{3},$$

if $3\delta \sum_{k=1}^{n} |b_k| < \varepsilon$. This yields an upper bound for those δ which are admissible in (6.24).

Theorem 6.12. (Kronecker). *Let* λ_k, θ_k, $k = 1, 2, \ldots, n$ *be real arbitrary numbers. In order that the system of diophantine inequalities*

(6.26) $|\lambda_k \tau - \theta_k| < \delta \pmod{2\pi}$, $k = 1, 2, \ldots, n$,

have solutions τ *for any* $\delta > 0$, *it is necessary and sufficient that any relation of form* $\sum_{k=1}^{n} m_k \lambda_k = 0$, *where* m_k *are integers will imply that* $\sum_{k=1}^{n} m_k \theta_k \equiv 0 \pmod{2\pi}$.

Proof. Let us first prove the necessity of the condition. If for any $\delta > 0$, the system (6.26) admits a solution, then for each of these solutions we may write

(6.27) $\lambda_k \tau - \theta_k - 2\pi l_k = \delta_k$, $k = 1, 2, \ldots, n$,

where $-\delta < \delta_k < \delta$, and l_k are integers. If we multiply (6.27) by m_k and add these relations, then considering that $\sum_{k=1}^{n} m_k \lambda_k = 0$, we obtain

$$\sum_{k=1}^{n} m_k \theta_k + 2\pi \sum_{k=1}^{n} m_k l_k = -\sum_{k=1}^{n} m_k \delta_k .$$

It follows that

(6.28)
$$\left| \sum_{k=1}^{n} m_k \theta_k + 2\pi \sum_{k=1}^{n} m_k l_k \right| < \delta \sum_{k=1}^{n} |m_k|.$$

Since $\delta > 0$ is arbitrary, we have that $\sum_{k=1}^{n} m_k \theta_k$ must be an integral multiple of 2π.

In order to prove the sufficiency of the condition, we shall show that the maximum value of the modulus of the function

$$f(t) = 1 + \sum_{k=1}^{n} e^{i(\lambda_k t - \theta_k)}$$

equals $n + 1$, that is, the value of the function

$$F(u_1, u_2, \ldots, u_n) = 1 + u_1 + u_2 + \cdots + u_n$$

at the point $(1, 1, \ldots, 1)$. If p is a natural number, we set

(6.29)
$$[f(t)]^p = \sum_j d_j e^{i\beta_j t},$$

(6.30)
$$[F(u_1, u_2, \ldots, u_n)]^p = \sum a_{n_1, n_2, \ldots, n_n} u_1^{n_1}, u_2^{n_2}, \ldots u_n^{n_n}.$$

We see that expansion (6.29) follows from expansion (6.30) if we make $u_k = e^{i(\lambda_k t - \theta_k)}$, $k = 1, 2, \ldots, n$, and cancel similar terms (for equal β_j). If

$$\sum_{k=1}^{n} m'_k \lambda_k = \sum_{k=1}^{n} m''_k \lambda_k$$

then from the conditions of the theorem follows that

$$\sum_{k=1}^{n} m'_k \theta_k = \sum_{k=1}^{n} m''_k \theta_k \quad (\text{mod } 2\pi).$$

Thus the coefficients of similar terms have equal arguments (mod 2π). Since the modulus of a sum of complex numbers equals the sum of moduli of the terms when the terms have equal arguments (mod 2π), it means that

(6.31)
$$\sum |d_j| = \sum a_{n_1, n_2, \ldots n_n} = [F(1, 1, \ldots, 1)]^p = (n + 1)^p.$$

If we admit that $\sup_{t \in R} |f(t)| = h < n + 1$, then

$$|\alpha_j| = |M\{[f(t)]^p e^{-i\beta_j t}\}| \leq h^p.$$

The number of terms in expansion (6.30) is smaller than $(p + 1)^n$. Thus we may write that $\sum|\alpha_j| < (p + 1)^n h^p$. From (6.31) it follows that $(n + 1)^p < (p + 1)^n h^p$, which is impossible if p is sufficiently large.

Hence $\sup_{t \in R} |f(t)| = n + 1$, showing that there exist values of t for which the quantities $\lambda_k t - \theta_k$, $k = 1, 2, \ldots, n$ are arbitrarily close to integral multiples of 2π.

Remark. If $\lambda_1, \lambda_2, \ldots, \lambda_n$ are such that there is no relation of the form $\sum_{k=1}^{n} m_k \lambda_k = 0$, where m_k are integers not all zero, then it means that system (6.26) has solutions for any $\delta > 0$ and for any $\theta_1, \theta_2, \ldots, \theta_n$.

Lemma 6.2. *Let*

$$\varphi(t) \sim \sum_{k=1}^{\infty} a_k e^{i\lambda_k t}$$

be an almost periodic function, and let λ be a real number which is not a linear combination with rational coefficients of the Fourier exponents λ_k. For any $\varepsilon > 0$, there exists a number δ, $0 < \delta < \pi/2$ and an integer n, such that $\varphi(t)$ has ε-translation numbers τ which satisfy the system

$$(6.32) \quad |\lambda_k \tau| < \delta, \quad k = 1, 2, \ldots, n, \quad |\lambda \tau - \pi| < \delta, (\text{mod } 2\pi).$$

Proof. It follows from Lemma 6.1 that there exist numbers δ and n depending on ε, such that any solution of system (6.24) is an ε-translation number of the function $\varphi(t)$. One should also prove that among these solutions, there exists at least one which satisfies the inequality $|\lambda \tau - \pi| < \delta$. This result follows immediately from the observation made with regard to Kronecker's theorem, since a relation of the form $\sum_{k=1}^{n} m_k \lambda_k + m\lambda = 0$, where m_k and m are integers, is impossible. One takes $\theta_1 = \theta_2 = \cdots = \theta_n = 0$, $\theta_{n+1} = \pi$.

Theorem 6.13. *Let $f(t)$ be an almost periodic function with values in the Banach space X. The quantity*

$$(6.33) \qquad a(\lambda) = M\{f(t)e^{-i\lambda t}\}$$

is different from the null element of X only for an at most countable set of complex values of λ.

Proof. Let us consider the numerical function

$$(6.34) \qquad \varphi(t) = \sup_{u \in R} \|f(u + t) - f(u)\|.$$

Since

$$|\varphi(t+\tau) - \varphi(t)| = \left| \sup_{u \in R} \|f(u+t+\tau) - f(u)\| - \sup_{u \in R} \|f(u+t) - f(u)\| \right|$$

$$\leqslant \sup_{u \in R} \left| \|f(u+t+\tau) - f(u)\| - \|f(u+t) - f(u)\| \right|$$

$$\leqslant \sup_{u \in R} \|f(u+t+\tau) - f(u+t)\| = \sup_{u \in R} \|f(u+\tau) - f(u)\|,$$

$\varphi(t)$ is continuous and almost periodic. That is, any ε-translation number of $f(t)$ is also an ε-translation number of $\varphi(t)$. Conversely, from $\varphi(0) = 0$ it follows that any ε-translation number of $\varphi(t)$ is also an ε-translation number of $f(t)$.

Let $\{\lambda_\alpha\}$ be the set of Fourier exponents of the function $\varphi(t)$. The set $\mathscr{M}$ of linear combinations of the form $r_1 \lambda_1 + r_2 \lambda_2 + \cdots + r_n \lambda_n$, $n = 1, 2, \ldots$, where r_i are rational numbers is a countable set.

We now show that for any $\lambda \bar{\in} \mathscr{M}$, $a(\lambda) = 0$. It follows that $a(\lambda) \neq 0$ for at most a countable set of values of λ.

If $\lambda \bar{\in} \mathscr{M}$ then according to Lemma 6.1 one can find for any $\varepsilon > 0$ an ε-translation number τ of the function $\varphi(t)$, such that

(6.35) $$|\lambda\tau - \pi| < \frac{\pi}{2} \quad (\text{mod } 2\pi).$$

From (6.35) we get that

(6.36) $$|1 - e^{-i\lambda\tau}| > 1.$$

Since

$$a(\lambda) = \lim_{T \to \infty} \frac{1}{T} \int_\tau^{\tau+T} f(t)e^{-i\lambda t}\, dt = e^{-i\lambda\tau} M\{f(t+\tau)e^{-i\lambda t}\}$$

$$= e^{-i\lambda\tau}a(\lambda) + e^{-i\lambda\tau}M\{[f(t+\tau) - f(t)]e^{-i\lambda t}\},$$

and τ is an ε-translation number for $f(t)$, too, it means that

$$\|a(\lambda)\| \cdot |1 - e^{-i\lambda\tau}| \leqslant \varepsilon.$$

Considering (6.35), we obtain that

$$\|a(\lambda)\| < \varepsilon.$$

The theorem is proved since ε is arbitrary.

Let $\lambda_1, \lambda_2, \ldots, \lambda_n, \ldots$ be those values of λ for which $a(\lambda) \neq 0$, and let $A_k = a(\lambda_k)$, $k = 1, 2, \ldots$.

The series $\sum_{k=1}^{\infty} A_k e^{i\lambda_k t}$ is called *Fourier series* associated to the function f. One may also write that

$$(6.37) \qquad\qquad f(t) \sim \sum_{k=1}^{n} A_k e^{i\lambda_k t}.$$

As in the case of numerical almost periodic functions, the Fourier series associated to an almost periodic function with values in a Banach space plays an important role in the study of the properties of this function. We shall not go into details of this problem.

However, it is an important fact that Bochner's summation methods remain valid in the case of almost periodic functions in Banach spaces, although it requires a different proof from that indicated in the case of numerical functions. In this way we shall obtain the approximation theorem.

The proof of the approximation theorem will be based on the uniqueness theorem.

Let ϕ be a continuous linear functional on the Banach space X, i.e., an element of the dual X^*.

Lemma 6.3. *If $f: R \to X$ is an almost periodic function and $\phi \in X^*$, then the composed function $(\phi \circ f)(t) = \phi(f(t))$, $t \in R$, is a numerical almost periodic function.*

Proof. Since ϕ is linear and bounded, the inequality

$$(6.38) \qquad |\phi(f(t+\tau)) - \phi(f(t))| \leqslant \|\Phi\| \cdot \|f(t+\tau) - f(t)\|,$$

is true and the lemma is proved.

Lemma 6.4. *If f is almost periodic from R to X and $\phi \in X^*$, then*

$$(6.39) \qquad\qquad \phi(M\{f\}) = M\{\phi \circ f\}.$$

Proof. Since ϕ is linear and continuous, it is commutative with the integration, so that

$$(6.40) \qquad \phi\left(\frac{1}{T}\int_0^T f(t)\, dt\right) = \frac{1}{T}\int_0^T \phi(f(t))\, dt,$$

where by taking the limit we get (6.39).

Theorem 6.14. *If f and g are almost periodic functions defined on R, with values in the Banach space X and have the same Fourier series, then $f(t) \equiv g(t)$, $t \in R$.*

Proof. According to the hypothesis, for any real λ we have

(6.41) $$M\{f(t)e^{-i\lambda t}\} = M\{g(t)e^{-i\lambda t}\},$$

i.e., the means of the functions $f(t)\,e^{-i\lambda t}$ and $g(t)^{-i\lambda t}$ are equal. Hence, from Lemma 6.4 for $\phi \in X^*$ we have

(6.42) $$M\{\phi(f(t))e^{-i\lambda t}\} = M\{\phi(g(t))e^{-i\lambda t}\},$$

which means that the almost periodic numerical functions $\phi \circ f$ and $\phi \circ g$ have the same Fourier series. Using the uniqueness theorem for the numerical almost periodic functions, we obtain that

(6.43) $$\phi(f(t)) = \phi(g(t)), \qquad t \in R.$$

The equality $f(t) = g(t)$, $t \in R$ follows from a corollary of the known Hahn–Banach theorem, which claims that from $\Phi(x) = \Phi(y)$ for any $\Phi \in X^*$ it follows that $x = y$.

Theorem 6.15. *Assume given an almost periodic function on R with values in the Banach space X,*

$$f(t) \sim \sum_{k=1}^{\infty} A_k e^{i\lambda_k t}.$$

There exists a sequence of trigonometric polynomials

$$\sigma_m(t) = \sum_{k=1}^{n} r_{k,m} A_k e^{i\lambda_k t}, \qquad n = n(m),$$

which converges uniformly to f on the whole real line in the topology of the norm of X, as $m \to \infty$. The numbers $r_{k,m}$ are rational and depend on λ_k and m, but not on A_k.

Proof. As we can see, Theorem 6.15 generalizes Theorem 1.24 in the case of numerical functions. Essentially, the proof is the same, so we shall present it briefly.

Consider the trigonometric polynomial

$$K_n(t) = \frac{1}{n} \frac{\sin^2 (nt/2)}{\sin^2 (t/2)} = \sum_{v=-n}^{n} \left(1 - \frac{|v|}{n}\right) e^{-ivt}.$$

If $\beta_1, \beta_2, \ldots, \beta_n, \ldots$ is a basis of the Fourier exponents $\lambda_1, \lambda_2, \ldots, \lambda_n, \ldots$

of the function f, then let us define

$$\mathcal{K}_m(t) = K_{(m!)^2}\left(\frac{\beta_1 t}{m!}\right) \cdots K_{(m!)^2}\left(\frac{\beta_m t}{m!}\right)$$

$$= \sum \left(1 - \frac{|v_1|}{(m!)^2}\right) \cdots \left(1 - \frac{|v_m|}{(m!)^2}\right) \exp\left(-i\left(\frac{v_1\beta_1}{m!} + \cdots + \frac{v_m\beta_m}{m!}\right)t\right),$$

the summation being extended to $|v_j| \leqslant (m!)^2$, $j = 1, 2, \ldots, m$.

Since $\mathcal{K}_m$ is a numerical almost periodic function and the translate of f, $f_t(u) = f(t + u)$ is almost periodic, the product $\mathcal{K}_m \cdot f_t$ is an almost periodic function with values in X.

Let

$$\sigma_m(t) = M\{\mathcal{K}_m f_t\} = \lim_{T \to \infty} \frac{1}{T} \int_0^T \mathcal{K}_m(u)f(u + t)\, du,$$

which after an elementary calculation becomes

$$(6.44) \quad \sigma_m(t) = \sum \left(1 - \frac{|v_1|}{(m!)^2}\right) \cdots \left(1 - \frac{|v_m|}{(m!)^2}\right)$$

$$\times a\left(\frac{v_1\beta_1}{m!} + \cdots + \frac{v_m\beta_m}{m!}\right) \exp\left(i\left(\frac{v_1\beta_1}{m!} + \cdots + \frac{v_m\beta_m}{m!}\right)t\right),$$

the summation being extended to $|v_j| \leqslant (m!)^2$, $j = 1, 2, \ldots, m$, and $a(\lambda)$ is given by (6.33).

The polynomials $\sigma_m(t)$ are called Bochner–Fejér polynomials of the function f. We can write for m sufficiently large

$$\sigma_m(t) = \sum_{k=1}^n r_{k,m} A_k e^{i\lambda_k t},$$

with

$$r_{k,m} = \left(1 - \frac{|v_1|}{(m!)^2}\right) \cdots \left(1 - \frac{|v_m|}{(m!)^2}\right).$$

$$\lambda_k = \frac{v_1\beta_1}{m!} + \cdots + \frac{v_m\beta_m}{m!} = r_1\beta_1 + \cdots + r_h\beta_h$$

which follows from the same argument as that used in the proof of Theorem 1.24.

Using the definition of the polynomials σ_m, after some upper estimates we obtain

$$(6.45) \qquad \|\sigma_m(t + \tau) - \sigma_m(t)\| \leqslant \sup_t \|f(t + \tau) - f(t)\|, \qquad t \in R$$

since

$$M\left\{ K_{(m!)^2}\left(\frac{\beta_1 t}{m!}\right) \cdots K_{(m!)^2}\left(\frac{\beta_m t}{m!}\right) \right\} = 1.$$

From (6.45) one sees that the sequence $\{\sigma_m(t)\}$ is an equi-continuous and equi-almost periodic sequence. We also show that for any fixed t, the set of values of the sequence $\{\sigma_m(t)\}$ is relatively compact in X. Then to the sequence $\{\sigma_m(t)\}$ one could apply Theorem 6.10. For this purpose we use a result by Phillips (547) (§3.1).

Let X be a Banach space, A a bounded set from X and X^* the dual of X. Let M_A be the Banach space of complex functions bounded on A with the topology of uniform convergence on A. Let T be the linear map from X^* into M_A defined by

$$(T\varphi)(x) = \varphi(x), \qquad x \in A.$$

The needed result can be stated as follows:

The necessary and sufficient condition that A be relatively compact in X is that T be compact.

In order that A be relatively compact it is sufficient that from any bounded sequence $\{\varphi_n\} \subset X^*$ one could extract a subsequence $\{\varphi_{k_n}\}$, uniformly convergent on A.

In our case, if $\{\varphi_n\} \subset X^*$ is a bounded sequence and $f: R \to X$ is almost periodic, then from

$$|\varphi_n(f(t))| \leqslant \|\varphi_n\| \cdot \|f(t)\|,$$

and

$$|\varphi_n(f(t + \tau)) - \varphi_n(f(t))| \leqslant \|\varphi_n\| \cdot \|f(t + \tau) - f(t)\|,$$

it follows that the sequence of numerical functions $\{\varphi_n \circ f\}$ is equi-bounded, equi-continuous, and equi-almost periodic. Based on Theorem 6.10, we may select a subsequence $\{\varphi_{k_n} \circ f\}$, uniformly convergent on R.

We now show that the sequence $\{\varphi_{k_n} \circ \sigma_p\}$ converges uniformly with respect to $t \in R$. We have

$$(6.46) \qquad \varphi_{k_n}(\sigma_p(t)) = \lim_{T \to \infty} \frac{1}{T} \int_t^{t+T} \mathcal{K}_p(u - t)\varphi_{k_n}(f(u)) \, du.$$

Let $\varepsilon > 0$ and $N(\varepsilon)$ be such that for any $m, n \geqslant N(\varepsilon)$,

(6.47)
$$\sup_{u \in R} |\varphi_{k_n}(f(u)) - \varphi_{k_m}(f(u))| < \varepsilon.$$

From (6.46) and (6.47) we get

$$|\varphi_{k_n}(\sigma_p(t)) - \varphi_{k_m}(\sigma_p(t))|$$

$$= \left| \lim_{T \to \infty} \frac{1}{T} \int_t^{t+T} \mathscr{K}_p(u - t) [\varphi_{k_n}(f(u)) - \varphi_{k_m}(f(u))] \, du \right|$$

$$\leqslant \varepsilon \lim_{T \to \infty} \frac{1}{T} \int_t^{t+T} \mathscr{K}_p(u - t) \, du = \varepsilon \lim_{T \to \infty} \frac{1}{T} \int_0^T \mathscr{K}_p(u) \, du = \varepsilon$$

for any $n, m \geqslant N(\varepsilon)$ and $p = 1, 2, \ldots$. We proved thus that for any $t \in R$, the set $\{\sigma_m(t)\}$ is relatively compact in X.

Applying Theorem 6.10 to the sequence $\{\sigma_m\}$, let us retain a subsequence which we also denote by $\{\sigma_m\}$, uniformly convergent on R to a function $g : R \to X$. Obviously, g is almost periodic (Theorem 6.8).

Finally, let us show that $f = g$ using the uniqueness theorem.

Let us compute the Fourier coefficients $\tilde{a}(\lambda)$ of the function g.

$$\tilde{a}(\lambda) = M\{g(t)e^{-i\lambda t}\} = \lim_{m \to \infty} M\{\sigma_m(t)e^{-i\lambda t}\}$$

since the mean is a linear functional on $AP(X)$.

If $\lambda \neq \lambda_n$, $n = 1, 2, \ldots$, then $\tilde{a}(\lambda) = 0$ based on the fact that

$$\lim_{T \to \infty} \frac{1}{T} \int_0^T e^{i\lambda t} e^{-i\mu t} \, dt = \begin{cases} 0 & \text{if } \lambda \neq \mu, \\ 1 & \text{if } \lambda = \mu. \end{cases}$$

Let now $\lambda = \lambda_k$. If m exceeds some m_0, then

(6.48)
$$M\{\sigma_m(t)e^{-i\lambda_k t}\} = \left(1 - \frac{|v_1|}{(m!)^2}\right) \cdots \left(1 - \frac{|v_m|}{(m!)^2}\right) A_k,$$

where $v_i = r_i m!$ for $i \leqslant h$ and $v_i = 0$ for $i > h$.

Since $\lim_{m \to \infty} r_{k,m} = 1$ (see 1.86), from (6.48) it follows that

$$\tilde{a}(\lambda_k) = A_k = a(\lambda_k), \qquad k = 1, 2, \ldots$$

and the theorem is completely proved.

Theorems 6.6, 6.8, and 6.15 together lead to the important result that: the set of almost periodic functions defined in R with values in the Banach space X coincides with the set of continuous normal functions and with the set of functions with approximation property.

4. Almost periodic functions in the Muckenhoupt sense and in the Stepanov sense

In this paragraph we shall apply the theory of almost periodic functions with values in Banach spaces to the study of almost periodic functions in the mean (of order $p \geq 1$; the case $p = 2$ corresponds to the functions studied in Chapter II, §2). These functions are an extension of those studied by C. Muckenhoupt, as revealed for the first time by Bochner.

Let us now suppose that $X = L_p[0, 1]$, $p \geq 1$, and the considered functions may take complex values. As known, $L_p[0, 1]$ is a Banach space, and in the case $p = 2$ it is a Hilbert space. First, we notice that there are no essential differences between $L_p[0, 1]$ and $L_p[a, b]$ or $L_p(\Omega)$, where Ω is a measurable set from an euclidean space. The following properties are valid for these cases, too.

Let $f(x, t)$ be a function defined for all $x \in [0, 1]$ and for any $t \in R$, such that $f(x, t) \in L_p[0, 1]$, for any $t \in R$. We shall say that $f(x, t)$ is almost periodic in the mean of order p, $p \geq 1$, if the function $t \to f(x, t) \in L_p[0, 1]$ is almost periodic.

In other words, $f(x, t)$ is almost periodic in the mean of order p, if there exist some functions $\delta(\varepsilon, t) > 0$ and $l(\varepsilon) > 0$, defined for $\varepsilon > 0$ and $t \in R$ with the following properties:

1) $|t_1 - t| < \delta$ implies $\int_0^1 |f(x, t_1) - f(x, t)|^p \, dx < \varepsilon^p$;

2) any interval of length l on the real line contains at least a point τ such that

$$\int_0^1 |f(x, t + \tau) - f(x, t)|^p \, dx < \varepsilon^p, \qquad t \in R.$$

The general theorems from §1 can be applied to this special case and we obtain the following results:

a) if $f(x, t)$ is an almost periodic function in the mean of order p, then there exists a number $M_f > 0$ such that

$$\left\{ \int_0^1 |f(x, t)|^p \, dx \right\}^{1/p} \leq M_f, \qquad t \in R.$$

b) if $f(x, t)$ is almost periodic in the mean of order p, then for any $\varepsilon > 0$ there is a $\delta(\varepsilon) > 0$ such that

$$|t_1 - t_2| < \delta, \quad t_1, t_2 \in R, \quad \text{implies} \quad \int_0^1 |f(x, t_1) - f(x, t_2)|^p \, dx < \varepsilon^p;$$

c) if $f(x, t)$ is almost periodic in the mean of order p, then from any sequence $\{f(x, t + h_n)\}$ one may extract a subsequence which converges in the mean, of order p, uniformly with respect to $t \in R$;

d) the sum of two almost periodic functions of order p is a function of the same kind.

Conversely, if $f(x, t) \in L_p[0, 1]$ for $t \in R$ and if the condition 1) is satisfied [i.e., if $f(x, t)$ is continuous in the mean of order p], then the normality in the sense of the convergence in the mean of order p, uniformly with respect to $t \in R$ implies the almost periodicity in the mean of order p for the function $f(x, t)$.

The theory of the Fourier series may also be applied, but we shall not dwell upon this. We only note that from the general theory presented in §3 follows the existence of the limit

$$\lim_{T \to \infty} \frac{1}{T} \int_0^T f(x, t)\, dt,$$

in the sense of the convergence in $L_p[0, 1]$.

Let us now turn to the notion of almost periodic function in the Stepanov sense. We shall indicate Stepanov's definition, and then we shall show how the theory of the almost periodic functions in the Stepanov sense is included in the theory of almost periodic functions with values in a Banach space.

Let $f(t)$ be a numerical function defined almost everywhere in R, such that $f \in L_p[a, b]$ for any bounded interval $[a, b] \in R$.

We shall say that the function $f(t)$ is S^p-almost periodic if for any $\varepsilon > 0$ there is a number $l(\varepsilon) > 0$, such that any interval of length l of the real line contains at least one point τ for which

(6.49)
$$\sup_{x \in R} \left\{ \int_x^{x+1} |f(t + \tau) - f(t)|^p \, dt \right\}^{1/p} < \varepsilon.$$

One may show that the same class of functions is obtained by replacing (6.49) with

(6.50)
$$\sup_{x \in R} \left\{ \frac{1}{\alpha} \int_x^{x+\alpha} |f(t + \tau) - f(t)|^p \, dt \right\}^{1/p} < \varepsilon,$$

where α is an arbitrary positive number.

Let now $f(t)$ be an S^p-almost periodic function. Consider the function of two variables $\varphi(x, t) \equiv f(x + t)$, defined for $0 \leqslant x \leqslant 1$ and $t \in R$. From (6.49) it follows that

(6.51)
$$\left\{ \int_0^1 |\varphi(x, t + \tau) - \varphi(x, t)|^p \, dx \right\}^{1/p} < \varepsilon, \qquad t \in R.$$

Since

$$\lim_{h \to 0} \int_0^1 |\varphi(x, t + h) - \varphi(x, t)|^p \, dx = 0$$

it means that $\varphi(x, t)$ is an almost periodic function in the mean of order p.

Properties a, b, and c stated above for almost periodic functions in the Muckenhoupt sense hold also for S^p-almost periodic functions.

Let us now give a condition which shows when an S^p-almost periodic function can be reduced to an almost periodic function in the Bohr sense.

Theorem 6.16. *If an S^p-almost periodic function $(p \geqslant 1)$ is uniformly continuous on the real line, then it is almost periodic in the Bohr sense.*

Proof. Hölder's inequality

$$\int_x^{x+1} |f(x+\tau) - f(x)|\, dx \leqslant \left\{ \int_x^{x+1} |f(x+\tau) - f(x)|^p\, dx \right\}^{1/p}$$

shows us that it suffices to consider only the case $p = 1$.

Consider the function

$$\varphi_h(x) = \frac{1}{h} \int_0^h f(x+t)\, dt, \qquad h > 0.$$

One finds out that

$$|\varphi_h(x+\tau) - \varphi_h(x)| \leqslant \frac{1}{h} \int_x^{x+h} |f(t+\tau) - f(t)|\, dt.$$

Thus $\varphi_h(x)$ is continuous and almost periodic in the Bohr sense, if one considers that $f(x)$ is S^1-almost periodic.

We now note that

$$|\varphi_h(x) - f(x)| = \frac{1}{h} \left| \int_0^h [f(x+t) - f(x)]\, dt \right| \leqslant \frac{1}{h} \int_0^h |f(x+t) - f(x)|\, dt.$$

But $0 \leqslant t \leqslant h$. The uniform continuity of the function $f(x)$ leads to $|f(x+t) - f(x)| < \varepsilon$, if $h < \delta(\varepsilon)$. Therefore,

$$|f_h(x) - f(x)| < \varepsilon, \qquad x \in R \qquad \text{if } h < \delta(\varepsilon),$$

which shows us that $f(x)$ is the limit in the sense of uniform convergence on the whole line of almost periodic functions in the Bohr sense.

According to Theorem 1.6, $f(x)$ is almost periodic in the Bohr sense.

Remark. This theorem shows that Stepanov's definition is more general than that given by Bohr only outside the class of functions which are uniformly continuous on the real line.

Another example of almost periodic function defined on R with values in a Banach space are the almost periodic functions which depend uniformly on parameters, when the set Ω is compact (Chapter II, §1). If Ω is a compact set from the n-dimensional complex space, let us denote by $C(\Omega)$ the Banach space of the complex functions which are continuous and defined on Ω.

Let $f(Z, t)$ be a complex valued function, continuous and defined in $\Omega \times R$. We define the function $\varphi : R \to C(\Omega)$ by $\varphi(t) = f(\cdot, t)$ with $\|\varphi(t)\| = \sup_{Z \in \Omega} \|f(Z, t)\|$.

From

$$\|\varphi(t + \tau) - \varphi(t)\| = \sup_{Z \in \Omega} |f(Z, t + \tau) - f(Z, t)|$$

it follows that the definition of almost periodicity of φ coincides with the definition of almost periodicity of f with respect to t, uniformly with respect to $Z \in \Omega$.

If Ω is an open set, then there exists an increasing sequence of compact sets $\{\Omega_n\}$ such that $\bigcup_{n=1}^{\infty} \Omega_n = \Omega$. The set of continuous functions on Ω with complex values forms a locally convex space, the topology being defined by the family of seminorms

$$\|f(Z)\|_n = \sup |f(Z)|, \quad Z \in \Omega_n.$$

Let $\mathscr{C}(\Omega)$ be this space.

It is natural to call the function $f(Z, t)$ which is continuous in $\Omega \times R$, *almost periodic* with respect to t, if the map $\varphi : R \to \mathscr{C}(\Omega)$ defined by the same formula as above is almost periodic. The definition of almost periodic functions with values in a local convex space is given in the following paragraph.

5. Weakly almost periodic functions. The primitive of an almost periodic function

The purpose of this paragraph is the introduction of the notion of weakly almost periodic function in order to give a new characterization of almost periodic functions with values in a Banach space, and to obtain some results concerning the primitive of an almost periodic function. Most of the results of this paragraph are due to L. Amerio.

Let X_{lc} be a separated (Hausdorff) locally convex space with respect to the field of complex numbers. The topology of such a space may be defined with a sufficient family of seminorms $\{p_\alpha ; \alpha \in A\}$. Any neighborhood U of the origin is determined by an $\varepsilon > 0$ and a finite number of seminorms, by

(6.63) $\quad U = U(0, p_{\alpha_1}, \ldots, p_{\alpha_n}, \varepsilon) = \{x ; p_{\alpha_i}(x) < \varepsilon, \quad i = 1, 2, \ldots, n\}.$

The neighborhoods U are symmetric, convex, balanced, and absorbing. A function $f : R \to X_{lc}$, continuous in R is called *almost periodic* if for any neighborhood U of the origin there is a number $l = l(U) > 0$, such that any interval of the real line of length l, contains at least one point of abscissa τ with the property

(6.54) $$f(t + \tau) - f(t) \in U, \qquad t \in R$$

Adapting the proofs given in §1 of this chapter, one may easily find the following results: the set of values of an almost periodic function $f: R \to X_{lc}$ is precompact in X_{lc} and thus bounded; an almost periodic function $f: R \to X_{lc}$ is uniformly continuous in R; the limit of a uniformly convergent sequence on R of almost periodic functions is almost periodic.

Let us now consider a Banach space X, and let us denote by X_w the set X with the weak topology associated with the structure of Banach space X. As known, X_w is a locally convex separated space, whose topology is given by the sufficient family of seminorms $\{p_\varphi ; \varphi \in X^*\}$ defined by

$$p_\varphi(x) = |\varphi(x)|, \qquad x \in X,$$

where X^* is the dual of X.

A function $f: R \to X$ is called *weakly almost periodic*, if the function $f: R \to X_w$ is almost periodic.

Theorem 6.17. *Let X be a Banach space. The necessary and sufficient condition that the function $f: R \to X$ be weakly almost periodic is that for any $\varphi \in X^*$ the numerical function $t \to \varphi(f(t))$, $t \in R$ be almost periodic.*

Proof. If f is weakly almost periodic (and thus weakly continuous), for any fixed φ, the numerical function $\varphi \circ f$ is continuous in R. Writing the definition of the weakly almost periodic function with the neighborhood $U(\theta, \varphi, \varepsilon)$, we get for the corresponding ε-translation numbers that

(6.56) $$|\varphi(f(t + \tau)) - \varphi(f(t))| < \varepsilon, \qquad t \in R,$$

which shows that $\varphi \circ f$ is almost periodic. The necessity is thus proven.

To prove the sufficiency, let $U = U(\theta, \varphi_1, \varphi_2, \ldots, \varphi_n, \varepsilon)$ be a neighborhood of the origin. Since $\varphi_i \circ f, i = 1, 2, \ldots, n$, are almost periodic numerical functions, they form an equi-almost periodic family (there exist thus common almost periods). Writing (6.56) for $\varphi_i, i = 1, 2, \ldots, n$, this is equivalent to $f(t + \tau) - f(t) \in U$, which shows us that f is weakly almost periodic.

Theorem 6.18. *If X is a Banach space, then the necessary and sufficient condition that a function $f: R \to X$ be almost periodic is that f be weakly almost periodic and the set of its values be relatively compact in X.*

Proof. The necessity follows from the fact that a strongly continuous function is also weakly continuous, from Lemma 6.3 and Theorem 6.5.

To prove the sufficiency, we shall use again Phillips' result [547] (§3.1), used also in the proof of Theorem 6.15.

Since the set $H_f = \{f(t); t \in R\}$ is relatively compact in X, Phillips' theorem implies that from any bounded sequence $\{\varphi_m\} \subset X^*$ we can retain a uniformly convergent subsequence on H_f; in other words, from any bounded sequence $\{\varphi_n\} \subset X^*$, may be extracted a subsequence $\{\varphi_{k_n}\}$ such that the sequence $\{\varphi_{k_n} \circ f\}$ is uniformly convergent on R. Thus the set of functions $\{\varphi \circ f\}$ is relatively compact in the topology of uniform convergence on R of the bounded functions, a topology of Banach space. Since the functions $\varphi \circ f$ are almost periodic it follows that the set $\{\varphi \circ f; \varphi \in X^*, \|\varphi\| \leqslant 1\}$ is equi-almost periodic and equicontinuous.

For t', t'' fixed from R, it follows that

$$\|f(t') - f(t'')\| = \sup_{\|\varphi\| \leqslant 1} |\varphi(f(t')) - \varphi(f(t''))|,$$

which together with the equi-continuity and equi-almost periodicity of the set $\{\varphi \circ f; \varphi \in X^*, \|\varphi\| \leqslant 1\}$, shows that the function $f: R \to X$ is continuous and almost periodic.

A classical result from the theory of almost periodic functions is that concerning the almost periodicity of the primitive of an almost periodic numerical function f:

$$(6.57) \qquad F(t) = \int_0^t f(u)\, du.$$

Theorem 4.1 clarifies this problem under the hypothesis of boundedness for F.

If f is almost periodic from R in the Banach space X, the boundedness of F does not generally imply the almost periodicity of F. The following theorem is true:

Theorem 6.19. *If f is almost periodic from R into the Banach space X, and the set of values of F, defined by (6.57) is relatively compact in X, then F is almost periodic.*

Proof. It follows from (6.57) that for any $\varphi \in X^*$

$$(6.58) \qquad \varphi(F(t)) = \int_0^t \varphi(f(u))\, du.$$

From (6.58) we get that $\varphi \circ F$ is almost periodic since it coincides with the bounded primitive of an almost periodic numerical function. Therefore F

is weakly almost periodic. Since the set of its values is relatively compact in X, Theorem 6.18 yields the almost periodicity of F.

The condition of relative compactness of F is only sufficient and thus one needs classes of Banach spaces in which the almost periodicity of the primitive is preserved under the hypothesis of boundedness. A remarkable result in this direction was obtained by L. Amerio. First, we shall establish two lemmas.

Lemma 6.5. *If an equi-continuous and equi-almost periodic sequence of almost periodic functions from R in the space X converges pointwise on a dense set from R, then the sequence converges uniformly in R.*

The proof of this lemma is included in the final part of the proof of Theorem 6.10.

Lemma 6.6. *Let F be a weakly almost periodic function from R in the Banach space X. If for a sequence of real numbers $\{h_n\}$ the sequence of functions $\{F(t + h_n)\}$ converges weakly pointwise in R to the function $\tilde{F}(t)$, then*

$$(6.59) \qquad \sup_{t \in R} \|F(t)\| = \sup_{t \in R} \|\tilde{F}(t)\|.$$

Proof. If we fix a $\varphi \in X^*$, then $\varphi \circ F$ is almost periodic, and the sequence $\{\varphi(F(t + h_n))\}$ is equi-continuous and equi-almost periodic. This sequence of numerical functions, converges pointwise on R. Thus, according to Lemma 6.5 this sequence will be uniformly convergent on R to $\varphi \circ \tilde{F}$. Since $\varphi \circ \tilde{F}$ is almost periodic for any $\varphi \in X^*$, it follows that F is weakly almost periodic.

Since for a given φ the sequence $\{\varphi(F(t + h_n))\}$ converges uniformly on R to $\varphi(\tilde{F}(t))$, the sequence $\{\varphi(\tilde{F}(t - h_n))\}$ converges uniformly on R to $\varphi(F(t))$. Thus

$$(6.60) \qquad \lim_{n \to \infty}{}^* F(t + h_n) = \tilde{F}(t), \qquad \text{uniformly on } R,$$

and

$$(6.61) \qquad \lim_{n \to \infty}{}^* \tilde{F}(t - h_n) = F(t), \qquad \text{uniformly on } R.$$

The sign * indicates the convergence in the weak topology of the Banach space X.

For $\varphi \in X^*$ with $\|\varphi\| \leqslant 1$ and an $\varepsilon > 0$, it follows from (6.60) for n sufficiently large that

$$(6.62) \qquad |\varphi(\tilde{F}(t))| < |\varphi(F(t + h_n))| + \varepsilon \leqslant \|\varphi\|.$$

$$\|F(t + h_n)\| + \varepsilon \leqslant \sup_{t \in R} \|F(t)\| + \varepsilon.$$

If we fix t in (6.62), and consider that $\|x\| = \sup|\varphi(x)|$, $\|\varphi\| \leqslant 1$ we obtain

$$\|\tilde{F}(t)\| \leqslant \sup_{t \in R}\|F(t)\| + \varepsilon,$$

whence

(6.63) $$\sup_{t \in R}\|\tilde{F}(t)\| \leqslant \sup_{t \in R}\|F(t)\|.$$

Making a similar argument but starting from (6.61), we find that

(6.64) $$\sup_{t \in R}\|F(t)\| \leqslant \sup_{t \in R}\|\tilde{F}(t)\|,$$

which together with (6.63) leads to (6.59).

Theorem 6.20. *Let f be an almost periodic function from R in the uniformly convex Banach space X. The primitive of f*

$$F(t) = \int_0^t f(u)\, du,$$

is almost periodic if and only if it is bounded in X.

Proof. The necessity follows from Theorem 6.1.

Before proving the sufficiency, let us recall that a Banach space X is called uniformly convex if for any number σ, $0 < \sigma \leqslant 2$, there exists a number $\delta(\sigma) > 0$ such that from $\|x\| \leqslant 1$, $\|y\| \leqslant 1$, $\|x - y\| \geqslant \sigma$ will follow that $\|x + y\| \leqslant 2(1 - \delta(\sigma))$.

If x and y are arbitrary (and not only in the unit sphere), then the condition of uniform convexity becomes:

(6.65) $$\|x - y\| \geqslant \sigma \max(\|x\|, \|y\|),$$

implies that

(6.66) $$\left\|\frac{x + y}{2}\right\| \leqslant (1 - \delta(\sigma)) \max(\|x\|, \|y\|).$$

To prove that F is almost periodic, we shall show that it is weakly almost periodic and the set of its values is relatively compact in X.

The fact that F is weakly almost periodic follows from

$$\varphi(F(t)) = \int_0^t \varphi(f(u))\, du$$

and

$$|\varphi(F(t))| \leqslant \|\varphi\| \cdot \|F(t)\|,$$

where φ is an arbitrary element of X^*. Let us now suppose that the set of values of F is not relatively compact in X. There exists then a $\sigma > 0$ and a sequence $\{h_n\} \subset R$, such that

(6.67) $$\|F(h_j) - F(h_k)\| \geq 2\sigma, \qquad j \neq k.$$

From the normality of f follows that from the sequence $\{h_n\}$ we can extract a subsequence which will be denoted also by $\{h_n\}$, such that

(6.68) $$\lim_{n \to \infty} f(t + h_n) = \tilde{f}(t), \qquad \text{uniformly on } R.$$

Since the sequence $\{F(h_n)\}$ is bounded in X, and X is reflexive (any uniformly convex Banach space is reflexive),† we can extract a subsequence denoted also by $\{F(h_n)\}$, which is weakly convergent

(6.69) $$\lim_{n \to \infty}{}^* F(h_n) = c \in X.$$

For any $t \in R$ we have

$$F(t + h_k) = F(h_k) + \int_{h_k}^{h_k + t} f(u)\, du = F(h_k) + \int_0^t f(u + h_k)\, du,$$

whence it follows that $\{F(t + h_k)\}$ is weakly uniformly convergent in any finite interval

(6.70) $$\lim_{k \to \infty}{}^* F(t + h_k) = c + \int_0^t \tilde{f}(u)\, du = \tilde{F}(u).$$

Since F is weakly almost periodic, from Lemma 6.6 it follows that

(6.71) $$\sup_{t \in R} \|\tilde{F}(t)\| = \sup_{t \in R} \|F(t)\| = M.$$

One finds that for a fixed t we have that

$$\|F(t + h_j) - F(t + h_k)\| = \left\| F(h_j) - F(h_k) + \int_0^t [f(u + h_j) - f(u + h_k)]\, du \right\|$$

$$\geq \|F(h_j) - F(h_k)\| - \left\| \int_0^t [f(u + h_j) - f(u + h_k)]\, du \right\|$$

$$\geq 2\sigma - \left\| \int_0^t [f(u + h_j) - f(u + h_k)]\, du \right\| \geq \sigma$$

$$\geq \sigma M^{-1} \max\{\|F(t + h_j)\|, \|F(t + h_k)\|\}, \qquad j, k > n_0(t).$$

From the last inequality and the uniform convexity of the space X, it

† M. M. Day: Normed linear spaces. Springer Verlag, Berlin, 1962.

follows that for j, $k > n_0(t)$,

$$\left\| \frac{F(t + h_j) + F(t + h_k)}{2} \right\| \leqslant (1 - \delta(\sigma M^{-1}))\max(\|F(t + h_j)\|, \|F(t + h_k)\|)$$

$$\leqslant (1 - \delta(\sigma M^{-1}))M.$$

For an arbitrary φ from X^* with $\|\varphi\| = 1$, we have

$$(6.72) \quad \left| \varphi\left(\frac{F(t + h_j) + F(t + h_k)}{2} \right) \right| \leqslant \|\varphi\| \cdot \left\| \frac{F(t + h_j) + F(t + h_k)}{2} \right\|$$

$$\leqslant (1 - \delta(\sigma M^{-1}))M.$$

Taking the limit in (6.72), we obtain by (6.70)

$$|\varphi(\tilde{F}(t))| \leqslant (1 - \delta(\sigma M^{-1}))M$$

whence

$$\|\tilde{F}(t)\| \leqslant (1 - \delta(\sigma M^{-1})M$$

and

$$\sup_{t \in R} \|\tilde{F}(t)\| \leqslant (1 - \delta(\sigma M^{-1}))M < M = \sup_{t \in R} \|F(t)\|,$$

which contradicts (6.59). The theorem is completely proved.

Corollary. If f is almost periodic from R in a Hilbert space H, and the primitive of f is bounded in H, then the primitive is almost periodic. The claim of the corollary is true, since a Hilbert space is uniformly convex. The results of this paragraph have numerous applications to the theory of partial differential equations.

In concluding this paragraph, we shall consider an example which shows that the boundedness of the primitive of an almost periodic function with values in a Banach space does not suffice (in general) for its almost periodicity.

Let c_0 be the space of real-valued sequences that converging to zero. The norm in c_0 is given by $|x| = \sup |x_n|$, $n = 1, 2, \ldots$. Let $f : R \to c_0$ be defined as follows: $f(t) = \{f_n(t)\}$, $f_n(t) = n^{-1}\phi(n^{-1}t)$, $n = 1, 2, \ldots$, where $\phi(t)$ is a real-valued almost periodic function with a bounded (hence, almost periodic) integral. For instance, $\phi(t)$ can be chosen to be any trigonometric polynomial for which zero is not among the Fourier exponents. Since $\phi(t)$ is bounded on R, it is obvious that f is a map from R into c_0. Let us denote by $\Phi(t)$ any primitive of $\phi(t)$.

First, let us note that $f(t)$ is almost periodic as a map from R into c_0. Indeed, any linear functional on c_0 can be represented as $L(x) = \Sigma\alpha_n x_n$, with $\Sigma|\alpha_n| < \infty$. This means that $L(f)(t)$ is almost periodic for any linear functional L. Therefore, $f(t)$ is weakly almost periodic. On the other hand, a subset $M \subset c_0$ is relatively compact if and only if it is uniformly bounded in c_0, and $x_n \to 0$ as $n \to \infty$, uniformly with respect to $x \in M$. By means of this criterion one can easily see that $f(t)$ has a relatively compact range in c_0. Using Theorem 6.18, we obtain the almost periodicity of $f(t)$.

Second, we notice that $F(t) = \int_0^t f(s)\, ds = \{\Phi(n^{-1}t) - \Phi(0)\}$ is weakly almost periodic. We can show that the range of $F(t)$ is not relatively compact in c_0. Indeed, if the range of $F(t)$ were relatively compact in c_0, then the convergence $\Phi(n^{-1}t) - \Phi(0) \to 0$ would be uniform with respect to $t \in R$ (as $n \to \infty$). Since $\Phi(t)$ is an almost periodic function which does not reduce to a constant, the above property cannot be true. For every n, one can get large values of t such that $\Phi(n^{-1}t)$ is close to the sup $\Phi(t)$ or to inf $\Phi(t)$.

The above example turns out to be very characteristic for the problem of the almost periodicity of the primitive of an almost periodic function with values in a Banach space. As shown by M. I. Kadets, the following result holds true.

Theorem 6.21. *Let X be a Banach space which does not possess a subspace isomorphic to the Banach space c_0. Then the primitive of an almost periodic function from R into X is almost periodic if and only if it is bounded.*

The complete proof of Theorem 6.21 can be found in the book by B. M. Levitan and V. V. Zhikov (745).

Bibliographical notes

Paragraphs 1 and 3 of this chapter have been written on the basis of the paper by S. Bochner (68) and a synthesis work by L. Amerio (23). The proof of Theorem 6.15 is close to that given by S. Zaidman (694).

The notion of weakly almost periodic function was introduced by L. Amerio (9), who gave the important Theorem 6.17. The proof of this theorem is that which was recently given by B. M. Levitan (467), using a result of R. S. Phillips (547) regarding the characterization of relatively compact sets in Banach spaces.

Regarding the almost periodicity of the primitive of an almost periodic function, we have included together with the classical result of S. Bochner (69) established here with the help of the notion of weakly almost periodic function, Theorem 6.20 of L. Amerio (7, 23). Another recent contribution regarding the almost periodicity of the primitive of an almost periodic function is that made by B. M. Levitan (467): Let $f(t)$ be an almost periodic function with values in a Banach space X, and let $F(t)$ be a primitive of $f(t)$. If the mean of $F(t)$ exists, uniformly with respect to $u \in R$,

$$M\{F\} = \lim_{T \to \infty} \frac{1}{T} \int_u^{u+T} F(t) \, dt,$$

and

$$\sup_{u \in R} \frac{1}{L} \int_u^{u+L} \|F(t)\| \, dt < \infty,$$

for any $L > 0$, then $F(t)$ is almost periodic.

The following problem formulated by L. Amerio (23) remains open: is the boundedness of $F(t)$ sufficient for its almost periodicity, if X is a reflexive space?

The fundamental theorems of the theory of almost periodic functions with values in a Banach space have been redemonstrated by J. Kopeć (396) by the methods of functional analysis.

Among numerous generalizations of the notion of almost periodic function we have chosen in this chapter that by Stepanov. A more comprehensive generalization was given by H. Weyl (661). The monograph of A. S. Besicovitch (49) studies in detail different generalizations of the notion of almost periodicity, based on the idea of completeness of a metric space. It is obvious that the space of almost periodic functions $AP(X)$ may be obtained by completing the metric space of trigonometric polynomials (with coefficients from X), with respect to the metric $d(f, g) = \sup \|f(t) - g(t)\|$, $t \in R$. If one chooses another metric, then one obtains another space of almost periodic functions. For example, to obtain the notion of almost periodic function in the Besicovitch sense, one starts from the metric

$$d(f, g) = \left\{ \lim_{T \to \infty} \sup \frac{1}{2T} \int_{-T}^{T} |f(t) - g(t)|^p \, dt \right\}^{1/p}$$

where $p \geqslant 1$. This leads to B^p-almost periodic functions. One shows that for any B^p-almost periodic function there exists a mean in the Bohr sense, and therefore with any function of this type one may associate a Fourier series. A remarkable fact from the theory of B^p-almost periodic function is the following: If $\{a_n\}$ is a sequence of complex numbers such that $\sum |a_n|^2 < \infty$, then there exists a B^2-almost periodic function whose Fourier coefficients are nothing else but the numbers a_n. For a detailed study of almost periodic functions in the Stepanov, Weyl, or Besicovitch sense, we refer the reader to the monograph by B. M. Levitan (465). A. S. Kovanko (401) has introduced and studied a class of almost periodic functions in the Besicovitch sense. In a series of papers (see References)

A. S. Kovanko studies the fundamental properties of different spaces of almost periodic functions. We shall present here the compactness criterion for the families of S^p-almost periodic functions.

Let $\mathscr{F}$ be a family of S^p-almost periodic functions. It is relatively compact in S^p if and only if the following conditions hold:

1) For any $h > 0$, the set of almost periodic functions in the Bohr sense

$$f_h(x) = \frac{1}{h} \int_x^{x+h} f(t)\, dt, \qquad f \in \mathscr{F},$$

is compact in the sense of uniform convergence on R.

2) For any $\varepsilon > 0$, one may determine a $\delta(\varepsilon)$ such that $h < \delta$ will imply

$$\sup_{x \in R} \left\{ \int_x^{x+1} |f(t) - f_h(t)|^p\, dt \right\}^{1/p} < \varepsilon.$$

We recall that the set of S^p-almost periodic functions is a metric space in which the function defining distance is

$$d(f, g) = \sup_{x \in R} \left\{ \int_x^{x+1} |f(t) - g(t)|^p\, dt \right\}^{1/p}.$$

H. Tornehave (626) has studied almost periodic function of real variable with values in a metric space. The definition of these functions is similar to the classical Bohr definition, this time the modulus being replaced by the distance. In the approximation theorem for these functions one requires the local connection by arcs of the space of values and its completeness. The functions by which one approximates are of the form $f(\alpha_1 t, \alpha_2 t, \ldots, \alpha_n t)$, $t \in R$, with values in the same space, so that $f(u_1, u_2, \ldots, u_n)$ is periodic with respect to each of the real arguments $u_1, u_2, \ldots, u_n$, of period $2n\pi$. The natural number n depends on the function which it approximated and on the accuracy of approximation, and $\alpha_1, \alpha_2, \ldots, \alpha_n$ are real numbers, linearly rational independent. One shows that the approximation theorem remains generally invalid if one does not require the completeness of the space of values.

Considering the generalized translation operators, B. M. Levitan (452) defined the notion of almost periodicity with respect to a family of such operators. That is if $\{T^s\}$; $s \in R$, is a family of operators of this type (operating in different spaces of continuous functions on the real line), then one says that the function f is almost periodic with respect to the family $\{T^s\}$, $s \in R$, if the family of functions $\{T^s f\}$; $s \in K$, is normal (in the sense of uniform convergence on the real line). Levitan has studied extensively such classes of operators, their connection with harmonical analysis, and has indicated numerous applications.

Almost periodic functions with values in Banach spaces, and even more generally, in locally convex spaces or metric spaces, have been investigated by many authors during the last two decades. We mention first the monographs by L. Amerio and G. Prouse (705), B. M. Levitan and V. V. Zhikov (745), A. A. Pankov (754), and S. Zaidman (775). Basic results, as well as the more specialized results required for applications to partial differential equations, are included in the above-quoted references. They also contain further valuable references on the subject. Indeed all these monographs, taken together, provide a rather complete description of the status of this theory.

Several authors have contributed to the theory in a considerable number of journal papers. While no claim of completeness of the list of references can be advanced, we shall quote here a few items in which significant contributions have been made to this theory. In his papers (706), (707), Bolis Basit is concerned with the problem of almost periodicity of the primitive of an almost periodic function, or the boundedness of the primitive when the given function is only uniformly continuous. He provides boundedness conditions by having recourse to spectral properties. In (721), C. Corduneanu and J. A. Goldstein obtain almost periodicity criteria for the solutions of abstract differential equations. Almost periodicity criteria for the solutions of evolution equations can be found in the monograph by A. Haraux (735) and in many papers published by him in recent years. In particular, he deals with some generalized concepts of almost periodicity. In (757) A. Precupanu investigates the properties of almost periodic functions with values in a Fréchet space, obtaining most of the classical types of almost periodic functions as special cases (including functions defined on abstract groups). Interesting contributions regarding the almost periodicity of solutions of abstract differential equations are contained in A. A. Pankov (754), S. Zaidman (772), (773), (778).

Various generalizations of almost periodic functions and properties of families of almost periodic functions can be found in references (733), (767), (768), (756). For instance, in (767) a characterization of those functions whose derivative (a.e.) is almost periodic in the sense of Stepanov is provided. In (756), compactness criteria are given for families of almost periodic functions.

Another proof of the approximation theorem for almost periodic functions with values in a Banach space, simpler than the proof given in paragraph 3, can be found in the book (744) by M. A. Krasnoselskii, et al.

VII

Almost periodic functions on groups

In this chapter we shall consider another generalization of the notion of almost periodicity, due to J. Von Neumann, concerning numerical functions defined on groups. The main results of the H. Bohr theory were extended to this class of functions. These results are: the existence of the mean value, the theory of the Fourier series, and the uniqueness and approximation theorems. Also, unexpected relations were revealed between the theory of almost periodic functions on groups and different chapters of the group theory.

If the group is topological, the theory of almost periodic functions is considerably enriched, becoming closer to the classical Bohr theory. In the last paragraph are presented some facts in this direction.

1. Elementary properties of almost periodic functions on groups. The existence of the mean value

In this chapter we shall consider functions defined on a group G, assuming numerical values.

Among the characteristic properties of almost periodic functions of a real variable, the normality is the property best suitable for the definition, when one turns from the additive group of real numbers to an arbitrary group G.

We shall say that *the function $f(x)$ is almost periodic on the right (left), if the family of functions $\{f(xa)\}$ ($\{f(ax)\}$), $a \in G$, is compact in the sense of uniform convergence on G.*

From this definition it follows that any function which is almost periodic *on the right (left) on G is bounded.*

Indeed, if $f(x)$ is not bounded then there will exist a sequence $\{a_n\} \subset G$ such that $f(a_n) \to \infty$ as $n \to \infty$. But then from the sequence $\{f(xa_n)\}$ ($\{f(a_n x)\}$) one cannot extract a subsequence convergent in $x = e$ (the unit of G).

In the construction of the theory of almost periodic functions on groups an important role is played by a theorem on the compactness of families of numerical functions defined on arbitrary sets (generally not organized as topological spaces).

Let X be an arbitrary set and let $\{f(x)\}$, $x \in X$ be a family of numerical functions defined on X. A finite decomposition $X = \bigcup_{i=1}^{m} A_i$ such that

$x, y \in A_i$, $i = 1, 2, \ldots, m$ implies $|f(x) - f(y)| < \varepsilon$ for any function of the family under consideration is called ε-*decomposition*.

Theorem 7.1. *Let $f(x)$ be a family of equi-bounded functions on X. This family is compact in the sense of uniform convergence on X, if and only if for any $\varepsilon > 0$, there exists an ε-decomposition of X corresponding to the family $\{f(x)\}$.*

Proof. Let us first admit that the family reduces to a single function $f(x) = g(x) + ih(x)$, where g and h are real-valued functions. If $-M < g(x) < M$ for $x \in X$, we divide the interval $(-M, M)$ as follows: $-M = l_0 < l_1 < \cdots l_n = M$ such that $l_i - l_{i-1} < \varepsilon/2$, $i = 1, 2, \ldots, m$.

Let $B_i = \{x; x \in X, l_{i-1} < g(x) < l_i\}$, $i = 1, 2, \ldots, m$. It is obvious that the sets B_i give us an $(\varepsilon/2)$-decomposition of X for $g(x)$. Similarly, we may construct an $(\varepsilon/2)$-decomposition of X for $h(x)$ formed from the sets C_j, $j = 1, 2, \ldots, n$. Let us consider the sets $B_i \cap C_j$, $i = 1, 2, \ldots, m$; $j = 1, 2, \ldots, n$. If we denote them by A_k in an arbitrary order, then we get an ε-decomposition of X for $f(x)$, consisting of a finite number of sets. If we consider two complex functions $f_1(x)$ and $f_2(x)$, then the intersections of two arbitrary sets from two ε-decompositions of X for these two functions form an ε-decomposition of X for the family consisting of $f_1(x)$ and $f_2(x)$. In this way one finds out that for any $\varepsilon > 0$, there is an ε-decomposition of X which corresponds to a finite family of functions $f_1(x), f_2(x), \ldots, f_p(x)$.

If $\{f(x)\}$ is a compact infinite family regarded as a part of the space of all bounded functions on X with the norm $\|f\| = \sup_{x \in X} |f(x)|$, then according to Hausdorff's theorem for any $\varepsilon > 0$ there is a finite family of functions $f_1(x), \ldots, f_p(x)$ forming an $(\varepsilon/3)$-grid. Let now $X = \bigcup_{i=1}^{m} A_i$ be an $(\varepsilon/3)$-decomposition of X for the family which forms the grid. Let us show that this decomposition is an ε-decomposition for the family $\{f(x)\}$. Indeed, for any $f(x)$ of the considered family, there exists a $f_k(x)$, $1 \leqslant k \leqslant p$, such that $\sup_{x \in X} |f(x) - f_k(x)| < \varepsilon/3$. If $x, y \in A_i$, $i = 1, 2, \ldots, m$, then $|f_k(x) - f_k(y)| < \varepsilon/3$. Thus $|f(x) - f(y)| \leqslant |f(x) - f_k(x)| + |f_k(x) - f_k(y)| + |f_k(y) - f(y)| < \varepsilon$, which proves the necessity of the condition.

To prove the sufficiency of the condition, let $\{\varepsilon_n\}$ be a sequence of positive numbers converging to zero, and let us consider for every ε_k an ε_k-decomposition of X: $X = \bigcup_{j=1}^{n_k} A_j^{(k)}$.

In every $A_j^{(k)}$ we choose a point $x_j^{(k)}$. It is obvious that the set $\{x_j^{(k)}\}$ is countable. Since the functions of the family under consideration are uniformly bounded, we may extract (using the diagonal method) from any sequence of functions belonging to the family a subsequence $\{f_n(x)\}$, which converges at any point $x_j^{(k)}$. Let us show that this subsequence is uniformly convergent on X. If $\varepsilon > 0$ is given, then we have $3\varepsilon_r < \varepsilon$, if r is sufficiently large. We may determine an $N(\varepsilon) > 0$ such that

(7.1) $$|f_n(x_j^{(k)}) - f_m(x_j^{(k)})| < \frac{\varepsilon}{3}, \qquad k \leqslant r, j \leqslant n_r,$$

as soon as $n, m \geqslant N(\varepsilon)$. If $x \in X$, then there exists an index j such that $x \in A_j^{(r)}$. It follows that

(7.2) $$|f_n(x) - f(x_j^{(r)})| < \frac{\varepsilon}{3}, \qquad |f_m(x) - f_m(x_j^{(r)})| < \frac{\varepsilon}{3},$$

From (7.1) and (7.2) it follows that

$$|f_n(x) - f_m(x)| < \varepsilon$$

if $n, m \geqslant N(\varepsilon)$, which proves the uniform convergence of the sequence $\{f_n(x)\}$.

Theorem 7.2. *Any function which is almost periodic on the right is also almost periodic on the left.*

Proof. According to Theorem 7.1 for any $\varepsilon > 0$ there exists an ε-decomposition of the group G, corresponding to the family $\{f(xa)\}$, $a \in G$, where $f(x)$ is an almost periodic function on the right. Let $G = \bigcup_{i=1}^{n} A_i$ be this decomposition. We choose an element $x_i \in A_i$. For any $x \in G$, there exists an $j \leqslant n$ such that $\sup_{a \in G} |f(xa) - f(x_j a)| < \varepsilon$. This means that the functions $f(x_1 a), f(x_2 a), \dots, f(x_n a)$ forms a finite ε-grid for the family $\{f(xa)\}$, but this time x is considered a parameter. Applying Hausdorff's theorem to the metric space of bounded functions on G (the distance is derived from the norm $\|f\| = \sup_{x \in G} |f(x)|$), it follows that the family $\{f(xa)\}$, where x is a parameter, is compact in the sense of uniform convergence on G. That is, $f(x)$ is almost periodic on the left.

Similarly one may show that an almost periodic function on the left is almost periodic on the right.

Remark. This theorem allows us to speak from now on about *functions* which are *almost periodic on groups*; i.e., functions which are almost periodic on the left and on the right (in the sense of the definition).

Theorem 7.3. *If $f(x)$ and $g(x)$ are almost periodic on G so are the functions $\bar{f}(x)$, $\alpha f(x)$, $f(x^{-1})$, $f(axb)$, $f(x) + g(x)$, and $f(x)g(x)$, where α is a complex number and $a, b \in G$.*

The limit of a uniformly convergent sequence (on G) of almost periodic functions is also an almost periodic function.

The proof of this theorem is not difficult.

Theorem 7.4. The family of functions $\{f(axb)\}$, a, $b \in G$ is compact if $f(x)$ is almost periodic on G.

Proof. Let $G = \bigcup_{i=1}^{n} A_i$ be an $(\varepsilon/3)$-decomposition for the family $\{f(xb)\}$. We choose an element b_i in every A_i. For any $b \in G$, there exists b_i such that

$$(7.3) \qquad\qquad |f(xb) - f(xb_i)| < \frac{\varepsilon}{3}.$$

For any i, $f(xb_i)$ is almost periodic. It follows that the family $\{f(axb_i)\}$, $a \in G$, $i = 1, 2, \ldots, n$, is compact. We consider for every i an $(\varepsilon/3)$-decomposition of G for the above indicated family. We shall have $G = \bigcup_{j=1}^{k_i} B_{ij}$. From $x, y \in B_{ij}$ it follows that

$$(7.4) \qquad\qquad |f(axb_i) - f(ayb_i)| < \frac{\varepsilon}{3}.$$

Let us consider now the decomposition of the group G whose elements are all the intersections of the forms $B_{1j_1} \cap B_{2j_2} \cap \cdots B_{nj_n}$, where $j_p = 1, 2, \ldots, k_p$. Obviously, there exists only a finite number of such intersections.

Let us show that this decomposition is an ε-decomposition of the group G for the family $\{f(axb)\}$, $a, b \in G$. Indeed, if x, y belong to an element of the decomposition, it means that they are in a B_{ij} for any $i = 1, 2, \ldots, n$, and for a certain j. From (7.3) and (7.4) it follows that

$$|f(axb) - f(ayb)| \leqslant |f(axb) - f(axb_i)| + |f(axb_i) - f(ayb_i)|$$

$$+ |f(ayb_i) - f(ayb)| < \varepsilon,$$

which proves the theorem.

The converse of this theorem is obviously true, a fact which shows that as a definition of the almost periodicity of the function $f(x)$ may be taken the normality of the family $f(axb)$, $a, b \in G$.

Following A. Weil [653], we shall prove that the study of functions which are almost periodic on an abstract group reduces to the study of uniformly continuous functions on a precompact topological group.

We assume that $f(x)$ is an almost periodic numerical function on the group G. We define for any pair of elements $x, y \in G$

$$(7.5) \qquad\qquad d(x, y) = \sup_{a,b \in G} |f(axb) - f(ayb)|$$

It is obvious that $d(x, y)$ is a semidistance.

From the properties of the group it follows for two arbitrary elements c, d that

(7.6) $$d(x, y) = d(cxd, cyd).$$

Let E be the set of elements $x \in G$ characterized by the condition $d(x, e) = 0$, where e is the unit element of the group G. From expression (7.6) and the triangular inequality for $d(x, y)$ one concludes that E is an invariant subgroup of G, i.e., $e_1, e_2 \in E$ implies that $e_1 e_2^{-1} \in E$ and $ae_1 a^{-1} \in G$ for any $a \in G$.

The equivalence relation $x \sim y$, if and only if $xy^{-1} \in E$, divides the group G in equivalence classes $\{\eta\}$. We denote by G/E the factor group.

G/E is a metric space with the distance defined by

(7.7) $$d(\eta_x, \eta_y) = d(x, y),$$

where η_x, η_y represent the equivalence classes which contain the elements x, y.

Since $x' \in \eta_x$ and $y' \in \eta_y$ implies that $d(x, y) = d(x', y')$, the function $d(\eta_x, \eta_y)$ defined by (7.6) is effectively a distance.

According to (7.5), $f(x) = f(y)$ if $x \sim y$, which permits us to consider the function $f(x)$ as defined on the group G/E. We denote $F(\eta_x) = f(x)$.

Theorem 7.5. *The group G/E with the metric given by (7.7) is precompact, and the function $F(\eta_x)$ is uniformly continuous on this group.*

Proof. From (7.6) and from the triangular inequality we have that

(7.8) $$d(xy, x'y') \leqslant d(xy, x'y) + d(x'y, x'y') = d(x, x') + d(y, y'),$$

$$d(x^{-1}, y^{-1}) = d(xx^{-1}y, xy^{-1}y) = d(y, x) = d(x, y).$$

The last two relations prove the continuity of the group operations of G/E in the metric (7.7).

Let $G = \bigcup_{i=1}^{n} A_i$ be an ε-decomposition of the group G for the family of functions $\{f(axb)\}$, $a, b \in G$. We choose an element x_i from every A_i. It is obvious that the system $\{\eta_{x_i}\}$, $i = 1, 2, \ldots, n$, represent an ε-grid for the group G/E with metric given by (7.7), which proves the precompactness of this group.

The uniform continuity of the function $F(\eta_x)$ follows from the following obvious expressions

(7.9) $$|F(\eta_x) - F(\eta_y)| = |f(x) - f(y)| \leqslant d(x, y) = d(\eta_x, \eta_y).$$

This property of almost periodic functions on groups could be utilized in order to prove, in a different way than below, some of the main results of this chapter.

In order to establish the existence of the mean value of an almost periodic function, we need the following lemma, which is also of interest in other problems.

Lemma 7.1. *Let $\{a_i\}$ and $\{b_i\}$ be two sets consisting of n distinct elements each. Suppose that for every element a_i there is a part of the set $\{b_i\}$, $\varphi(a_i) \in \{b_i\}$ such that $\bigcup_{j=1}^{k} \varphi(a_{i_j})$, $1 \leqslant k \leqslant n$ contains at least k elements.*

Then there is a one-to-one correspondence between the sets $\{a_i\}$ and $\{b_i\}$ such that to any a_i there corresponds an element from $\varphi(a_i)$.

Proof. The lemma is obvious in the case $n = 1$. Let us now assume that it is true for sets consisting of at most $n - 1$ elements, and let us show that it remains true in the case of n elements.

Let us admit that for any group of r elements a_i, $1 \leqslant r < n$, there are at least $r + 1$ elements b_i. We fix an element a_{i_0} and we choose a $b_{j_0} \in \varphi(a_{i_0})$. There remain $n - 1$ elements a_i and $n - 1$ elements b_i, and for any group of r elements a_i there exist a group of at least r elements b_i. According to the induction hypothesis, among the $n - 1$ elements a_i and the $n - 1$ elements b_i, one can construct a one-to-one correspondence so that to any a_i there corresponds an element from $\varphi(a_i)$. Completing this correspondence with $a_{i_0} \to b_{j_0}$ we get the statement in the case of n elements.

Let us now admit that at least to one group of r elements a_i, $1 \leqslant r \leqslant n$ there corresponds a group of r elements b_i. According to the induction hypothesis between the elements of these groups one may construct a one-to-one correspondence with the property mentioned in the assertion. We must show that between the other two groups of $n - r$ elements one may set up a one-to-one correspondence of the indicated type. Indeed, if we admit that there are among the $n - r$ elements a_i a group of elements for which there are less than h elements b_i, to this group together with the r elements a_i will correspond a group of elements b_i which contains less than $h + r$ elements, which is impossible.

Hence the lemma is proved.

Corollary. If the set X admits two decompositions $X = \bigcup_{i=1}^{n} A_i$, $X = \bigcup_{i=1}^{n} B_i$ such that the union of any group of r parts B_i, $1 \leqslant r < n$ contains no more than r parts A_i, then one may set up a one-to-one correspondence between the set A_i and B_i such that the sets in correspondence have a nonvoid intersection.

Indeed, with every A_i we associate those B_j for which $A_i \cap B_j \neq \phi$. If to a group of k elements A_i would correspond a group of $l < k$ elements B_i, this would mean that the reunion of these elements B_i contains $k > l$ elements A_i, which is impossible. Lemma 7.1 may thus be applied.

Theorem 7.6. *If $f(x)$ is an almost periodic function on G, there exists a number $M\{f\}$ called the mean value of the function $f(x)$, with the following property: for any $\varepsilon > 0$, we may determine $n = n(\varepsilon)$ elements $a_1, a_2, \ldots, a_n$ of the group G, such that*

(7.10)
$$\left| M\{f\} - \frac{1}{n} \sum_{i=1}^{n} f(ba_i c) \right| < \varepsilon, \qquad b, c \in G.$$

The proof of this theorem will be made in several steps.

a) If $R = \{a_1, a_2, \ldots, a_n\}$ is a finite system of elements of G, then we denote by

(7.11)
$$M(R) = \frac{1}{n} \sum_{i=1}^{n} f(a_i).$$

Let $G = \bigcup_{i=1}^{n} A_i$ be a minimal ε-decomposition of G for $f(x)$.

By minimal ε-decomposition we understand an ε-decomposition containing the smallest number of components. If from every A_i we choose a representative a_i, then we can form the arithmetic mean (7.11), which depends on the decomposition as well as on the choice of the system of representatives $R = \{a_1, a_2, \ldots, a_n\}$.

However, if $R' = \{a'_1, a'_2, \ldots, a'_n\}$ is another system of representatives for the decomposition under consideration, then the following relation holds true

(7.12)
$$|M(R) - M(R')| < \varepsilon,$$

which follows immediately from $|f(a_i) - f(a'_i)| < \varepsilon$, $i = 1, 2, \ldots, n$.

b) Let $G = \bigcup_{i=1}^{n} B_i$ be another minimal ε-decomposition of G for $f(x)$. If $S = \{b_1, b_2, \ldots, b_n\}$ is a system of representatives for this decomposition, then

(7.13)
$$|M(R) - M(S)| < 2\varepsilon.$$

We note that the decompositions $G = \bigcup_{i=1}^{n} A_i$ and $G = \bigcup_{i=1}^{n} B_i$ satisfy the conditions from the corollary of Lemma 7.1. Indeed, if a number of r sets B_i, $r < n$ would cover s, $s > r$, sets A_i, then n would not be the minimal number of terms, admitted in an ε-decomposition. Therefore, for these, two decompositions one may choose a common system of representatives say $T = \{c_1, c_2, \ldots, c_n\}$. We have however

$$|M(R) - M(S)| \leqslant |M(R) - M(T)| + |M(T) - M(S)| < 2\varepsilon,$$

considering (7.12).

c) If $R = \{a_1, a_2, \ldots, a_n\}$ is a system of representatives for a minimal ε-decomposition $G = \bigcup_{i=1}^{n} A_i$, then

$$(7.14) \qquad \left| M(R) - \frac{1}{n} \sum_{i=1}^{n} f(ba_i c) \right| < 2\varepsilon, \qquad b, c \in G.$$

Indeed, $G = \bigcup_{i=1}^{n} bA_i c$ is a minimal ε-decomposition of G, for any $b, c \in G$, and we only have to apply (7.13).

d) Let us call ε-*approximate mean value* of $f(x)$ and let us denote it by $M\{f(x), \varepsilon\}$, any number which satisfies the following condition: there exists a system of elements $\{a_1, a_2, \ldots, a_n\}$ of G such that

$$(7.15) \qquad \left| M\{f, \varepsilon\} - \frac{1}{m} \sum_{j=1}^{m} f(ba_j c) \right| < 2\varepsilon, \qquad b, c \in G.$$

From what has been proved in c) it follows that such approximate means exist for any $\varepsilon > 0$.

If $M\{f(x), \varepsilon\}$ and $M\{f(x), \varepsilon'\}$ are two approximate means of $f(x)$, then

$$(7.16) \qquad |M\{f, \varepsilon\} - M\{f(x), \varepsilon'\}| < 2(\varepsilon + \varepsilon').$$

For $M\{f(x), \varepsilon\}$ (7.15) holds, and for $M \in \{f(x), \varepsilon'\}$ we shall have an inequality of the form

$$(7.17) \qquad \left| M\{f, \varepsilon'\} - \frac{1}{n} \sum_{i=1}^{n} f(ba'_i c) \right| < 2\varepsilon', \qquad b, c \in G.$$

If in (7.15) we make $c = e$, and give to b the "values" $a'_1, a'_2, \ldots, a'_n$, then we obtain n inequalities. If we take the arithmetic mean of the first members and second members, then we get

$$(7.18) \qquad \left| M\{f, \varepsilon\} - \frac{1}{nm} \sum_{i,j} f(a'_i a_j) \right| < 2\varepsilon,$$

Let us make $b = e$ in (7.17), and give to c the "values" $a_1, a_2, \ldots, a_m$. As above, it follows that

$$(7.19) \qquad \left| M\{f, \varepsilon'\} - \frac{1}{mn} \sum_{i,j} f(a'_i a_j) \right| < 2\varepsilon'.$$

From (7.18) and (7.19) we obtain (7.16).

e) Let us assign to ε' a sequence of values approaching zero. $M\{f, \varepsilon'\}$ is bounded on this sequence, as follows from (7.16). In other words, there exists a sequence $\{\varepsilon'_k\}$, $\lim_{k \to \infty} \varepsilon'_k = 0$, such that

(7.20) $$\lim_{k \to \infty} M\{f, \varepsilon'_k\} = M\{f\}.$$

From (7.16) we get that

(7.21) $$|M\{f\} - M\{f, \varepsilon\}| \leqslant 2\varepsilon.$$

The number $M\{f\}$ is nothing else but the mean value which we wanted to construct, since from (7.15) and (7.21) we have

(7.22) $$\left| M\{f\} - \frac{1}{m} \sum_{j=1}^{m} f(ba_j c) \right| < 4\varepsilon, \qquad b, c \in G,$$

and from (7.21) we have

$$\lim_{\varepsilon \to 0} M\{f, \varepsilon\} = M\{f\},$$

which proves the uniqueness of the mean value.

The mean value of an almost periodic function on a group G has the same properties as those encountered in the case of almost periodic functions of a real variable.

Theorem 7.7. *The mean value defined in the preceding theorem has the following properties:*

1) $M\{\alpha f\} = \alpha M\{f\}$, for any α complex;
2) $M\{f + g\} = M\{f\} + M\{g\}$;
3) $M\{1\} = 1$;
4) *if* $f(x) \geqslant 0$ *for* $x \in G$, *then* $M\{f\} \geqslant 0$;
5) *if* $f(x) \geqslant 0$ *for* $x \in G$, *and* $f(x) \not\equiv 0$, *then* $M\{f\} > 0$;
6) $M\{f\} \leqslant M\{|f|\}$;
7) $M\{\bar{f}\} = \overline{M\{f\}}$;
8) $M\{f(xa)\} = M\{f(ax)\} = M\{f(x)\}$, $a \in G$;
9) $M\{f(x^{-1})\} = M\{f(x)\}$;
10) *if* $\lim_{n \to \infty} f_n(x) = f(x)$, *uniformly on* G, *then* $\lim_{n \to \infty} M\{f_n\} = M\{f\}$.

Proof. Some of these properties follow immediately from the definition of the mean value, and the others may be proved on the basis of the preceding properties. We shall only prove properties 2 and 5.

Let us consider a minimal ε-decomposition for the functions of the families $\{f(axb)\}$ and $\{g(axb)\}$. Such a decomposition exists, as follows from Theorems 7.1 and 7.4. If we choose a system of representatives $\{x_1, x_2, \ldots, x_n\}$ for this decomposition, and consider the fact that the arithmetic mean of the values of $f + g$ for $x = ax_i b$, $i = 1, 2, \ldots, n$ is the sum of the arithmetic means of f and g, then property 2 follows from the fact that $M\{f\}$ may be approximated with any degree of accuracy by arithmetic means of the form

$$\frac{1}{n} \sum_{i=1}^{n} f(ax_i b),$$

uniformly with respect to $a, b \in G$.

To prove property 5, we note that we shall have $f(x_0) > \alpha > 0$, at least for one $x_0 \in G$. Let $\{f(xa_i)\}$, $i = 1, 2, \ldots, n$, be a finite $(\varepsilon/2)$-grid for the family $\{f(xa)\}$. For any $a \in G$, there is an a_i, $1 \leqslant i \leqslant n$, such that

$$(7.22) \qquad |f(xa) - f(xa_i)| < \frac{\alpha}{2}.$$

Making $x = x_0 a_i^{-1}$ in (7.22), we obtain

$$(7.23) \qquad |f(x_0 a_i^{-1} a) - f(x_0)| < \frac{\alpha}{2},$$

whence it follows that $f(x_0 a_i^{-1} a) > \alpha/2$. Therefore, for any $a \in G$,

$$(7.24) \qquad \sum_{i=1}^{n} f(x_0 a_i^{-1} a) > \frac{\alpha}{2}.$$

If in (7.24) we take the mean with respect to a and consider 2 and 4, then we have

$$M\{f\} \geqslant \frac{\alpha}{2n} > 0,$$

which proves property 5.

Corollary. The set of almost periodic functions defined on the group G is a prehilbertian space, if the addition and multiplication by scalars are defined as usual, and the scalar product by $(f, g) = M\{f\bar{g}\}$.

We shall denote by U_G the space of almost periodic functions on G with the norm $\|f\| = (M\{f\bar{f}\})^{1/2}$.

Theorem 7.8. *If a linear functional $M'\{f\}$, defined in the space of almost periodic functions on G, assuming complex values, satisfies conditions* 1, 2, 3, 4, *and* 8, *then*

$$(7.25) \qquad M'\{f\} \equiv M\{f\}.$$

Proof. If we consider the fact that $M'\{f\}$ satisfies conditions 1 and 2, then we can restrict ourselves to the case of real-valued functions. Let $\varepsilon > 0$ be arbitrary. There exist n elements $a_1, a_2, \ldots, a_n$ such that

$$M\{f\} - \varepsilon < \frac{1}{n} \sum_{i=1}^{n} f(ba_i) < M\{f\} + \varepsilon, \qquad b \in G.$$

Considering b as argument and applying M', we obtain

$$M\{f\} - \varepsilon < M'\{f\} < M\{f\} + \varepsilon,$$

which proves (7.25).

Remark. According to this theorem, if G is the additive group of real numbers, then the mean value defined in this paragraph coincides with the integral mean (in the Bohr sense).

It is interesting to notice that the existence of the mean value was established in this paragraph without assuming that G is a topological group and that the function is continuous.

2.   *Unitary representations of groups. The Fourier series associated with an almost periodic function*

In order to construct the Fourier series associated with an almost periodic function on a group, we need some notions referring to the linear representations of the groups.

We call *linear representation* of order r of the group G a matrix $D(x) = (d_{ij}(x))$, $i, j = 1, 3, \ldots, r$, whose elements are numerical functions defined on G, such that:

1) $D(e) = E_r$, where e is the unit of G, and E_r is the unit matrix of order r;

2) $D(xy) = D(x)D(y)$, $x, y \in G$.

The representation $D(x)$ is called bounded (almost periodic), if the functions $d_{ij}(x)$, $i, j = 1, 2, \ldots, n$, are bounded (almost periodic) on G.

If $D(x)$ is a unitary matrix, for any $x \in G$, then the representation $D(x)$ is called *unitary*.

Let us also recall that two representations of G, $C(x)$ and $D(x)$, are said to be *equivalent* if they have the same order, and if there is a constant non-singular matrix A, such that

$$D(x) = A^{-1}C(x)A, \qquad x \in G.$$

The relation between the bounded representations, those almost periodic and those unitary, is expressed in

Theorem 7.9. *The following three conditions for a representation $D(x)$ of the group G are equivalent:*
1) $D(x)$ *is bounded;*
2) $D(x)$ *is almost periodic;*
3) $D(x)$ *is equivalent with a unitary representation.*

Proof. Since 2) → 1) and 3) → 1), it is sufficient to show that any bounded representation is almost periodic and equivalent with a unitary representation.

Indeed, if $D(x) = (d_{ij}(x))$ is bounded, then from

$$d_{ij}(ax) = \sum_{k=1}^{r} d_{ik}(a)\, d_{kj}(x),$$

it follows the normality of the functions $d_{ij}(x)$. We can convince ourselves that this is true, if we apply the Bolzano–Weierstrass theorem to every numerical set $\{d_{ij}(a)\}$, $a \in G$, $i, j = 1, 2, \ldots, r$.

Let us now show that 1) → 3). We consider the constant matrix $A = (a_{ij})$, $i, j = 1, 2, \ldots, r$, where

(7.25) $$a_{ij} = M\left\{ \sum_{k=1}^{r} d_{ik}(x)\, \overline{d_{jk}(x)} \right\}, \qquad i, j, = 1, 2, \ldots, r.$$

These mean values exist since 1) → 2). One finds out that $\overline{a_{ij}} = a_{ji}$, which shows that A is a Hermitian matrix. If $z_1, z_2, \ldots, z_r$ are complex numbers not all zero, then

$$\sum_{i,j=1}^{r} \left[\sum_{k=1}^{r} d_{ik}(x)\, \overline{d_{jk}(x)} \right] z_i\, \bar{z}_j = \sum_{k=1}^{r} \left| \sum_{i=1}^{r} d_{ik}(x) z_i \right|^2 > 0.$$

Taking the mean with respect to x, and considering (7.25), we get

$$\sum_{i,j=1}^{r} a_{ij} z_i\, \bar{z}_j > 0,$$

if $\sum_{i=1}^{r} |z_i|^2 > 0$. In other words, A is a positively defined Hermitian matrix. There exists then a unitary matrix U such that $U^{-1}AU$ is a diagonal matrix whose elements λ_i, $i = 1, 2, \ldots, r$, on the main diagonal are positive. Consider the diagonal matrix $V = (U^{-1}AU)^{1/2}$. One establishes that

$$A = UV^2U^{-1} = (UVU^{-1})(UVU^{-1}) = B^2.$$

If one considers the invariance property of the mean value with respect to translations, then

$$a_{ij} = M\left\{\sum_{k=1}^{r} d_{ik}(ax)\, \overline{d_{jk}(ax)}\right\}$$

$$= M\left\{\sum_{k=1}^{r}\left(\sum_{l=1}^{r} d_{il}(a)\, d_{lk}(x)\right)\cdot\left(\sum_{m=1}^{r}\overline{d_{jm}(a)\, d_{mk}(x)}\right)\right\}$$

$$= \sum_{l,m=1}^{r} d_{il}(a)\, \overline{d_{jm}(a)}\, a_{lm}.$$

This relation says that $A = D(a)AD^*(a)$, where $D^*(a) = (\overline{d_{ij}(a)})$, $i, j = 1, 2, \ldots, r$ is the adjoint matrix of $D(a)$. Therefore $B^2 = D(a)B^2D^*(a)$ or $B^{-1}D(a)BBD^*(a)B^{-1} = E_r$. Since B is Hermitian, $B = B^*$, $B^{-1} = (B^{-1})^*$, which permits us to write

(7.26) $$[B^{-1}D(a)B]\cdot[B^{-1}D(a)B]^* = E_r.$$

Equality (7.26) shows that $B^{-1}D(a)B$ is a unitary matrix, so that the theorem is proved.

Let us recall now some notions from the theory of vector spaces of finite dimension. These notions will be necessary in order to state a lemma of Schur with numerous applications in the theory of linear representations of groups.

Let E be a vector space of dimension r on the field of complex numbers. Let $\{e_1, e_2, \ldots, e_r\}$ be a basis for E, so that any element $\alpha \in E$ can be uniquely represented in the form $\alpha = \sum_{i=1}^{r} \alpha_i e_i$ where α_i are complex numbers. Any linear transformation $\alpha' = A\alpha$ of the space E into itself is of the form $\alpha'_i = \sum_{j=1}^{r} a_{ij}\alpha_j$. The transformation A is completely characterized by the matrix (a_{ij}), and in the sequel we shall not make distinction in notation between the transformation A and the matrix (a_{ij}). Let us also notice that under a change of the basis $e'_i = \sum_{j=1}^{r} t_{ij}e_j$, the transformation which in the old basis had the matrix A, will have now the matrix $TAT^{-1} = A'$, where $T = (t_{ij})$ must be nonsingular.

If the linear transformation A admits an invariant space E_1 of dimension s, $0 < s < r$, i.e., if $AE_1 \subset E_1$, then the matrix A will be of the form

(7.27) $$A = \begin{pmatrix} B & C \\ 0 & D \end{pmatrix},$$

if in E we choose a basis whose first s vectors belong to E_r. In (7.27) B and D are square matrices of order s, $r - s$, respectively, and C is a rectangular matrix. By 0 we also denote a rectangular matrix whose elements are all zero.

One sees without difficulty that the linear transformation given by the matrix A in the same basis in which (7.27) is true, admits also an invariant space of dimension $r - s$, generated by the last $r - s$ vectors of the basis.

Let Σ be a family of linear transformations of E into itself. We say that Σ is *reducible* if there exists an s-dimensional subspace E_1 of E, $0 < s < r$ invariant under all the transformations of the family Σ.

In the contrary case, Σ is called *irreducible*.

If in E we fix a certain basis, then we can speak about *reducible* or *irreducible* families of matrices, the meaning of these notions being obvious.

Suppose now that Σ is a reducible family of matrices, and let us show that the family Σ^*, formed from the adjoint matrices of those from Σ, is also reducible.

We mentioned above that there exists a nonsingular matrix T such that any matrix of the family $T \Sigma T^{-1}$ will have the special form (7.27). According to a preceding observation, all the transformations which in the new basis correspond to the matrices $(A')^* = (TAT^{-1})^*$, $A \in \Sigma$, admit an invariant subspace $E'_1 \subset E$, the same for all the transformations. But $A^* = T^*(A')^*(T^*)^{-1}$, which proves the assertion.

Lemma 7.2. *Let Σ be an irreducible family of square matrices of order m and Ω an irreducible family of square matrices of order n. If A is a matrix of type (m, n), such that*

$$(7.28) \qquad\qquad \Sigma A = A\Omega,$$

then there are possible only two cases: either all the elements of A are zero or $m = n$ and A is nonsingular.

Proof.† Let us first mention that (7.28) must be interpreted in the following manner: for any $B \in \Sigma$, there exists $C \in \Omega$ such that $BA = AC$ and conversely.

Let E be a vector space of dimension m with a determined basis. The matrices of the family Σ may be interpreted as linear transformations of E into itself. Let $A = (a_{ij})$, and let a_k be the vector of E with the components a_{ik}, $i = 1, 2, \ldots, m$. Let E_1 be the linear subspace of E, generated by the vectors a_k, $k = 1, 2, \ldots, n$. Let us show that E_1 is invariant for all the transformations from Σ.

If $B = (b_{ij}) \in \Sigma$ and $C = (c_{ij}) \in \Omega$, such that $BA = AC$, then by applying the transformation B to a_k we shall obtain a vector a'_k whose components are $a'_{ik} = \sum_{j=1}^{m} b_{ij} a_{jk}$, $i = 1, 2, \ldots, m$. From $BA = AC$ it follows that

† L. S. Pontrjagin, Topological groups, Princeton, 1948.

$a'_{ik} = \sum_{j=1}^{n} a_{ij} c_{jk}$, $i = 1, 2, \ldots, m$. Thus, the vectors a'_k may be linearly expressed in terms of the vectors a_k, which proves that $a'_k \in E_1$, $k = 1, 2, \ldots, n$, i.e., E_1 is invariant under all the transformations from Σ.

Since Σ is irreducible, the dimension of E_1 is 0 or m. In the first case, a_k are all zero, i.e., A have all the elements zero. In the second case, among the vectors a_k, $k = 1, 2, \ldots, n$, we may determine m linearly independent vectors so that

$$(7.29) \qquad\qquad\qquad n \geqslant m.$$

Consider now the families Σ^* and Ω^* consisting of the adjoint matrices to those from Σ and Ω. As remarked above, the families Σ^* and Ω^* are irreducible. Since $\Omega^* A^* = A^* \Sigma^*$, we may apply to the families Σ^* and Ω^* the previous considerations. Thus only the following cases are possible: either all the elements of A^* are zero, which means that all the elements of A are zero or there are n linearly independent columns of A which implies $m \geqslant n$. Taking also into account (7.29), we get $m = n$ and hence A^*, and thus A, is nonsingular.

Corollary 1. If Σ is a irreducible family of square matrices of order r, and A is also a matrix of the same kind, permutable with any matrix from Σ, then $A = \lambda E_r$, λ being a complex number.

Indeed, let λ be a characteristic root of the matrix A. It is obvious that $A - \lambda E_r$ is permutable with any matrix from Σ, and therefore we may apply Schur's lemma. Since $\det(A - \lambda E_r) = 0$, $A = \lambda E_r$.

Corollary 2. If Σ is an irreducible family of matrices pairwise permutable, then all the matrices from Σ are of first order (i.e., complex numbers).

From the previous corollary it follows that any $A \in \Sigma$ is of the form $A = \lambda E_r$. But such a family may be reducible only if $r = 1$.

Now we may turn to prove the orthogonality relations for the unitary representations.

Theorem 7.10. *If $D(x)$ and $E(x)$ are two unitary representations of the group G, irreducible and nonequivalent of orders m and n, then*

$$(7.30) \qquad M\{d_{ij}(x)\overline{e_{kl}(x)}\} = 0, \; 1 \leqslant i, j \leqslant m, \; 1 \leqslant k, l \leqslant n.$$

Proof. Let B be an arbitrary constant $m \times n$ matrix, and let $A(x) = D(x)BE(x^{-1})$, $A = M\{A(x)\}$. Since the mean value is invariant under translations, then

$$D(y)AE(y^{-1}) = M_x\{D(y)D(x)BE(x^{-1})E(y^{-1})\}$$
$$= M_x\{D(yx)BE((yx)^{-1})\} = M_x\{A(yx)\} = A.$$

Thus $D(y)A = AE(y)$, and according to Schur's lemma we get the following: either $m = n$ and A is a nonsingular matrix or A is zero. In the first case we would have $E(y) = A^{-1}D(y)A$, i.e., D and E would be equivalent. This is a contradiction. In the second case we have

(7.31) $$M_x\{D(x)BE(x^{-1})\} = 0.$$

Choosing B such that the element in column l and row j is 1 and all the others zero, from (7.31) will result (7.30).

Theorem 7.11. *If $D(x)$ is an irreducible unitary representation of order r of the group G, then*

(7.32) $$M\{d_{ij}(x)\,\overline{d_{kl}(x)}\} = \begin{cases} 1/r, & \text{if } i = k, j = l, \\ 0, & \text{if } i \neq k \text{ or } j \neq l. \end{cases}$$

Proof. Let B be a square matrix of order r. As in the preceding theorem, we consider the matrix $A(x) = D(x)BD(x^{-1})$, $A = M\{A(x)\}$. We obtain $D(y)AD(y^{-1}) = A$ since the mean value is invariant under translations, which shows that $D(y)A = AD(y)$. According to corollary 1 of Schur's lemma, $A = \alpha E_r$, where α is a complex number. Therefore

'(7.33) $$M_x\{D(x)BD(x^{-1})\} = \alpha E_r.$$

Taking the *trace* of both sides and considering that two equivalent matrices have the same trace, we get $SpB = \delta_{jl}$ (the Kronecker symbol). From (7.33) it follows that

$$M\{d_{ij}(x)\,\overline{d_{kl}(x)}\} = \frac{1}{r}\,\delta_{ik}\,\delta_{jl},$$

which is equivalent to (7.32).

The construction of the Fourier series associated with an almost periodic function on the group G becomes now a simple problem.

Consider the set of unitary irreducible representations of the group G. The equivalence relation $D(x) \sim E(x)$, if there exists a constant matrix A for which $D(x) = A^{-1}E(x)A$, divides the set of these representations into disjoint classes (equivalence classes). We choose a representative of each class. Let $\mathcal{M}$ be the set of these representatives.

If $f(x)$ is an almost periodic function on G, and $D(x) \in \mathcal{M}$ is of order r, then we call *Fourier matrix* of the function $f(x)$ the matrix

(7.34) $$A(D) = rM\{f(x)D'(x^{-1})\},$$

D' being the *transpose* of D. From (7.34) it follows that the elements of $A(D)$ are

$$a_{ij}(D) = rM\{f(x)\,\overline{d_{ij}(x)}\}, \qquad 1 \leqslant i, j \leqslant r.$$

According to Theorems 7.10 and 7.11, two different representations from $\mathcal{M}$ satisfy conditions (7.30), and an arbitrary representation from the same family satisfies conditions (7.32). Thus the Bessel inequality holds for any finite system of representations from $\mathcal{M}$:

$$(7.35) \qquad \sum_{n=1}^{N} \frac{1}{r_n} \sum_{i,j=1}^{r_n} |a_{ij}(D_n)|^2 \leqslant M\{|f(x)|^2\}.$$

This inequality is found by the same method as that used in Chapter I, §3.

According to (7.35), the set of representations from $\mathcal{M}$ for which $A(D) \neq 0$ is at most countable. Let $D_n(x) = (d_{ij}^{(n)}(x))$, $1 \leqslant i, j \leqslant r_n$ be these representations, and let $A_n = A(D_n) = (a_{ij}^{(n)})$, $1 \leqslant i, j \leqslant r_n$.

The series

$$(7.36) \qquad \sum_{n=1}^{\infty} \sum_{i,j=1}^{r_n} a_{ij}^{(n)}\, d_{ij}^{(n)}(x),$$

is called the *Fourier series* associated with the function $f(x)$. This can also be denoted by:

$$(7.37) \qquad f(x) \sim \sum_{n=1}^{\infty} \sum_{i,j=1}^{r_n} a_{ij}^{(n)}\, d_{ij}^{(n)}(x).$$

Since (7.35) is valid for any N, the inequality

$$(7.38) \qquad \sum_{n=1}^{\infty} \frac{1}{r_n} \sum_{i,j=1}^{r_n} |a_{ij}^{(n)}|^2 \leqslant M\{|f(x)|^2\}$$

is satisfied. It is useful to note that

$$\sum_{i,j=1}^{r_n} a_{ij}^{(n)}\, d_{ij}^{(n)}(x) = \mathrm{Sp}(A_n D'_n(x)),$$

$$\sum_{i,j=1}^{r_n} |a_{ij}^{(n)}|^2 = \mathrm{Sp}(A_n A_n^*).$$

These relations allow us to write (7.37) in the form

$$(7.39) \qquad f(x) \sim \sum_{n=1}^{\infty} \mathrm{Sp}(A_n D'_n(x)),$$

and (7.38) in the form

$$(7.40) \qquad \sum_{n=1}^{\infty} \frac{1}{r_n} (\mathrm{Sp}\, A_n A_n^*) \leqslant M\{|f(x)|^2\}.$$

Remark. In the case in which G is an abelian group, the unitary irreducible representations of G are of order 1, according to corollary 2 of the Schur lemma.

Therefore the Fourier series of the almost periodic function $f(x)$ is of the form

(7.41) $$f(x) \sim \sum_{n=1}^{\infty} a_n \varphi_n(x), \qquad a_n = M\{f\bar{\varphi}_n\},$$

where $\varphi_n(x)$ are numerical functions called characters of the group G. If we assume that G is topological and $f(x)$ is continuous, then $\varphi_n(x)$ will be continuous (Corollary of Theorem 7.16).

In particular, if G is the additive group of real numbers R, then $\varphi_n(x)$ must satisfy the following conditions: a) $\varphi_n(0) = 1$; b) $|\varphi_n(x)| = 1$; $x \in R$; c) $\varphi_n(x + y) = \varphi_n(x)\varphi_n(y)$, x, $y \in R$. But the continuous functions on the real line which satisfy a, b, and c are $\varphi_n(x) = e^{i\lambda_n x}$ with λ_n real. The Fourier series attached to the function $f(x)$ achieves the well known form in the case of numerical almost periodic functions.

3. The uniqueness and approximation theorems

We shall present several problems from the theory of almost periodic functions of two variables and we shall dwell especially upon a class of functional equations defined on an abstract group G.

If $f(x, y)$ is a numerical function defined on $G \times G$, we shall say that it is almost periodic if the family $\{f(ax, by)\}$, a, $b \in G$, is compact in the uniform convergence topology on $G \times G$. Since $G \times G$ is also a group, the theory of almost periodic functions of two variables is not essentially different from the above theory. One may verify that if $f(x)$ is almost periodic on G, then $f(xy)$ and $f(xy^{-1})$ are almost periodic functions on $G \times G$.

If $f(x, y)$ is almost periodic on $G \times G$, then $g(y) = M_x\{f(x, y)\}$ is almost periodic on G, and we may therefore consider for $f(x, y)$ the iterated mean values $M_y\{M_x\{f(x, y)\}\}$ and $M_x\{M_y\{f(x, y)\}\}$. Since each of these iterated mean values satisfies the properties given in Theorem 7.7, the following relation holds:

(7.42) $$M_x\{M_y\{f(x, y)\}\} = M_y\{M_x\{f(x, y)\}\} = M\{f(x, y)\}.$$

Consider the space U_G of almost periodic functions on G, organized as a prehilbertian space in terms of the scalar product $(f, g) = M\{f\bar{g}\}$ (the corollary of Theorem 7.7). We denote by $\tilde{U}_G$ the space obtained by completing U_G in the norm $\|f\| = (M\{f\bar{f}\})^{1/2}$.

$\tilde{U}_G$ is a Hilbert space containing U_G as a dense part of it.

Let $f(x)$ be an almost periodic function on G such that $f(xy^{-1}) = f(yx^{-1})$, x, $y \in G$. We denote by T the operator

(7.43) $$T\varphi(x) = M_y\{f(xy^{-1})\varphi(y)\}, \qquad \varphi \in U_G.$$

Since $f(xy^{-1})\varphi(y)$ is an almost periodic function on $G \times G$, according to the above observations, it easily follows that $T: U_G \to U_G$.

Lemma 7.3. *The operator T may be extended to a completely continuous self-adjoint operator, $\tilde{T}: \tilde{U}_G \to \tilde{U}_G$.*

Proof. Let us notice that T operates continuously on U_G. Indeed, from the properties of the mean value and from the Schwartz inequality, for any $\varphi \in U_G$ it follows that

$$M\{|T\varphi|^2\} \leqslant M_x\{|M_y\{f(xy^{-1})\varphi(y)\}|^2\} \leqslant M_x\{M_y\{|f(xy^{-1})|^2\}M\{|\varphi|^2\}.$$

Thus $\|T\varphi\| \leqslant M_1\|\varphi\|$. Since U_G is dense in $\tilde{U}_G$, T may be extended by continuity to a linear continuous operator $\tilde{T}$, defined on the space $\tilde{U}_G$.

Let X_i, $(i = 1, 2, \ldots, n)$ be an ε-decomposition of the group G for the family of functions $\{f_y(x) = f(xy^{-1}); y \in G\}$.

From the relation

$$|T\varphi(x') - T\varphi(x'')|^2 \leqslant M_y\{|f(x'y^{-1}) - f(x''y^{-1})|^2\}M\{|\varphi|^2\}, \qquad \varphi \in U_G,$$

and from the fact that $f(x)$ is almost periodic, from $x', x'' \in X_i$ and $\|\varphi\| \leqslant A$, it follows $|T\varphi(x') - T\varphi(x'')| < M_1\varepsilon$. Thus according to Theorem 7.1, the family of functions $\{T\varphi(x) : \|\varphi\| \leqslant A\}$ is compact in the topology of uniform convergence on G. Since on the space U_G this topology is stronger than the topology induced by the norm $\|f\| = (M\{f\bar{f}\})^{1/2}$, it follows that $T: U_G \to U_G$ is compact. This implies the compactness of the operator $\tilde{T}$.

Since $f(xy^{-1}) = f(yx^{-1})$, according to (7.42) it follows that $\tilde{T}$ is self-adjoint.

Corollary. If $f(x) \not\equiv 0$ is an almost periodic function on G, and $f(xy^{-1}) = f(yx^{-1})$, the functional equation

$$(7.44) \qquad \lambda\varphi(x) = M_y\{f(xy^{-1})\varphi(y)\}$$

admits at least one eigenvalue $\lambda_0 \neq 0$ and for every eigenvalue a finite dimensional space of eigenfunctions $\mathcal{N} \subset \tilde{U}_G$.

Indeed, equation (7.44) may also be written in the form $\lambda\varphi = \tilde{T}\varphi$, where $\tilde{T}$ is a linear self-adjoint compact operator, defined on the Hilbert space $\tilde{U}_G$. The corollary follows then from the Hilbert–Schmidt theory.

We now may turn to the proof of the uniqueness theorem. As noticed in Chapter I, §3, it suffices to prove the following

Theorem 7.12. *If the Fourier series for the almost periodic function $f(x)$ is the identically zero series, then $f(x) \equiv 0$ on G.*

The proof will be made in two steps.

a) Let us show first that from the hypothesis we made it results that the Fourier matrix of the function $f(x)$, with respect to any unitary representation of G, is the zero matrix.

For the time being we know only that the Fourier matrix of $f(x)$ vanishes for any representation $D(x) \in \mathcal{M}$. Since any other irreducible unitary representation of G may be expressed in the form $C(x) = U^{-1} D(x) U$, where $D(x) \in \mathcal{M}$, and U is unitary, the Fourier matrix is zero with respect to any irreducible unitary representation.

Let $C(x)$ be a unitary reducible representation of G. If $C(x)$ is of order r, then we choose in the r-dimensional vector space E an orthonormal basis, such that the first s vectors of this basis, $0 < s < r$ generate the invariant subspace $E_1 \subset E$. The orthogonal complement of E_1 in E, say E_2, is also invariant, since a unitary transformation preserves the orthogonality of vectors. These considerations show us that there exists a unitary matrix V such that

$$(7.45) \qquad V^{-1} C(x) V = \begin{pmatrix} C_1(x) & 0 \\ 0 & C_2(x) \end{pmatrix},$$

where $C_1(x)$ and $C_2(x)$ are unitary representations of G of order s and $r - s$. If $C_1(x)$ and $C_2(x)$ are irreducible, then we have obtained a decomposition of $C(x)$ into irreducible representations. In the contrary case we apply to $C_1(x)$ and $C_2(x)$ the same method of decomposition. Finally, we shall obtain a relation of the form

$$(7.46) \qquad C(x) = U^{-1} \begin{pmatrix} C_1(x) & 0 & \cdots & 0 \\ 0 & C_2(x) & \cdots & 0 \\ 0 & 0 & \cdots & 0 \\ 0 & 0 & \cdots & C_n(x) \end{pmatrix} U,$$

where U is a unitary matrix, and $C_1(x)$, $C_2(x)$, $\ldots$, $C_n(x)$ are irreducible unitary representations of G.

From (7.46) follows a), if we consider that the Fourier matrix of $f(x)$ with respect to any unitary irreducible representation is zero.

b) Let $f(x)$ be an almost periodic function on G such that the Fourier matrix with respect to any unitary representation is zero. Let us now show that $f(x) \equiv 0$ on G.

Consider the function

$$(7.47) \qquad g(x) = M_y\{f(xy)\overline{f(y)}\}.$$

One may easily establish that $g(x)$ is almost periodic and that $g(xy^{-1}) =$

$g(yx^{-1})$. Then, according to the corollary of Lemma 7.3 the functional equation

(7.48) $$\lambda\varphi(x) = M_z\{g(xz^{-1})\varphi(z)\}$$

admits at least one eigenvalue $\lambda_0 \neq 0$. Let $\varphi_1(x), \ldots, \varphi_n(x)$ be an ortho-normal basis in the linear space of the eigenfunctions which are associated with λ_0. One also finds out that the functions $\varphi_1(xy^{-1}), \varphi_2(xy^{-1}), \ldots, \varphi_n(xy^{-1})$ are also eigenfunctions of equation (7.48) corresponding to λ_0. Indeed, $\lambda_0 \varphi_i(xy^{-1}) = M_z\{g(xy^{-1}z^{-1})\varphi_i(z)\} = M_t\{g(xt^{-1})\varphi_i(ty^{-1})\}$, $i = 1, 2, \ldots, n$, since the mean value is invariant under translations. It follows that

(7.49) $$\varphi_i(xy^{-1}) = \sum_{j=1}^{n} e_{ij}(y)\varphi_j(x), \qquad i = 1, 2, \ldots, n.$$

On the other hand, one can easily see that the system $\varphi_1(xy^{-1})$, $\varphi_2(xy^{-1}), \ldots, \varphi_n(xy^{-1})$ is an orthonormal system. Thus, the matrix $E(y) = (e_{ij}(y))$, $i, j = 1, 2, \ldots, n$, transforms an orthonormal system into another orthonormal system. Thus $E(y)$ is unitary for any $y \in G$.

Let us now show that $E(y)$ is a linear representation of G. Indeed formulas (7.49) allow us to write

$$\varphi_i(x(yz)^{-1}) = \varphi_i(xz^{-1}y^{-1}) = \sum_{j=1}^{n} e_{ij}(yz)\varphi_j(x) = \sum_{j=1}^{n} e_{ij}(y)\varphi_j(xz^{-1})$$

$$= \sum_{j=1}^{n} e_{ij}(y) \sum_{k=1}^{n} e_{jk}(z)\varphi_k(x), \qquad i = 1, 2, \ldots, n.$$

Since $\varphi_1(x), \varphi_2(x), \ldots, \varphi_n(x)$ are linearly independent, we have the following relations: $e_{ij}(yz) = \sum_{k=1}^{n} e_{ik}(y)e_{kj}(z)$, $i, j = 1, 2, \ldots, n$. This means that $E(yz) = E(y)E(z)$. If in (7.49) we take $y = e$, then we get that $e_{ij}(e) = \delta_{ij}$, $i, j = 1, 2, \ldots, n$, i.e., $E(e) = E_n$. Therefore, $E(y)$ is a unitary representation of the group G.

Because the Fourier matrix of $f(x)$ with respect to any unitary representation of G is zero, the same thing can be said about $g(x)$. Indeed, if $D(x)$ is a unitary representation of G of order r, then the Fourier matrix of $g(x)$ with respect to $D(x)$ is

$$B = (b_{ij}) = (rM\{g(x)\,\overline{d_{ij}(x)}\}) = rM\{g(x)\overline{D'(x^{-1})}\}$$
$$= rM_x\{\overline{D'(x^{-1})}M_y\{f(xy)\overline{f(y)}\}\} = rM_y\{\overline{f(y)}M_x\{f(xy)\overline{D'(x^{-1})}\}\}$$
$$= rM_y\{\overline{f(y)}M_z\{f(z)\overline{D'(yz^{-1})}\}\} = rM_z\{f(z)\overline{D'(z^{-1})}\} \cdot M_y\{\overline{f(y)}\overline{D'(y)}\}\}$$
$$= \frac{1}{r}AA^*,$$

A being the Fourier matrix of the function $f(x)$ with respect to $D(x)$.

Therefore,

(7.50) $M\{g(x)\overline{e_{ij}(x^{-1})}\} = M\{g(x)e_{ji}(x)\} = 0,$ $1 \leqslant i, j \leqslant n.$

We shall show that these relations are in contradiction with the above-mentioned fact that equation (7.48) admits nontrivial solutions. From (7.50) we get

(7.51) $M_y\{g(xy^{-1})e_{ij}(y^{-1})\} = M_z\{g(z)e_{ij}(x^{-1}z)\}$

$$= \sum_{k=1}^n e_{ik}(x^{-1})M_z\{g(z)e_{kj}(z)\} = 0.$$

If we set $x = e$ in (7.49) and change y into y^{-1}, then we obtain

$$\varphi_i(y) = \sum_{j=1}^n e_{ij}(y^{-1})\varphi_j(e).$$

From (7.51) we get for $i = 1, 2, \ldots, n$

$$M_y\{g(xy^{-1})\varphi_i(y)\} = \sum_{j=1}^n \varphi_j(e)M_y\{g(xy^{-1})e_{ij}(y^{-1})\} = 0,$$

which contradicts $M_y\{g(xy^{-1})\varphi_i(y)\} = \lambda_0 \varphi_i(y)$, $i = 1, 2, \ldots, n$, $\lambda_0 \neq 0$, $\varphi_i(y) \not\equiv 0$ on G.

Thus $g(x) \equiv 0$ on G. Making $x = e$ in (7.47) and using the property 5 of the mean value, we get that $f(x) \equiv 0$ on G.

Theorem 7.13. *For any almost periodic function on G the Parseval inequality*

(7.52) $$\sum_{n=1}^{\infty} \frac{1}{r_n} \mathrm{Sp}(A_n A_n^*) = M\{|f(x)|^2\}$$

holds true.

Proof. We consider the almost periodic function $g(x)$ defined by (7.47). As seen in the proof of the preceding theorem, the Fourier matrices of the functions $g(x)$ and $f(x)$ with respect to the same unitary representation, are related by $B = (1/r)AA^*$. Thus we may write

(7.53) $$g(x) \sim \sum_{n=1}^{\infty} \frac{1}{r_n} \mathrm{Sp}(A_n A_n^* D'(x)).$$

If $E(x) = U^{-1}D(x)U$ is a representation of G, unitarily equivalent with $D(x)$, then $C = rM\{f(x)E'(x^{-1})\} = V^{-1}AV$, where V is unitary and A is the Fourier matrix of $f(x)$ with respect to the unitary representation $D(x)$.

If A is Hermitian, then V may be determined so that C be diagonal. It follows that

$$\mathrm{Sp}(CE'(x)) = \mathrm{Sp}(V^{-1}AVV^{-1}D'(x)V) = \mathrm{Sp}(AD'(x)).$$

Since the Fourier matrices of $g(x)$, $B_n(x) = (1/r_n)A_n A_n{}^*$ are Hermitian, we may assume that they are diagonal: $B_n = (1/r_n)\lambda_i \delta_{ij}$, $i, j = 1, 2, \ldots, n$. Since B_n is positive definite, $\lambda_i \geqslant 0$, $i = 1, 2, \ldots, n$.

Let us now show that the Fourier series (7.53) is uniformly convergent on G. Indeed,

$$\mathrm{Sp}(B_n D'_n(x)) = \frac{1}{r_n} \sum_{i=1}^{r_n} \lambda_i \, d_{ii}^{(n)}(x).$$

Since $|d_{ii}^{(n)}(x)| \leqslant 1$, $i = 1, 2, \ldots, r_n$,

(7.54) $$|\mathrm{Sp}(B_n D'_n(x))| \leqslant \frac{1}{r_n} \sum_{i=1}^{n} \lambda_i = \frac{1}{r_n} \mathrm{Sp}(A_n A_n{}^*).$$

But in the last part of (7.54) there appears a term of a convergent numerical series (according to the Bessel inequality (7.40)). Its sum is an almost periodic function on G, whose Fourier series is also the series (7.53). According to Theorem 7.11,

$$g(x) = \sum_{n=1}^{\infty} \frac{1}{r_n} \mathrm{Sp}(A_n A_n{}^* D'_n(x)).$$

For $x = e$ one obtains the Parseval inequality (7.52).

Remark. Theorem 7.13 is equivalent with the uniqueness theorem. Indeed, if one admits Parseval's inequality, this means that for an almost periodic function whose Fourier series is identically zero, the relation $M\{|f(x)|^2\} = 0$ holds true. According to the properties of the mean value $f(x) \equiv 0$ on G.

Theorem 7.14. *Let $f(x)$ be an almost periodic function on G. For any $\varepsilon > 0$, there exists a polynomial*

$$P_n(x) = \sum_{k=1}^{n} \mathrm{Sp}(B_k E'_k(x)),$$

where $E_k(x)$, $k = 1, 2, \ldots, n$, are irreducible unitary representations of G, and B_k, $k = 1, 2, \ldots, n$, are constant matrices such that

(7.55) $$|f(x) - P_n(x)| < \varepsilon, \qquad x \in G.$$

Proof. It is obvious that the representations $E_k(x)$, as well as the matrices B_k, depend on ε.

Consider the function $g(x) = \sup_{t \in G} |f(xt) - f(t)|$. It is easy to show that this function is almost periodic on G.

According to Wiener, let us define the real function

$$\varphi_\varepsilon(v) = \begin{cases} 1 - \dfrac{v}{\varepsilon}, & 0 \leqslant v \leqslant \varepsilon, \\ 0, & v > \varepsilon. \end{cases}$$

Consider now the function

$$\psi_\varepsilon(x) = \frac{\varphi_\varepsilon(g(x))}{M\{\varphi_\varepsilon(g(x))\}},$$

and notice that $\varphi_\varepsilon(g(x))$ is a non-negative almost periodic function. This function is not identically zero since $\varphi_\varepsilon(g(e)) = \varphi_\varepsilon(0) = 1$. Therefore, the mean is positive so that $\psi_\varepsilon(x)$ is almost periodic.

The function $f_\varepsilon(x) = M_y\{f(y)\psi_\varepsilon(xy^{-1})\}$ possesses the property

(7.56) $$|f(x) - f_\varepsilon(x)| < \varepsilon, \qquad x \in G.$$

Indeed, if $\psi_\varepsilon(xt^{-1}) \neq 0$, then $\varphi_\varepsilon(g(xt^{-1})) \neq 0$, i.e., $g(xt^{-1}) < \varepsilon$. This means that $\sup_{y \in G} |f(xt^{-1}y) - f(y)| < \varepsilon$ or $\sup_{y \in G} |f(xy) - f(ty)| < \varepsilon$. For $y = e$ one get $|f(x) - f(t)| < \varepsilon$. Since $M_t\{\psi_\varepsilon(xt^{-1})\} = 1$,

$$f(x) = M_t\{f(x)\psi_\varepsilon(xt^{-1})\}.$$

From $f(x)$ one subtracts $f_\varepsilon(x)$ and considering the fact that $\psi_\varepsilon(xt^{-1}) \neq 0$, this implies that $|f(x) - f(t)| < \varepsilon$. One obtains (7.56).

Now suppose that

$$f(x) \sim \sum_{h=1}^{\infty} \frac{1}{r_n} \operatorname{Sp}(A_n D'_n(x)).$$

If $\eta > 0$, then there exists $N(\eta)$ such that

(7.57) $$\sum_{n=N+1}^{\infty} \frac{1}{r_n} \operatorname{Sp}(A_n A^*_n) < \eta.$$

Let us denote now

$$S_N(x) = \sum_{n=1}^{N} \frac{1}{r_n} \operatorname{Sp}(A_n D'_n(x)), \qquad r_N(x) = f(x) - S_N(x).$$

The Parseval inequality leads to

$$M\{|r_N(x)|^2\} = \sum_{n=N+1}^{\infty} \frac{1}{r_n} \operatorname{Sp}(A_n A_n^*) < \eta.$$

Since

$$f_\varepsilon(x) = M_t\{f(t)\psi_\varepsilon(xt^{-1})\}$$
$$= M_t\{S_N(t)\psi_\varepsilon(xt^{-1})\} + M_t\{r_N(t)\psi_\varepsilon(xt^{-1})\}$$
$$= \sum_{n=1}^{N} \frac{1}{r_n} \text{Sp}(A_n M\{D'_n(t^{-1})\psi_\varepsilon(t)\}D'_n(x)) + M_t\{r_N(t)\psi_\varepsilon(xt^{-1})\},$$

and, according to the Cauchy inequality,

$$M_t\{r_N(t)\psi_\varepsilon(xt^{-1})\}| \leqslant (M_t\{|r_N(t)|^2\})^{1/2}(M_t\{\psi_\varepsilon^2(xt^{-1})\})^{1/2}$$
$$< \sqrt{\eta}(M\{\psi_\varepsilon^2(t)\})^{1/2},$$

it means that we shall have

(7.58) $$\left| f_\varepsilon(x) - \sum_{n=1}^{N} \frac{1}{r_n} \text{Sp}(A_n M\{D'_n(t^{-1})\psi_\varepsilon(t)\}D'_n(x)) \right| < \varepsilon, \qquad x \in G,$$

if

(7.59) $$\eta < \frac{\varepsilon^2}{M\{\psi_\varepsilon^2(t)\}}.$$

The theorem now follows from (7.56) and (7.58), if we use the fact that

$$\sum_{n=1}^{N} \frac{1}{r_n} \text{Sp}(A_n M\{D'_n(t^{-1})\psi_\varepsilon(t)\}D'_n(x))$$

is a polynomial of the form specified in the statement.

4. Almost periodic functions on topological groups

The results obtained in the preceding paragraphs did not assume any restriction of topological nature on the group on which the almost periodic function is defined. In this paragraph we shall study some properties which appear when we assume that the almost periodic function is continuous on a topological group.

Theorem 7.15. *An almost periodic function continuous on a topological group is uniformly continuous.*

Proof. We must prove that for any $\varepsilon > 0$ there exists a neighborhood $U(e)$ of the unit such that $\sup |f(xa) - f(xb)| < \varepsilon$ for any $a, b \in G$ and $a^{-1}b \in U$. Consider an $(\varepsilon/3)$-decomposition of the group G, $G = \bigcup_{i=1}^{n} A_i$ corresponding to the family $\{f(xa)\}$, $a \in G$. Let us choose for any A_i an element $x_i \in A_i$. For any $x \in G$ there exists j, $1 \leqslant j \leqslant n$, such that $\sup_{a \in G} |f(xa) - f(x_j a)| < \varepsilon/3$. Since $f(x)$ is continuous, for any $\varepsilon > 0$ and x_j, $j = 1, 2, \ldots, n$, there exists an $U_j(e)$ such that $|f(x_j a) - f(x_j b)| < \varepsilon/3$ for $a, b \in G$, $a^{-1}b \subset U_j(e)$. Let $U(e) = \bigcap_{j=1}^{u} U_j(e)$. From the inequality

$$\sup_{x \in G} |f(xa) - f(xb)| \leqslant \sup_{x \in G} |f(xa) - f(x_j a)| + \sup_{x \in G} |f(xb) - f(x_j b)|$$
$$+ \sup_{x \in G} |f(x_j a) - f(x_j b)|,$$

it follows that $U(e)$ is the sought neighborhood.

Theorem 7.16. *If $f(x)$ is an almost periodic function continuous on the topological group G, then the irreducible unitary representations of G which appear in the Fourier series attached to the function $f(x)$ are continuous.*

Proof. Suppose that

$$f(x) \sim \sum_{n=1}^{\infty} \operatorname{Sp}(A_n D'_n(x)) = \sum_{n=1}^{\infty} \sum_{i,j=1}^{r_n} a_{ij}^{(n)} d_{ij}^{(n)}(x),$$

where the matrices A_n and $D_n(x)$ have the meaning indicated in §2.

Consider the function $g(x) = M_y\{f(xy)\overline{f(y)}\}$. As remarked above, $g(x)$ is almost periodic on G, and its Fourier series is given by (7.53),

$$g(x) \sim \sum_{n=1}^{\infty} \frac{1}{r_n} \operatorname{Sp}(A_n A_n^* D'_n(x)).$$

We shall prove that $g(x)$ is continuous on G. Indeed, since $f(x)$ is continuous and almost periodic, it is uniformly continuous (Theorem 7.15). Thus for any $\varepsilon > 0$, there exists a neighborhood of the unit $U(e)$ such that $\sup_{y \in G} |f(x_1 y) - f(x_2 y)| < \varepsilon$ for x_1, $x_2 \in G$; $x_1 x_2^{-1} \in U(e)$. It follows then from the Schwarz inequality that:

$$(7.60) \qquad |g(x_1) - g(x_2)| \leqslant \varepsilon (M\{|f|^2\})^{1/2},$$

which proves that the function $g(x)$ is continuous.

To prove the theorem we consider the system of functions $\{\varphi_i(x)\}$ defined by

$$(7.61) \qquad \varphi_i(x) = \sum_{j=1}^{r_n} \beta_j d_{ij}^{(n)}(x), \qquad (i = 1, 2, \ldots, n)$$

where the numbers β_j are chosen such that φ_i are the eigenfunctions of the functional equations:

$$(7.62) \qquad \lambda_0 \varphi(x) = M_z\{g(xz^{-1})\varphi(z)\}, \qquad \lambda_0 \neq 0.$$

Replacing the expressions for the functions $\varphi_i(x)$ given by formula (7.61) in (7.62), one concludes that the numbers β_j and λ_0 must verify the following system of equations:

$$(7.63) \qquad \sum_{k=1}^{r_n} \gamma_{jk}^{(n)} \beta_k = \lambda_0 \beta_j, \qquad (j = 1, 2, \ldots, n),$$

where $\gamma_{jk}^{(n)}$ are the elements of the matrix $A_n A_n{}^*$.

Since the matrix $A_n A_n{}^*$ is Hermitian, the system (7.63) admits a non-trivial solution $(\beta_1, \beta_2, \ldots, \beta_n)$ where λ_0 is a non-zero eigenvalue of $A_n A_n{}^*$. Employing the same argument used for $g(x)$, one shows that the functions $\varphi_i(x)$ are continuous on G.

From the relation

$$\varphi_i(xy) = \sum_{j=1}^{r_n} \varphi_j(x) \, d_{ij}^{(n)}(y)$$

and from Theorem 7.11, one concludes that

$$(7.65) \qquad M_x\{\varphi_i(xy)\overline{\varphi_j}(x)\} = \left[\frac{1}{r_n} \sum_{j=1}^{r_n} |\beta_j|^2\right] d_{ij}^{(n)}(y).$$

Since the left-hand side represents a continuous function, $d_{ij}^{(n)}$ are continuous functions. Thus the theorem is proved.

Corollary. If G is an abelian topological group and $f(x)$ is almost periodic and continuous on G, then the characters $\varphi_n(x)$ of the group G in the Fourier series of the function $f(x)$ are continuous numerical functions.

In general, on an arbitrary topological group, the set of almost periodic functions is distinct from the set of continuous functions. However, on a compact topological group the following result holds:

Theorem 7.17. *A function $f(x)$, continuous on a compact topological group, is almost periodic.*

Proof. Since $f(x)$ is continuous on the compact group G it is also uniformly continuous, i.e., for any $\varepsilon > 0$ there exists a neighborhood U of the unit element such that $|f(x) - f(y)| < \varepsilon$ if $xy^{-1} \in U(e)$. From the compactness of the group G one concludes that there exists a finite covering $G = \bigcup_{i=1}^n A_i$, where $A_i = U(e)a_i$, $(a_i \in G; i = 1, 2, \ldots, n)$.

If $x, y \in A_i$, then for any $a \in G$, $(ax)^{-1}(ay) \in U(e)$ and thus

$$\sup_{a \in G} |f(ax) - f(ay)| < \varepsilon.$$

Therefore the system A_i, $(i = 1, 2, \ldots, n)$ is an ε-decomposition of the group G for the family of functions $\{f(ax)\}$. According to Theorem 7.1 $f(x)$ is almost periodic.

Corollary. Any continuous function on a topological compact group G can be uniformly approximated by linear combinations of irreducible unitary representations of G.

Remark. An almost periodic function on a compact topological group is not necessarily continuous. This fact follows from the following example of an almost periodic function on the group of rational numbers given by H. D. Ursell [637]. Consider a continuous function $f(x)$ on the real line, of period 2π, and let us denote by $f_q(n + \alpha/q!) = f(n + \alpha n_q)$, where α, n_q, q are integers and $0 \leqslant \alpha \leqslant q!$. Let $v(x) = \sup_{y \in R} |f(y + x) - f(y)|$ and let $\varepsilon_q = \max_{0 \leqslant \alpha < (q-1)!} v(\alpha v_q)$, where $v_q = q n_q - n_{q-1}$. It is obvious that $|f_q(x) - f_{q-1}(x)| \leqslant \varepsilon$ for any rational x. Choosing the sequence n_q so that the series $\sum_1^\infty \varepsilon_q$ is convergent, it follows that $f_q(x)$ is convergent. If we denote by $\varphi(x)$ its limit, one may show that $\varphi(x)$ is normal on the group of rational numbers but it is not continuous.

Bibliographical notes

In writing this chapter, I have used especially the paper by J. von Neumann (524) and the monographs of W. Maak (489) and B. M. Levitan (465).

The existence theorem of the mean value in the form in which it is presented in §1 belongs to W. Maak. Another proof of this fundamental theorem based on the ergodic theory is found in K. Yosida's monograph (686). In the case of almost periodic functions on topological groups, there are other proofs for the theorem of existence of the mean given by A. Weil (654), H. Nikaidô (528), and in the case of abelian groups by E. Følner (265).

In the monograph by E. Hewitt and K. A. Ross [355] are obtained several integral representations for the mean in the case of almost periodic functions on locally compact groups. For this purpose one uses the Haar integral. We mention the following result which reminds us by its form of the classical definition:

Let G be an abelian topological group, locally compact and countable at infinity. If λ is a Haar measure on G, then there exists an increasing sequence H_n of open sets with compact closure, such that

$$M\{f\} = \lim_{n \to \infty} \frac{1}{\lambda(H_n)} \int_{H_n} f(x)\, dx,$$

for any continuous almost periodic function on G.

The possibility of reducing the study of almost periodic functions defined on an abstract group to the study of uniformly continuous functions defined on an abstract group was indicated by A. Weil (653). With regard to this we also mention E. R. van Kampen's paper (385).

Using the known Gelfand theorem of representation of Banach algebras, applied to the algebra of almost periodic functions on an abstract group G, one may establish the existence of a homomorphism between the algebra of almost periodic functions on G and that of continuous (complex) functions on a certain compact topological space S. If G is an abelian locally compact group, then one may establish an isomorphism between the algebra of almost periodic functions on G and the algebra of continuous functions on S. It is natural to call the space S the *Bohr compactification* of G. For this kind of result we refer the reader to Weil's monograph (654). Other ways of compactification, starting from almost periodic functions, are found in the works of H. Anzai and S. Kakutani (24).

The tight connection which exists between the almost periodic functions on groups and the ergodic theory has been revealed by W. Eberlein (225). Other remarkable results in this direction have been obtained by K. Yosida (685, 686).

The almost periodic functions on groups with values in Banach spaces were studied by S. Bochner and J. von Neumann (83). The definition and the fundamental properties are not essentially different from the case of complex-valued functions.

J. D. Weston (656) considered a more general class of almost periodic functions, i.e., the case in which the functions are defined on a topological group, and the values belong to a complete metric space X. Let $B(G, X)$ be the set of bounded mappings from G to X. In the set $B(G, X)$ one may define a distance function by

$$d(f, g) = \sup_{t \in G} \rho(f(t), g(t)), f, g \in B(G, X).$$

The continuous function $f \in B(G, X)$ is called almost periodic on G, if the set of functions $\{f(utv); t \in G\}$ is relatively compact in $B(G \times G, X)$.

E. Følner (268, 273) has extended the notion of almost periodic function in the Besicovitch sense to the functions defined on groups. The almost periodic functions in the Besicovitch sense on a local compact group have been discussed in R. Hirschfeld's paper (356).

The almost periodic functions on groups have interesting applications in the theory of group representation. The Weil–Neumann theory was successfully extended to the representations of a compact group. For problems of this kind we refer the reader to the monographs by E. Hewitt and K. A. Ross (355), L. H. Loomis (477), W. Maak (489), and K. Maurin (510).

Lately, numerous papers deal with the almost periodic functions on semigroups. The definition of almost periodicity on a semigroup may be formulated with the help of normality (Bochner). We mention the work of H. Günzler (300) in which are given many properties of almost periodic functions on semigroups, and also a rich bibliography with comments. The article by I. Glicksberg and K. de Leeuw (290) also contains interesting results in this field such as:

Any almost periodic (continuous) function on an abelian topological semigroup can be approximated by linear combinations of semicharacters of the semigroup.

By semicharacter of a semigroup H we mean a continuous map φ of H in the field of complex numbers, such that $|\varphi(a)| \leqslant 1$, and $\varphi(a + b) = \varphi(a)\varphi(b)$, $a, b \in H$.

The progress in the theory of almost periodic functions defined on groups has been steady during the last two decades. The monographs by J. F. Berglund, H. D. Junghenn and P. Miles (708), R. B. Burckel (709), C. F. Dunkel and D. E. Ramirez (725), B. M. Levitan and V. V. Zhikov (745), and A. A. Pankov (754) contain a good deal of information regarding some of the recent results and trends in the theory of almost periodic functions defined on groups.

There are many contributions in the form of journal papers. We will comment here on those due to T.-S. Wu (769), and P. Milnes (749), (750), (751). In (769) it is shown that, in general, right and left almost periodic functions on a group do not necessarily coincide if the Bohr (adapted) definition is used. This kind of almost periodicity is defined as follows: let f be a complex-valued function defined on the topological group G; assume that for every $\epsilon > 0$, there is a compact set $K_\epsilon \subset G$, such that $A_\epsilon K_\epsilon = G$, where the set A_ϵ is defined by $A_\epsilon = \{s \in G; |f(ts) - f(t)| < \epsilon \text{ for all } t \in G\}$. More precisely, this is the concept of right almost periodicity in the sense of Bohr. The kind of definition we have used in this chapter is the one based on normality, due to Bochner. It follows that Bohr's definition does not lead to the same class of almost periodic functions as the one due to Bochner. Moreover, as shown in the paragraph 3 of this chapter, the almost periodicity concept based on normality implies the approximation property. In the papers (750), (751), P. Milnes analyses these differences in depth. In particular, he shows that almost periodic functions in the Bohr sense do not form a linear space. But a function which is Bohr almost periodic and almost automorphic must be almost periodic in the meaning adopted in this chapter. In (749), the concept of weak almost periodicity in the sense of W. F. Eberlein is investigated.

Appendix

The Weierstrass theorem

We had the opportunity several times to use the Weierstrass theorem on approximation of continuous functions by polynomials. Since the proof of this theorem for functions of several variables is not usually found in texts of mathematical analysis, we consider that it is necessary to present it in this appendix. We select one of the simplest demonstrations.*

Before stating the theorem, let us specify some of the notations.

We shall denote by $X = (x_1, x_2, \ldots, x_s)$ the generic point of the s-dimensional euclidean space R^s, and by $|X - Y|$ the euclidean distance between the points X and Y:

$$|X - Y| = \sqrt{(x_1 - y_1)^2 + \cdots + (x_s - y_s)^2}.$$

By $X_1, X_2, \ldots, X_q$ we shall denote certain points of R^s.

Theorem. *If $f(x)$ is a real-valued function, continuous on a set $\mathcal{M}$ which is bounded and closed in R^s, then for any $\varepsilon > 0$ one can determine a polynomial $P_\varepsilon(X)$ such that*

(1) $$|f(X) - P_\varepsilon(X)| < \varepsilon, \qquad X \in \mathcal{M}.$$

Proof. Let D be the diameter of the set $\mathcal{M}$, i.e., $D = \sup |X - Y|$, X, $Y \in \mathcal{M}$. Let us set $M = \sup |f(X)|$, $X \in \mathcal{M}$. Given $\varepsilon > 0$, from the uniform continuity of $f(X)$ on $\mathcal{M}$ it follows that one can determine a $\delta(\varepsilon) > 0$ such that

(2) $$|f(X) - f(X')| < \frac{\varepsilon}{4},$$

if $|X - X'| < \delta$, X, $X' \in \mathcal{M}$.

Let $m = m(\delta)$ be the smallest natural number with the property

(3) $$\frac{1}{2^m} \leqslant \frac{\delta}{2s}.$$

---

* G. Viglino, Riv. Mat. Univ. Parma, **2**, 1951.

We divide R^s into hypercubes by the hyperplanes

$$x_k = \frac{l}{2^m}, \qquad k = 1, 2, \ldots, s; \qquad |l| = 0, 1, 2, \ldots .$$

We retain only the hypercubes which intersect the set $\mathcal{M}$. Since $\mathcal{M}$ is bounded and closed, there exists only a finite number of such hypercubes. Let us choose one point in each of the hypercubes. We obtain thus the points $X_1, X_2, \ldots, X_\mu$. If one of these points is situated on the boundary of one of these hypercubes, it could belong to more than one of the hypercubes of the above chosen family. We make the convention of considering this point only once, i.e., $X_i \neq X_j$, if $i \neq j$.

Set

(4) $$\varphi_r(X) = \frac{\delta^2 - |X - X_r|^2}{D^2}, \qquad X \in \mathcal{M}, \qquad r = 1, \ldots, \mu.$$

Every $\varphi_r(x)$ is a polynomial in $x_1, x_2, \ldots, x_s$. Since $0 \leqslant |X - X_r| \leqslant D$ for $X \in \mathcal{M}, r = 1, 2, \ldots, \mu$, it follows that

(5) $$-1 < \frac{\delta^2 - D^2}{D^2} \leqslant \varphi_r(X) \leqslant \frac{\delta^2}{D^2} < 1, \qquad X \in \mathcal{M}, \qquad r = 1, \ldots, \mu.$$

Let us denote by n the smallest natural number with the property

(6) $$\frac{8M\mu}{(1 + 3\delta^2/4D^2)^n} < \varepsilon.$$

The function

(7) $$\Phi(X) = \frac{\sum_{r=1}^{\mu} [1 + \varphi_r(X)]^n f(X_r)}{\sum_{r=1}^{\mu} [1 + \varphi_r(X)]^n}, \qquad X \in \mathcal{M},$$

is a rational function in $x_1, x_2, \ldots, x_s$. We have

(8) $$f(X) - \Phi(X) = \frac{\sum_{r=1}^{\mu} [1 + \varphi_r(X)]^n [f(X) - f(X_r)]}{\sum_{r=1}^{\mu} [1 + \varphi_r(X)]^n}.$$

Let us now fix X. The subscripts $1, 2, \ldots, \mu$ are divided into two categories: the first category includes those for which $|X - X_r| \leqslant \delta$, and the second category—those for which $|X - X_r| > \delta$. We divide the sum in the numerator of the fraction in (8) into two parts corresponding to the division with respect to the subscripts $1, \ldots, \mu$. By $\sum'$ we denote the summation over the subscripts of the first category, and by $\sum''$ we denote the summation over the subscripts of the second category.

Obviously we have

(9)
$$\left| \frac{\sum'[1 + \varphi_r(X)]^n[f(X) - f(X_r)]}{\sum\limits_{r=1}^{\mu}[1 + \varphi_r(X)]^n} \right| < \frac{\varepsilon}{4}, \qquad X \in \mathcal{M}.$$

In order to estimate the second sum we notice that for the subscripts r over which this summation is extended we have $\delta < |X - X_r| \leqslant D$. Thus $-1 < \varphi_r(X) < 0$ or $0 < 1 + \varphi_r(X) < 1$, $X \in \mathcal{M}$. This leads to

$$0 < [1 + \varphi_r(x)]^n < 1$$

and consequently,

(10)
$$\sum''[1 + \varphi_r(X)]^n < \mu, \qquad X \in \mathcal{M}.$$

The point X belongs to one of the hypercubes which cover $\mathcal{M}$. Therefore there exists an r_0 such that

(11)
$$|X - X_{r_0}| < \frac{\sqrt{s}}{2^m},$$

the right-hand side of inequality (11) representing the length of the diagonal of one of the considered hypercubes.

According to (3) and (11) we obtain

(12)
$$|X - X_{r_0}| < \frac{\delta}{2\sqrt{s}} \leqslant \frac{\delta}{2}.$$

Inequality (12) shows that r_0 belongs to $\sum'$, and we shall have

$$1 + \varphi_{r_0}(X) = 1 + \frac{\delta^2 - |X - X_{r_0}|^2}{D^2} \geqslant 1 + \frac{\delta^2 - \delta^2/4}{D^2} = 1 + \frac{3\delta^2}{4D^2}.$$

Thus

$$\sum_{r=1}^{\mu}[1 + \varphi_r(X)]^n > \left(1 + \frac{3\delta^2}{4D^2}\right)^n,$$

which permits us to write for $X \in \mathcal{M}$:

(13)
$$\left| \frac{\sum''[1 + \varphi_r(X)]^n[f(X) - f(X_r)]}{\sum\limits_{r=1}^{\mu}[1 + \varphi_r(X)]^n} \right| < \frac{2M\mu}{(1 + 3\delta^2/4D^2)^n} < \frac{\varepsilon}{4}.$$

From (8), (9), and (13) it follows that

(14) $$|f(X) - \Phi(X)| < \frac{\varepsilon}{2}, \qquad X \in \mathcal{M}.$$

Up to now we proved that the function $f(X)$ can be uniformly approximated on $\mathcal{M}$ by rational functions of the form (7). It remains to show now that any function of the form (7) can be approximated by polynomials.

We have

(15) $$\Phi(X) = \frac{S(X)}{R(X)}, \qquad X \in \mathcal{M},$$

$S(x)$ and $R(x)$ being polynomials in $x_1, x_2, \ldots, x_s$ of degree $2n$. Moreover, there exist two numbers λ and Λ such that $0 < \lambda \leqslant R(X) \leqslant \Lambda$, $X \in \mathcal{M}$. The existence of these numbers follows from the continuity of $R(X)$ on $\mathcal{M}$, and from the inequality $1 + \varphi_r(x) > 0$ for $x \in \mathcal{M}$, $r = 1, 2, \ldots, \mu$.

If $L > \Lambda$, then for $X \in \mathcal{M}$ we have

(16) $$\Phi(X) = \frac{S(X)}{L - [L - R(X)]} = \frac{S(X)}{L} \sum_{k=0}^{\infty} \left\{ \frac{L - R(X)}{L} \right\}^k,$$

since

$$0 < \frac{L - R(X)}{L} \leqslant \frac{L - \lambda}{L} < 1.$$

The convergence is therefore uniform. There exists a sufficiently large number N, such that if we set

(17) $$P_\varepsilon(X) = \frac{S(X)}{L} \sum_{k=0}^{N} \left\{ \frac{L - R(X)}{L} \right\}^k,$$

we shall have

(18) $$|\Phi(X) - P_\varepsilon(X)| < \frac{\varepsilon}{2}, \qquad X \in \mathcal{M}.$$

From (14) and (18) we obtain

$$|f(X) - P_\varepsilon(X)| < \varepsilon, \qquad X \in \mathcal{M},$$

i.e., what we wanted to prove.

We can now easily prove the Weierstrass theorem in the complex space.

Theorem. *If $f(z_1, z_2, \ldots, z_n)$ is a continuous function on a bounded, closed* *set in the n-dimensional complex space, then for any $\varepsilon > 0$ one can determine a polynomial $P_\varepsilon(z_1, \ldots, z_n; \bar{z}_1, \ldots, \bar{z}_n)$ such that*

(19) $$|f(z_1, \ldots, z_n) - P_\varepsilon(z_1, \ldots, z_n; \bar{z}_1, \ldots, \bar{z}_n)| < \varepsilon,$$

at any point of the considered set.

Proof. Consider $z_k = x_k + iy_k$, $k = 1, 2, \ldots, n$. The functions $\operatorname{Re} f = u(x_1, \ldots, x_n; y_1, \ldots, y_n)$ and $\operatorname{Im} f = v(x_1, x_2, \ldots, x_n; y_1, \ldots, y_n)$ are continuous on a bounded closed set in the $2n$-dimensional Euclidean space. According to the preceding theorem, for any $\varepsilon > 0$ there are some polynomials $Q(x_1, \ldots, x_n; y_1, \ldots, y_n)$ and $R(x_1, \ldots, x_n; y_1, \ldots, y_n)$ such that

(20) $$|u - Q| < \frac{\varepsilon}{2}, \qquad |v - R| < \frac{\varepsilon}{2},$$

in the considered set.

Let

(21) $$P(z_1, \ldots, z_n; \bar{z}_1, \ldots, \bar{z}_n) = Q + iR.$$

Since

$$x_k = \frac{z_k + \bar{z}_k}{2}, \qquad y_k = \frac{z_k - \bar{z}_k}{2i},$$

this means that $Q + iR$ is a polynomial in $z_1, \ldots, z_n, \bar{z}_1, \ldots, \bar{z}_n$ which allows us to write relation (21).

From (20) and (21) it follows immediately (19), which proves the theorem.

Remark. The approximation of complex functions by polynomials in $z_1, z_2, \ldots, z_n$ only is a very interesting and difficult problem. See, for example, W. Rudin, *Real and Complex Analysis*, McGraw-Hill, New York, 1966, Chapter 20.

References

1. ADRIANOVA, L. J. *The reducibility of systems of n linear differential equations with quasi-periodic coefficients.* (Russian). Vestnik Leningrad. Univ., **17** (1962), no. 7, 14–24.

2. ALBRYCHT, J., *Some remarks on the Marcinkiewicz-Orlicz space.* I, II, III. Bull. Acad. Polon. des Sciences, Cl.III, **4** (1956), 1–3; **7** (1959), 11–12, 55–56.

3. ————*The theory of Marcinkiewicz-Orlicz spaces.* Rozprawy Matematyczne, XXVII (1962), 1–55.

4. AMERIO, LUIGI. *Sulla convergenza in media delle serie* $\sum_{n=0}^{\infty} a_n e^{i\lambda_n x}$. Annali Scuola Norm. Sup. Pisa, **10** (1941), 191–198.

5. ————*Soluzioni quasi-periodiche, o limitate, di sistemi differenziali non lineari quasi-periodici, o limitati.* Annali mat. pura ed appl., **39**, (1955), 97–119.

6. ————*Funzioni quasi-periodiche ed equazioni differenziali.* Atti VI Congr. Un. Mat. Ital. (Naples, 1959), p. 119–144. Edizioni Cremonese, Rome, 1960.

7. ————*Sull'integrazione delle funzioni quasi-periodiche a valori in uno spazio hilbertiano.* Rend. Accad. Naz. Lincei (8), **28** (1960), 600–603.

8. ————*Quasi-periodicità degli integrali ad energia limitata dell'equazione delle onde, con termine noto quasi-periodico.* I, II, III. Rend. Accad. Naz. Lincei (8), **28** (1960), 147–152, 322–327, 461–466.

9. ————*Funzioni debolmente quasi-periodiche.* Rend. Sem. Mat, Padova, **30** (1960), 288–301.

10. ————*Sull'equazione delle onde con termine noto quasi-periodico.* Rend. Mat. Appl. (5), **19** (1960), 333–346.

11. ————*Problema misto e quasi-periodicità per l'equazione delle onde non omogenea.* Ann. Mat. pura ed appl. (4), **49** (1960), 393–417.

12. ————*Sulle equazioni differenziali quasi-periodiche astratte.* Ric. Mat., **9** (1960). 256–274.

13. ————*Ancora sulle equazioni differenziali quasi-periodiche.* Ric. Mat., **10** (1961), 31–32.

14. ————*Funzioni quasi-periodiche astratte e problemi di propagazione.* Conferenze del Seminario di Matematica dell'Università di Bari, No. 57 (1961), 1–14.

15. ————*Sulle equazioni lineari quasi-periodiche negli spazi hilbertiani.* I, II. Rend. Accad. Naz. Lincei, **31** (1961), 110–117, 197–205.

16. ————*Soluzioni quasi-periodiche delle equazioni lineari iperboliche quasi-periodiche.* Rend. Accad. Naz. Lincei (8), **33** (1962), 179–186.

17. ————*Soluzioni quasi-periodiche di equazioni funzionali negli spazi di Hilbert.* Seminario dell'Istituto Nazionale di Alta Matematica (1962–1963), 787–796.

18. ————*Soluzioni quasi-periodiche di equazioni quasi-periodiche negli spazi hilbertiani.* Ann. Mat. pura ed appl., **61** (1963), 259–278.

19. ————*Sul teorema di approssimazione delle funzioni quasi-periodiche.* Rend. Accad. Naz. Lincei (8), **34** (1963), 97–104.

20. ————*Su un teorema di minimax per le equazioni differenziali astratte.* Rend. Accad. Naz. Lincei, **35** (1963), 409–416.

21. ————*Ancora sul teorema di approssimazione per le funzioni quasi-periodiche.* Rend. Accad. Naz. Lincei, **36** (1964), 101–106.

22. ———*Solutions presque-périodiques d'équations fonctionnelles dans les espaces de Hilbert.* Deuxième Colloque sur l'Analyse fonctionnelle, Liège, 1964, 11–35.

23. ———*Abstract almost-periodic functions and functional equations.* Boll. U.M.I., **21** (1966), 287–334.

24. ANZAI, H., KAKUTANI, S., *Bohr compactifications of a locally compact abelian group.* I, II. Proc. Imp. Acad. Tokyo, **19** (1943), 476–480, 533–539.

25. ARTEMENKO, A. *On positive functionals in the space of Bohr's almost periodic functions.* (Russian). Zap. Inst. Math. Mech. Kharkovsk. Gos. Univ., **16** (1940), 111–114.

26. ARTJUŠENKO, L. M. *The application of Fourier series for finding almost periodic solutions of equations with mean values.* (Russian). Izv. Vysš. Učebn. Zaved. Matematika, no. 4 (5) (1968), 21–27.

27. ARZUMANJAN, G. S. *On some extensions of the class of Bohr's almost periodic functions.* (Russian). Trudy Nauč. Isled. Inst. Mat. Phys. Azerb. Univ., **I** (1949), 27–49.

28. AVAKIAN, ARRA STEVE. *Almost periodic functions and the vibrating membrane.* J. Math. Phys., **XIV** (1935), 350–378.

29. BAILLETTE, AIMÉE. *Sur la transformée de Fourier-Carleman d'une fonction presque-périodique de spectre donné.* C.R. Acad. Sci. Paris, **258** (1964), 6049–6052.

30. BANG, T. *Om spaltning af naestenperiodiske Funktioners Fourierraekker.* Mat. Tidsskrift B (1941), 53–58.

31. ———*Une inégalité de Kolmogoroff et les fonctions presque-périodiques.* Mat.-fysiske Medd., **19**, 4 (1941), 1–28.

32. BARBĂLAT, I. *Application du principe topologique de T. Ważewski aux équations différentielles du second ordre.* Annales Polonici Mathematici, **V** (1958), 303–317.

33. ———*Solutions presque-périodiques des équations différentielles non-linéaires.* Com. Acad. R.S. România, **11** (1961), 155–159.

34. ———*Solutions presque-périodiques de l'équation de Riccati.* Com. Acad. R.S. România, **11** (1961), 161–165.

35. BARBU, V. *Solutions presque-périodiques pour un système d'équations linéaires aux dérivées partielles.* Ric. Mat., **XV** (1966), 207–222.

36. BELLMAN, RICHARD. *Almost periodic gap series.* Duke Math. J., **10** (1943), 641–642.

37. ———*Stability theory of differential equations.* McGraw-Hill, New York, 1953.

38. BEREZANSKIJ, YU. M. *On the theory of Levitan's almost periodic sequences.* (Russian). Doklady Akad. Nauk SSSR, **81** (1951), 493–496.

39. ———*On the theory of almost periodic functions with respect to translations in hypercomplex systems.* (Russian). Doklady Akad. Nauk SSSR, **85** (1952), 9–12.

40. ———*On generalized almost periodic functions and sequences, related with the difference-differential equations.* (Russian). Mat. Sb., **32** (1953), 157–194.

41. BERMAN, D. L. *Linear trigonometric polynomial operations in spaces of almost-periodic functions.* (Russian). Mat. Sb. (N.S.), **49** (91) (1959), 267–280.

42. BERTRANDIAS, JEAN-PAUL. *Functions pseudo-aléatoires et fonctions presque-périodiques.* C.R. Acad. Sci. Paris, **255** (1962), 2226–2228.

43. BESICOVITCH, A. S. *Sur quelques points de la théorie des fonctions presque-périodiques.* C.R. Acad. Sci. Paris, **180** (1925), 394–397.

44. ———*Über die Parsevalsche Gleichung für analytische fastperiodische Funktionen.* Acta. Math., **47** (1926), 283–295.

45. ———*On generalized almost periodic functions.* Proc. London Math. Soc., **25** (1926), 495–512.

46. ———*On Parseval's theorem for Dirichlet series.* Proc. London Math. Soc., **26** (1927), 25–34.

47. ———*On mean values of functions of a complex and of a real variable.* Proc. London Math. Soc., **27** (1928), 373–388.

48. ———*Analysis of conditions of generalized almost periodicity.* Acta. Math., **58** (1932), 217–230.

49. ———*Almost periodic functions.* Cambridge University Press, 1932.

50. BESICOVITCH, A. S., BOHR, H. *Some remarks on generalizations of almost periodic functions.* Mat.-fysiske Medd., **8**, 5 (1927), 1–33.

51. ———*On almost periodic properties of translation numbers.* J. London Math. Soc., **3** (1928), 172–176.

52. ———*Almost periodicity and general trigonometric series.* Acta Math., **57** (1931), 203–292.

53. BEURLING, A. *Un théorème sur les fonctions bornées et uniformément continues sur l'axe réel.* Acta Math., **77** (1945), 127–136.

54. ———*Sur une classe de fonctions presque-périodiques.* C.R. Acad. Sci. Paris, **225** (1947), 326–328.

55. BIRJUK, G. I. *On a theorem concerning the existence of almost periodic solutions for certain non linear differential systems with a small parameter.* (Russian). Doklady Akad. Nauk SSSR, **96** (1954), 5–7.

56. ———*On the problem of the existence of almost periodic solutions for systems with a small parameter in a singular case.* (Russian). Doklady Akad. Nauk SSSR, **97** (1954), 577–579.

57. BOCHNER, S. *Sur les fonctions presque-périodiques de Bohr.* C.R. Acad. Sci. Paris, **180** (1925), 1156–1158.

58. ———*Über Fouerierreihen fastperiodischen Funktionen,* Sitzb. Berliner Math. Ges., **26** (1927), 49–65.

59. ———*Properties of Fourier series of almost periodic functions.* Proc. London Math. Soc., **26** (1927), 433–452.

60. ———*Konvergenzsätze für Fourierreihen grenz periodischer Funktionen.* Math. Z., **27** (1927), 187–211.

61. ———*Beiträge zur Theorie der fastperiodischen Funktionen. I. Funktionen einer Variablen.* Math. Ann., **96** (1927), 119–147.

62. ———*Beiträge zur Theorie der fastperiodischen Funktionen. II. Funktionen mehrerer Variablen.* Math. Ann., **96** (1927), 383–409.

63. ———*Über die Struktur von Fourierreihen fastperiodischer Funktionen.* Sitzb. Math.-Naturwiss. Abt. Bayer. Akad. Wiss. (1928), 181–190.

64. ———*Über gewisse Differential- und allgemeinere Gleichungen deren Lösungen fastperiodisch sind. I. Teil. Der Existenzsatz.* Math. Ann., **102** (1929), 489–504.

65. ———*Über gewisse Differential- und allgemeinere Gleichungen deren Lösungen fastperiod sind. II. Teil. Der Beschränktheitssatz.* Math. Ann., **103** (1929), 588–597.

66. ———*Beitrag zur absoluten Konvergenz fastperiodischer Funktionen.* Jahresb. Deutschen Math. Vereinigung, **39** (1930), 52–54.

67. ———*Über gewisse Differential- und allgemeinere Gleichungen deren Lösungen fastperiodisch sind. III. Teil. Systeme von Gleichungen.* Math. Ann., **104** (1931), 579–587.

68. ——*Abstrakte fastperiodische Funktionen.* Acta Math. **61** (1933), 149–184.

69. ——*Remark on the integration of almost periodic functions.* J. London Math Soc., **8** (1933), 250–254.

70. ——*Homogeneous systems of differential equations with almost periodic coefficients.* J. London Math. Soc., **8** (1933), 283–288.

71. ——*Fastperiodische Lösungen der Wellengleichung.* Acta Math.,**62**(1934),227–237.

72. ——*On general Fourier series with gaps.* Prace Mat.-fiz., **43** (1936), 63–79.

73. ——*Additive set functions on groups.* Proc. Nat. Acad. Sci. U.S., **25** (1939), 158–160.

74. ——*A uniqueness theorem for analytic almost-periodic functions.* Duke Math. J **5** (1939), 937–940.

75. ——*Almost periodic solutions of the inhomogeneous wave equations.* Proc. Nat. Acad. Sci. U.S., **46** (1960), 1233–1236.

76. ——*Uniform convergence of monotone sequences of functions.* Proc. Nat. Acad. Sci. U.S., **47** (1961), 582–585.

77. ——*A new approach to almost periodicity.* Proc. Nat. Acad. Sci. U.S., **48** (1962), 2039–2043.

78. ——*Potential-theoretic approach to analytic Bohr almost periodic functions in half-planes.* Math. Ann., **150** (1963), 150–155.

79. ——*Continuous mappings of almost automorphic and almost periodic functions.* Proc. Nat. Acad. Sci. U.S., **52** (1964), 907–910.

80. BOCHNER, S., BOHNENBLUST, F. *Analytic functions with almost periodic coefficients.* Ann. Math., **35** (1934), 152–161.

81. BOCHNER, S., IZUMI, S. *Strong law of large numbers for sequences of almost periodic functions.* Ann. Math., **69** (1959), 718–732.

82. BOCHNER, S., JESSEN, B. *Distribution functions and positive-definite functions.* Ann. Math., **35** (1934), 252–257.

83. BOCHNER, S., NEUMANN, J. von. *On compact solutions of operational-differential equations.* I. Ann. Math., **36** (1935), 255–290.

84. ——*Almost periodic functions in groups.* II. Trans. Am. Math. Soc., **37** (1935), 21–50.

85. BOGDANOWICZ, WITOLD. *On the existence of almost periodic solutions for systems of ordinary differential equations in Banach spaces.* Arch. Rational Mech. Anal., **13** (1963), 364–370.

86. BOGOLIUBOV, N. N. *Sur l'approximation des fonctions par les sommes trigonométriques.* Comptes Rendus (Doklady) de l'Acad. Sci. de l'URSS (A) (1930), 147–152.

87. ——*Sur l'approximation trigonométrique des fonctions dans l'intervalle infini.* I. Bulletin (Izvestija) de l'Acad. Sci. de l'URSS, **1** (1931), 23–54.

88. ——*Sur l'approximation trigonométrique des fonctions dans l'intervalle infini.* II. Bulletin (Izvestija) de l'Acad. Sci. de l'URSS, **1** (1931), 149–160.

89. ——*On some arithmetic properties of almost periods.* (Russian). Zap. Kafedry Math. Phys. Akad. Nauk USSR, **4** (1939), 185–206.

90. ——*On some statistical methods in the mathematical physics.* (Russian). Lwow, 1945.

91. BOGOLIUBOV, N. N., KRYLOFF, N. M. *New methods in nonlinear mechanics.* (Russian). Moscow, 1934.

92. ——*Sur les solutions quasi-périodiques des équations de la mécanique non-linéaire.* C.R. Acad. Sci. Paris, **199** (1934), 1592–1593.

93. BOGOLIUBOV, N. N., MITROPOLSKIĬ YU. A. *Asymptotic methods in the theory of non-linear vibrations.* (Russian). Moscow, 1955.

94. BOHL, P. *Ueber die Dartsellung von Funktionen einer Variablen durch trigonometrische Reihen mite mehrer einer Variablen proportionalen Argumenten.* Dorpat, 1893.

95. ———*Ueber eine Differentialgleichung der Störungstheorie.* J. f. reine u. angew. Math., **131** (1906), 268–321.

96. BOHR, H. *Über eine quasi-periodische Eigenschaft Dirichletschen Reihen mit Anwendung auf die Dirichletschen L-Funktionen.* Math. Ann., **85** (1922), 115–122.

97. ———*Sur les fonctions presque-périodiques.* C.R. Acad. Sci. Paris, **177** (1923), 737–739.

98. ———*Sur l'approximation des fonctions presque-pérodiques par des sommes trigonométriques.* C.R. Acad. Sci. Paris, **177** (1923), 1090–1092.

99. ———*Zur Theorie der fastperiodischen Funktionen. I. Eine Verallgemeinerung der Theorie der Fourierreihen.* Acta Math., **45** (1925), 29–127.

100. ———*Zur Theorie der fastperiodischen Funktionen. II. Zusammenhang der fastperiodischen Funktionen mit Funktionen von unendlich vielen Variablen; gleichmässige Approximation durch trigonometrische Summen.* Acta. Math., **46** (1925), 101–214.

101. ———*En Saetning om Fourierraekker for naestenperiodiske Funktioner.* Mat. Tidsskrift (1925), 31–37.

102. ———*Einige Sätze über Fourierreihen fastperiodischer Funktionen.* Math. Z., **23** (1925), 38–44.

103. ———*Fastperiodische Funktionen.* Jahresb. Deutsch. Math. Ver., **34** (1925), 25–41.

104. ———*Über allgemeine Fourier- und Dirichlet-Entwicklungen.* Den Sjette Skandinaviske Matematikerkongress, København, 1925, 173–190.

105. ———*Sur les fonctions presque-périodiques d'une variable complexe.* C.R. Acad. Sci. Paris, **180** (1925), 645–647.

106. ———*Zur Theorie der fastperiodischen Funktionen. III. Dirichletentwicklung analytischer Funktionen.* Acta. Math., **47** (1926), 237–281.

107. ———*On the explicit determination of the upper limit of an almost periodic function.* J. London Math. Soc., **1** (1926), 109–112.

108. ———*Sur le théorème d'unicité dans la théorie des fonctions presque-périodiques.* Bull. Sci. Math., **50** (1926), 1–7.

109. ———*En Klasse hele transcendente Funktioner.* Mat. Tidsskrift (1926), 41–45.

110. ———*Allgemeine Fourier- und Dirichlet-Entwicklungen.* Abh. Math. Sem. Hambg. Univ., **4** (1926), 366–374.

111. ———*Ein Satz über analytische Fortsetzung fastperiodischer Funktionen.* J. f. reine u. angew. Math., **157** (1927), 61–65.

112. ———*Über die Verallgemeinerungen fastperiodischer Funktionen.* Math. Ann., **100** (1928), 357–366.

113. ———*Bericht über die Theorie der fastperiodischen Funktionen.* Atti del Congresso Int. dei Matematici, Bologna, **2** (1928), 283–288.

114. ———*Grenzperiodische Funktionen.* Acta Math., **52** (1928), 127–133.

115. ———*Über analytische fastperiodische Funktionen.* Math. Ann., **103** (1930), 1–14.

116. ———*Kleinere Beiträge zur Theorie der fastperiodischen Funktionen.* I, II. Mat.-fysiske Medd., **10**, 10 (1931), 1–17.

117. ———*Kleinere Beiträge zur Theorie der fastperiodischen Funktionen.* III, IV. Mat.-fysiske Medd., **10**, 12 (1931), 1–15.

118. ———*On the inverse function of an analytic almost periodic function.* Ann. Math., **32** (1931), 247–260.

119. ———*Almost periodicity and general trigonometric series.* Acta Math., **57** (1931), 203–292.

120. ———*Ueber fastperiodische ebene Bewegungen.* Comm. Math. Helvetici, **4** (1932), 51–64.

121. ———*Fastperiodische Funktionen.* Springer-Verlag, Berlin, 1932.

122. ———*Stabilitet og Naestenperiodicitet.* Mat. Tidsskrift (1933), 21–25.

123. ———*Un théorème général sur l'intégration d'un polynôme trigonométrique.* C.R. Acad. Sci. Paris, **200** (1935), 1276–1277.

124. ———*Kleinere Beiträge zur Theorie der fastperiodischen Funktionen.* V. Mat.-fysiske Medd., **13**, 8 (1935), 1–13.

125. ———*Kleinere Beiträge zur Theorie der fastperiodischen Funktionen.* VI. Mat.-fysiske Medd., **14**, 2 (1936), 1–8.

126. ———*Kleinere Beiträge zur Theorie der fastperiodischen Funktionen.* VII, VIII. Mat.-fysiske Medd., **14**, 7 (1936), 1–24.

127. ———*Ein allgemeiner Satz über die Integration eines trigonometrischen Polynoms* Prace matematyczno-fizyczne, **43** (1936), 273–288.

128. ———*Über fastperiodische Bewegungen.* Neuvième Congrès des Mathématiciens Scandinaves, Helsingfors, 1938, 39–61.

129. ———*Contributions to the theory of analytic almost periodic functions. On the behavior of an analytic almost periodic function in the neighborhood of a boundary for its almost periodicity.* Mat.-fysiske Medd., **20**, 18 (1943), 1–37.

130. ———*On some functional spaces.* Comptes Rendus du dixième Congrès des Math. Scandinaves, København, 1947, 313–319.

131. ———*On almost periodic functions and the theory of groups.* Am. Math. Monthly, **56** (1949), 595–609.

132. ———*A survey of the different proofs of the main theorems in the theory of almost periodic functions.* Proc. Intern. Congr. Math., Cambridge, **1** (1950), 339–348.

133. ———*On limit periodic functions of infinitely many variables.* Acta Sci. Math. Univ. Szegediensis, **12** (1950), 145–149.

134. ———*On the definition of almost periodicity.* J. d'Analyse Mathématique, **1** (1951), 11–27.

135. ———*Collected Mathematical Works*, vol. I, II, III. Dansk Mat. Forening. København, 1952.

136. BOHR, H., FLANDERS, D. A. *Algebraic equations with almost-periodic coefficients.* Mat.-fysiske Medd., **15**, 12 (1937), 1–49.

137. ———*Algebraic functions of analytic almost periodic functions.* Duke Math. J., **4** (1938), 779–787.

138. BOHR, H., FØLNER, E. *On some types of functional spaces. A contribution to the theory of almost periodic functions.* Acta Math., **76** (1944), 31–155.

139. BOHR, H., JESSEN, B. *Über fastperiodische Bewegungen auf einem Kreise.* Annali Scuola Norm. Sup. Pisa, Scienze fisiche e matematiche, **1** (1932), 385–398.

140. ———*Mean motions and almost periodic functions.* Colloques Internationaux du C.N.R.S., Analyse Harmonique, Paris, 1949, 75–84.

141. BOHR, H., NEUGEBAUER, O. *Über lineare Differentialgleichungen mit konstanten Koeffizienten und fastperiodischer rechter Seite.* Nachr. Ges. Wiss. Göttingen, Math.-Phys. Klasse (1926), 8–22.

142. BORUHOV, L. E. *A linear integral equation with almost periodic kernel and free term.* (Russian). Doklady Akad. Nauk SSSR, **57** (1947), 647–649.

143. ——— *On periodic and almost periodic solutions of the equations $y' + q(x)y = f(x)$.* (Russian). Nauč. Ež. Saratovsk. Univ. (1954), 656–657.

144. ——— *On a paper of V. V. Bystrenin.* (Russian). Nauč. Ež. Saratovsk. Univ. (1954), 658.

145. ——— *On almost periodicity of solutions of some linear differential systems with almost periodic coefficients.* (Russian). Nauč. Ež. Saratovsk. Univ. (1954), 659–660.

146. ——— *On almost periodic solutions of certain differential equations.* (Russian). Nauč. Ež. Saratovsk. Univ. (1954), 660–661.

147. ——— *On periodicity and almost periodicity of solutions of certain second order linear differential equations.* (Russian). Nauč. Ež. Saratovsk. Univ. (1954), 661–663.

148. BOURION, GEORGES. *Sur le cas classiques de convergence des séries de Fourier associées aux fonctions presque-périodiques de Bohr.* Bull Math. Soc. Sci. Math. Phys. R.P. Roumaine (N.S.), **2** (50) (1958), 1–4.

149. BRAUERS, N. *Différentiation et intégration des fonctions presque-périodiques de plusieurs variables réelles.* Acta Univ. Latviensis (1939), 235–263.

150. ——— *Sur l'integration des fonctions presque-périodiques de deux variables.* Comm. Math. Helvetici, **11** (1939), 330–335.

151. BRĀZMA, N. *Sur les fonctions presque-périodiques de plusieurs variables complexes.* Acta Univ. Latviensis, **20** (1941), 431–455.

152. BREDIHINA, E. A. *Some estimations for the best approximations of almost periodic functions.* (Russian). Doklady Akad. Nauk SSSR, **103** (1955), 731–754.

153. ——— *On absolute convergence of Fourier series of almost periodic functions.* (Russian). Doklady Akad, Nauk SSSR, **111** (1956), 1163–1166.

154. ——— *Two criteria for absolute convergence of Fourier series of almost periodic functions.* (Russian). Trudy Av. Inst. Kuibyshev, **3** (1957), 43–48.

155. ——— *On best approximations of almost periodic functions by entire functions of finite degree.* (Russian). Doklady Akad. Nauk SSSR, **117** (1957), 17–20.

156. ——— *Fourier series as a device for approximation of almost periodic functions.* (Russian). Doklady Akad. Nauk SSSR, **123** (1958), 219–222.

157. ——— *Some problems concerning the best approximation of almost periodic functions with a bounded spectrum.* (Russian). Doklady Akad. Nauk SSSR, **131** (1960), 721–724.

158. ——— *The summation of Fourier series of almost periodic functions* (Russian). Izv. Vysš. Učebn. Zaved. Matematika, no. **15** (18) (1960), 33–39.

159. ——— *Some estimates of the deviation of partial sums of Fourier series of almost-periodic functions.* (Russian). Mat. Sb. (N.S.), **50** (92) (1960), 369–382.

160. ——— *On the approximation of almost-periodic functions with bounded spectrum.* (Russian). Mat. Sb. (N.S.), **56** (98) (1962), 59–76.

161. ——— *Simultaneous approximation of almost-periodic functions and their derivatives.* (Russian). Doklady Akad. Nauk SSSR, **145** (1962), 17–20.

162. ——— *On the convergence of Fourier series of the almost periodic functions of Stepanov.* (Russian). Uspehi Mat. Nauk, **19** No. 6 (120) (1964), 133–137.

163. ———On the approximation of almost periodic functions. (Russian). Sirbisk. Mat. Ž., **5** (1964), 768–773.

164. ———On the convergence of Fourier series of almost periodic functions. (Russian). Doklady Akad. Nauk SSSR, **160** (1965), 259–262.

165. ———On approximation of almost periodic functions of class Q_ε. (Russian). Izv. Vysš. Učebn. Zaved. Matematika, No. **4** (1965), 17–23.

166. ———Approximation of Stepanov's almost periodic functions. (Russian). Doklady Akad. Nauk SSSR, **164** (1965), 255–258.

167. DE BRUIJN, N. G. Bijnaperiodicke multiplicatieve functies. Nieuw. Arch. Wisk., **22**, (1943), 81–95.

168. BURD, V. Š. On the branching of almost periodic solutions of non linear ordinary differential equations. (Russian). Doklady Akad. Nauk SSSR, **159** (1964), 239–242.

169. BURKHILL, H. Almost periodicity and nonabsolutely integrable functions. Proc. London Math. Soc., **53** (1951), 32–42.

170. ———Cesàro-Perron almost periodic functions. Proc. London Math. Soc., **2** (1952), 150–174.

171. ———The Cesàro-Perron scale of almost periodicity. Proc. London Math. Soc., **7** (1957), 481–497.

172. ———A note on mean values. J. London Math. Soc., **34** (1959), 1–4.

173. BYLOV, B. F. On stability from above of the greatest characteristic index of a system of linear differential equations with almost-periodic coefficients. (Russian). Mat. Sb. (N.S.), **48** (90) (1959), 117–128.

174. ———The structure of the solutions of a system of linear differential equations with almost periodic coefficients. (Russian). Mat. Sb. (N.S.), **66** (108) (1965), 215–229.

175. BYSTRENIN, V. V. On almost periodic solutions of certain ordinary differential equations. (Russian). Doklady Akad. Nauk SSSR, **33** (1941), 387–389.

176. ———On the approximation theorem in the theory of almost periodic functions. (Russian). Doklady Akad. Nauk SSSR, **33** (1941), 390–392.

177. CAMERON, R. H. Implicit functions of almost periodic functions. Bull. Am. Math. Soc., **40** (1934), 895–904.

178. ———Almost periodic transformations. Trans. Am. Math. Soc., **36** (1934), 276–291.

179. ———Linear differential equations with almost periodic coefficients. Ann. Math., **37** (1936), 29–42.

180. ———Almost periodic properties of bounded solutions of linear differential equations with almost periodic coefficients. J. Math. Phys., **15** (1936), 73–81.

181. ———Linear differential equations with almost periodic coefficients. Acta Math., **69** (1938), 21–56.

182. ———Quadratures involving trigonometric sums. J. Math. Phys., **19** (1940), 161–166.

183. CARACOSTA, GEORGES, DOSS, RAOUF. Sur l'intégrale d'une fonction presque-périodique. C.R. Acad. Sci. Paris, **246** (1958), 3207–3208.

184. CARTWRIGHT, M. L. Almost periodic solutions of certain second order differential equations. Rend. Sem. Mat. Fis. Milano, **31** (1961) ,100–110.

185. CASTELLANA-RIZZONELLI, PIERANITA. Sulle funzioni limitate, a valori vettoriali, con differenze quasi periodiche. Tamburini Editore, Milano, 1963, 14 p.

186. ČERESIZ, V. M. *On almost periodic solutions of non linear systems.* (Russian). Doklady Akad. Nauk SSSR, **165** (1965), 281–284.

187. CERNEAU, SIMONNE. *Sur la construction de solutions presque-périodiques de problèmes de perturbation singulière.* C.R. Acad. Sci. Paris, **260** (1965), 768–771.

188. CHENG, NAI-DANG. *On the constructive property of a class of uniformly almostperiodic functions.* (Chinese). Acta Math. Sinica, **13** (1963), 441–446.

189. CIGLER, JOHANN. *Einige Bemerkungen zur Theorie der fastperiodischen Funktionen.* **15** (1964), 155–160.

190. CINQUINI, SILVIO. *Sopra i polinomi di Bochner-Féjer e le funzioni quasi-periodiche secondo Stepanoff.* Ist. Lombardo Sci. Lett. Rendiconti, **9** (1945), 381–400.

191. ——*Funzioni quasi-periodiche.* Scuola Norm. Sup. Pisa, 1950.

192. ——*Sopra il problema dell'approssimazione delle funzioni quasi-periodiche.* Ann. Scuola Norm. Sup. Pisa, **5** (1951), 245–267.

193. ——*Funzioni quasi-periodiche ed equazioni differenziali.* Rend. Sem. Mat. Torino, **11** (1951–1952), 47–74.

194. ——*Sopra qualche concetto della teoria delle funzioni quasi-periodiche.* Matematiche (Catania), **12** (1957), 1–17.

195. ——*Sopra una definizione di funzione quasi-periodica.* Ist. Lombardo Sci. Lett. Rend. Cl. Sci. Mat. Nat., **91** (1957), 547–564.

196. CONTI, R., SANSONE, G. *Non-linear differential equations.* Pergamon Press, New York, 1964.

197. CORDUNEANU, C. *Soluţii aproape-periodice ale ecuaţiilor diferenţiale neliniare de ordinul al doilea.* Comunicările Acad. R.P.R., **5** (1955), 21–26.

198. ——*Soluţii asimptotic aproape-periodice ale ecuaţiilor diferenţiale neliniare de ordinul al doilea.* Studii şi cercetări ştiinţifice. Matematică (Iaşi), **6** (1955), 1–4.

199. ——*Cîteva consideraţii în legătură cu unele sisteme neliniare de ecuaţii diferenţiale.* Studii ši cerc. štiinţ. Matematică (Iaši), **VII** (1956), fasc. 2, 13–23.

200. ——*Funcţii aproape-periodice.* Editura Acad. R.S. România, Bucharest, 1961.

201. COSTINESCU, A., GHEORGHIU, N. *Observations sur les fonctions et les suites presquepériodiques avec des valeurs dans un espace de Banach.* Rev. Roumaine de Math. pures et appl., **XI** (1966), 421–424.

202. CRONIN, JANE. *Almost-periodic solutions and critical roots.* Duke Math. J., **29** (1962), 663–669.

203. DELSARTE, J. *Les fonctions moyenne-périodiques.* J. Math. pures et appl., **14** (1935), 403–453.

204. ——*Sur une extension nouvelle de la notion de presque-périodicité.* C.R. Acad. Sci. Paris, **206** (1938), 573–575.

205. ——*Une extension nouvelle de la théorie des fonctions presque-périodiques de Bohr.* Acta Math., **69** (1938), 259–317.

206. ——*Essai sur l'application de la théorie des fonctions presque-périodiques à l'arithmétique.* Ann. Sci. École Norm. Supér. Paris, **62** (1945), 185–204.

207. DEMIDOVICH, B. P. *On a case of almost periodicity of solutions of an ordinary differential equation of first order.* (Russian). Usp. Mat. Nauk, **8**, f.6 (1953), 103–106.

208. ——*On bounded solutions of a certain nonlinear system of ordinary differential equations.* (Russian). Mat. Sb., **40** (82) (1956), 73–94.

209. ——*On bounded solutions for some quasi-linear systems.* Doklady Akad. Nauk SSSR, **138** (1961), 1273–1275.
210. DEYSACH, LAWRENCE G., SELL, GEORGE R. *On the existence of almost periodic motions.* Michigan Math. J., **12** (1965), 87–95.
211. DOLCHER, MARIO. *Su un criterio di convergenza uniforme per le successioni monotone di funzioni quasi-periodiche.* Rend. Sem. Mat. Univ. Padova, **34** (1964), 191–199.
212. DOSS, RAOUF. *Contribution to the theory of almost periodic functions.* Ann. Math., **46** (1945), 196–219.
213. ——*Some theorems on almost periodic functions.* Am. J. Math., **72** (1950), 81–92.
214. ——*Groupes compacts et fonctions presque-périodiques généralisées.* Bull. Sci. Math., **77** (1953), 186–194.
215. ——*On generalized almost periodic functions.* Ann. Math., **59** (1954), 477–489.
216. ——*Sur une nouvelle classe de fonctions presque-périodiques.* C.R. Acad. Sci. Paris, **238** (1954), 317–318.
217. ——*On Riemann integrability and almost periodic functions.* Compositio Mathematica, **12** (1956), 271–283.
218. ——*On bounded functions with almost periodic differences.* Proc. Am. Math. Soc., **12** (1961), 488–489.
219. ——*On generalized almost periodic functions,* II. J. London Math. Soc., **37** (1962), 133–140.
220. ——*The integral of a generalized almost periodic function.* Duke Math J., **30** (1963), 39–46.
221. ——*On de la Vallée-Poussin's proof of the fundamental theorem in the theory of almost-periodic functions.* Quart. J. Math., **16** (1965), 188–196.
222. ——*On the almost-periodic solutions of a class of integro-differential-difference equations.* Ann. Math., (2), **81** (1965), 117–123.
223. DUNFORD, N., SCHWARTZ, J. T. *Linear operators.* Part I. Interscience, New York, 1958.
224. DZJADYK, V. K. *On a property of almost periodic polynomials.* (Russian). Ukrain. Mat. Ž., **13** (1961), 96–98.

225. EBERLEIN, W. F. *Abstract ergodic theorems and weak almost-periodic functions.* Trans. Am. Math. Soc., **67** (1949), 217–240.
226. ——*Spectral theory and harmonic analysis.* Proc. Symp. on Spectral Theory and Diff. Problems. Oklahoma Agr. Mech. College (1951), 209–219.
227. ERDÖS, P., KAC, M., VAN KAMPEN, E. R., WINTNER, A. *Ramanujan sums and almost-periodic functions.* Studia Mathematica, **9** (1940), 43–53.
228. ERDÖS, P., WINTNER, A. *Additive functions and almost periodicity (B^2).* Am. J. Math., **62** (1940), 635–645.
229. ESCLANGON, E. *Sur une extension de la notion de périodicité.* C.R. Acad. Sci. Paris, **135** (1902), 891–894.
230. ——*Sur les fonctions quasi-périodiques.* C.R. Acad. Sci. Paris, **137** (1903), 305–307.
231. ——*Les fonctions quasi-périodiques.* Thèse, Paris, 1904.
232. ——*Sur les fonctions quasi-périodiques moyennes, déduites d'une fonction quasi-périodique.* C.R. Acad. Sci. Paris, **157** (1913), 1389–1392.
233. ——*Sur les intégrales quasi-périodiques d'équations différentielles linéaires.* C.R. Acad. Sci. Paris, **158** (1914), 1254–1256.

234. ———Sur les intégrales bornées d'une équation différentielle linéaire. C.R. Acad. Sci. Paris, **160** (1915), 475–478.

235. ———Sur les intégrales quasi-périodiques d'une équation différentielle. C.R. Acad. Sci. Paris, **160** (1915), 652–653.

236. ———Sur les intégrales quasi-périodiques d'une équation différentielle. C.R. Acad. Sci. Paris, **161** (1915), 488–489.

237. ———Nouvelles recherches sur les fonctions quasi-périodiques. Annales Obs. Bordeaux, **16** (1919), 1–174.

238. ESSEN, MATS. On homomorphism and orthogonal systems. Arkiv Mat., **3** (1958), No. 6, 505–510.

239. EYMARD, PIERRE. Applications laissant stable l'ensemble des fonctions presque-périodiques sur certains groupes discrets. C.R. Acad. Sci. Paris, **249** (1959), 2459–2461.

240. EZEILO, J. O. C. On the existence of an almost periodic solution of a non-linear system of differential equations. Contributions to Differential Equations, **3** (1964), 337–349.

241. FAN, KY. Les fonctions asymptotiquement presque-périodiques d'une variable entière et leur application à l'étude de l'itération des transformations continues. Math. Z., **48** (1942/43), 685–711.

242. FAST, H., HARTMAN, S., STEINHAUS, H. Sur les presque-périodes des fonctions périodiques. Coll. Math., **1** (1948), 297–304.

243. FAVARD, J. Sur les fonctions harmoniques presque-périodiques. C.R. Acad. Sci. Paris, **182** (1926), 757–759.

244. ———Sur les fonctions harmoniques presque-périodiques. J. Math pures et appl., **6** (1927), 229–336.

245. ———Sur les équations différentielles à coefficients presque-périodiques. Acta Math., **51** (1927), 31–81.

246. ———Sur la répartition des points où une fonction presque-périodique prend une valeur donnée. C.R. Acad. Sci. Paris, **194** (1932), 1714–1716.

247. ———Leçons sur les fonctions presque-périodiques. Gauthier-Villars, Paris, 1933.

248. ———Sur la fonction conjuguée d'une fonction presque-périodique. Mat. Tidsskrift (1934), 57–60.

249. ———Application de la formule sommatoire d'Euler à la démonstration de quelques propriétés extrémales des intégrales des fonctions périodiques ou presque-périodiques. Mat. Tidsskrift (1936), 81–94.

250. ———Note sur les fonctions presque-périodiques. Mat. Tidsskrift (1936), 71–75.

251. ———Remarque sur les polynômes trigonométriques. C.R. Acad. Sci. Paris, **209** (1939), 746–748.

252. ———Sur certains systèmes différentiels scalaires linéaires et homogènes à coefficients presque-périodiques. Ann. Mat. pura ed appl. (4), **61** (1963), 297–316.

253. ———Sur les équations différentielles scalaires presque-périodiques. J. Math. pures appl. (9), **43** (1964), 87–97.

254. FEKETE, M. On generalized Fourier Series with non-negative coefficients. Proc. London Math. Soc., **39** (1935), 321–333.

255. FENCHEL, WERNER. Om Bevaegelser, der er naestenperiodiske paaner Flytninger. Mat. Tidsskrift B (1937), 75–80.

256. FENCHEL, WERNER, JESSEN, B. Über fastperiodische Bewegungen in ebenen Bereichen und auf Flächen. Mat.-fysiske Medd., **13**, 6 (1935), 1–18.

257. FOIAŞ, C. *Essais dans l'étude des solutions des équations de Navier-Stokes dans l'espace. L'unicité et la presque-périodicité des solutions " petites "*. Rend. Sem. Mat. Univ. Padova, **32** (1962), 261–264.

258. FOIAŞ, C., ZAIDMAN, S. *Almost periodic-solutions of parabolic systems*. Ann. Scuola Norm. Sup. Pisa (3), **15** (1961), 247–262.

259. FØLNER, E. *Om Nulpunktsmaengder for naestperiodiske Funktioner*. Mat. Tidsskrift (1942), 54–62.

260. ———*Bidrage til de generaliserede naestenperiodiske Funktioners Teori*. Univ. of København, 1944.

261. ———*Bemaerking om Naestenperiodicitetens Definition*. Mat. Tidsskrift (1944), 24–27.

262. ———*On the structure of generalized almost periodic functions*. Mat.-fysiske Medd., **21**, 11 (1945), 1–30.

263. ———*W-naestenperiodiske Funktioner i vilkaarlige Grupper*. Mat. Tidsskrift, (1946), 153–162.

264. ———*Almost periodic functions on Abelian groups*. Comptes Rendus du dixième Congrès des Mathématiciens Scandinaves (1947), 356–362.

265. ———*A proof of the main theorem for almost periodic functions in an Abelian group*. Ann. Math., **50** (1949), 559–569.

266. ———*Note on the definition of almost periodic functions in groups*. Mat. Tidsskrift (1950), 58–62.

267. ———*A theorem on almost periodic functions of infinitely many variables*. Mat.-fysiske Medd., **25**, 14 (1950), 1–15.

268. ———*On the dual spaces of the Besicovich almost periodic spaces*. Mat.-fysiske Medd., **29**, 1 (1954), 1–27.

269. ———*Generalization of a theorem of Bogoliouboff to topological Abelian groups*. Math. Scand., **2** (1954), 5–18.

270. ———*Note on a generalization of a theorem of Bogoliouboff*. Math. Scand., **2** (1954), 224–226.

271. ———*On groups with full Banach mean value*. Math. Scand., **3** (1955), 243–254.

272. ———*Note on groups with and without full Banach mean value*. Math. Scand., **5** (1957), 5–11.

273. ———*Besicovitch almost periodic functions in arbitrary groups*. Math. Scand., **5** (1957), 47–53.

274. FRANKLIN, P. *The elementary theory of almost periodic functions of two variables*. J. Math. Phys., **5** (1925), 40–55.

275. ———*The fundamental theorem of almost periodic functions of two variables*. J. Math. Phys., **5** (1925), 201–237.

276. ———*Approximation theorems for generalized almost periodic functions*. Math. Z., **29** (1928), 70–87.

277. ———*Almost periodic recurrent motions*. Math. Z., **30** (1929), 325–331.

278. ———*Classes of functions orthogonal on an infinite interval, having the power of the continuum*. J. Math. Phys., **8** (1929), 74–79.

279. FRÉCHET, M., *Les fonctions asymptotiquement presque-périodiques continues*. C.R. Acad. Sci. Paris, **213** (1941), 520–522.

280. ———*Sur le théorème ergodique de Birkhoff*. C.R. Acad. Sci. Paris, **213** (1941), 607–609.

281. ———*Les fonctions asymptotiquement presque-périodiques*. Rev. Sci., **79** (1941), 341–354.

282. ———*Une application des fonctions asymptotiquement presque-périodiques à l'étude des familles de transformations ponctuelles et au problème ergodique.* Rev. Sci., **79** (1941), 407–417.

283. ———*Les transformations asymptotiquement presque-périodiques discontinues et le lemme ergodique.* Proc. Roy. Soc. Edinburgh, **63** (1950), 61–68.

284. FREUDENTHAL, H. *Topologische Gruppen mit genügend vielen fastperiodischen Funktionen.* Ann. Math., **37** (1936), 57–77.

285. FROLOV, I. S. *On the theory of generalized almost periodic sequences.* (Russian). Doklady Akad. Nauk SSSR, **165** (1965), 493–496.

286. GELFAND, I. *To the theory of normed rings. III. On the ring of almost periodic functions.* Comptes Rendus (Doklady) de l'Acad. Sci. de l'URSS, **25** (1939), 573–574.

287. GENUYS, F. *Sur les fonctions presque-périodiques dans une bande.* C.R. Acad. Sci. Paris, **234** (1921), 1939–1941.

288. GHEORGHIU, N. *Asupra soluţiilor aproape-periodice şi asimptotic aproape-periodice ale ecuaţiilor diferenţiale neliniare de primul ordin.* Analele şt. Univ. Iaşi (s.n.), **1** (1955), 17–20.

289. ———*Funcţii aproape-periodice de variabilă continuă şi de variabilă întreagă.* Analele şt. Univ. Iaşi (s.n.), **2** (1956), 29–31.

290. GLICKSBERG, IRVING, DE LEEUW, KAREL. *Almost periodic compactifications.* Bull. Am. Math. Soc., **65** (1959), 134–139.

291. ———*Applications of almost periodic compactifications.* Acta Math., **105** (1961), 63–97.

292. ———*Almost periodic functions on semigroups.* Acta Math., **105** (1961), 99–140.

293. GODEMENT, R. *Les fonctions de type positif et la théorie des groupes.* Trans. Am. Math. Soc., **63** (1948), 1–84.

294. GOLDBERG, RICHARD R. *Convolutions transforms of almost periodic functions.* Riv. Mat. Univ. Parma, **8** (1957), 307–312.

295. GOLDMAN, O. *Analytic almost-periodic functions. I.* Proc. Nat. Acad. Sci. U.S., **40** (1954), 294–296.

296. GREVE, W. *Dini's theorem for almost periodic functions.* J. Australian Math. Soc., **2** (1961/1962), 143–146.

297. GUINNAUD, A. P. *Concordance and the harmonic analysis of sequence.* Acta. Math., **101** (1959), 235–271.

298. GÜNZLER, HANS. *Fastperiodische Lösungen linearer hyperbolischer Differentialgleichungen.* Math. Z., **71** (1959), 223–250.

299. ———*Eine Characterisierung harmonischer Funktionen.* Math. Ann., **141** (1960), 68–86.

300. ———*Fourierreihen verktorwertiger fastperiodischer Funktionen auf Halbgruppen,* Mathematisches Institut Göttingen (Seminar W. Maak), Bericht No. 5, 1961.

301. ———*Eigenfunktion-Entwicklungen und allgemeine Fastperiodizität.* Math. Ann., **146** (1962), 146–179.

302. ———*Entwicklungen nach Eigenfunktionen und allgemeine Fastperiodizität. II.* Math. Z., **79** (1962), 69–94.

303. ———*Entwicklungen verallgemeinerter fastperiodischer Funktionen nach Eigenfunktionen.* Math. Ann., **146** (1962), 287–313.

304. ———*Hyperbolic differential equations and almost-periodic functions.* Rend. Sem. Mat. Fis. Milano, **34** (1964), 165–201.

305. HALANAY, A. *Soluții aproape-periodice ale ecuației Riccati.* Studii și cercetări matematice, **4** (1953), 245–354.

306. ———*Soluții aproape-periodice ale unor sisteme neliniare.* Gazeta mat. și fiz., A, **7** (1955), 396–399.

307. ———*Soluții aproape-periodice pentru sistemele de ecuații diferențiale neliniare.* Comunicările Acad. R.P.R., **6** (1956), 13–17.

308. ———*Introducere în teoria calitativă a ecuațiilor diferențiale.* București, 1956.

309. ———*Solutions presque-périodiques des systèmes d'équations différentielles à argument retardé contenant un petit paramètre.* Comunicările Acad. R.P.R., **9** (1959), 1237–1242.

310. ———*Periodic and almost-periodic solutions of systems of differential equations with lagging argument.* (Russian). Rev. Math. pures et appl., **4** (1959), 685–691.

311. ———*Almost periodic-solutions of systems of differential equations with a lagging argument and small parameter.* (Russian). Rev. Math. pures et appl., **5** (1960), 75–79.

312. ———*Solutions presque-périodiques des systèmes linéaires héréditaires.* C.R. Acad. Sci. Paris, **257** (1963), 827–829.

313. ———*Periodic and almost-periodic solutions of certain singularly perturbed systems with retarded argument.* (Russian). Rev. Math. pures et appl., **8** (1963), 285–292.

314. ———*Teoria calitativă a ecuațiilor diferențiale.* Editura Acad. R.S. România, Bucharest, 1963. (English translation, Academic Press, 1966).

315. ———*Almost-periodic solutions of linear delay-systems.* (Russian). Rev. Roumaine de Math. pures et appl., **9** (1964), 71–79.

316. HALE, J. K. *Oscillations in nonlinear Systems.* McGraw-Hill, New York, 1963.

317. ———*Periodic and almost-periodic solutions of functional differential equations.* Arch. Rational Mech. Anal., **15** (1964), 289–304.

318. HALE, J. K., SEIFERT, G. *Bounded and almost-periodic solutions of singularly perturbed equations.* J. Math. Anal. Appl., **3** (1961), 18–24.

319. ———*Bounded and almost-periodic solutions of singularly perturbed equations. Qualitative methods in the theory of non-linear vibrations.* Proc. Intern. Symp. Non-linear Vibrations, Kiev, vol. II, 1961, 427–432.

320. HALMOS, P., VAUGHAM, U. *The marriage problem.* Am. J. Math., **72** (1950), 214–215.

321. HAMMERSTEIN, A. *Ueber die Vollständigkeitsrelation in der Theorie der fastperiodischen Funktionen.* Sitzungsb. Preuss. Akad. Wiss. (1928), 17–20.

322. HARASAHAL, V. H. *Almost-periodic solutions of non-linear systems of differential equations.* (Russian). Prikl. Mat. Meh., **24** (1960), 565–567.

323. ———*The structure of solutions and the correctness of linear systems of differential equations with quasi-periodic coefficients.* (Russian). Doklady Akad. Nauk SSSR, **146** (1962), 1290–1293.

324. ———*On quasi-periodic solutions of systems of ordinary differential equations.* (Russian). Prikl. Mat. Meh., **27** (1963), 672–682.

325. ———*On quasi-periodic solutions of differential equations.* (Russian). Izv. Vysš. Učebn. Zaved. Matematika, No. 2 (39), (1964), 152–164.

326. HARTMAN, Ph. *Mean motions and almost periodic functions.* Trans. Am. Math. Soc., **46** (1939), 66–81.

327. HARTMAN, Ph., VAN KAMPEN, E. R., WINTNER, A. *Mean motions and distribution functions.* Am. J. Math., **59** (1937), 261–269.

328. ———*On the distribution functions of almost periodic functions.* Am. J. Math., **60** (1938), 491–500.

329. HARTMAN, Ph., WINTNER, A. *On the secular constants of almost-periodic functions.* Travaux Inst. Math. Tbilissi, **2** (1937), 37–41.

330. ———*On the almost-periodicity of additive number-theoretical functions.* Am. J. Math., **62** (1940), 753–758.

331. ———*On the convexity of averages of analytic almost-periodic functions.* Am. J. Math., **63** (1941), 581–583.

332. ———*Additive functions and almost-periodicity.* Duke Math. J., **9** (1942), 112–119.

333. ———*The L^2-space of relative measures.* Proc. Nat. Acad. Sci. U.S., **33** (1947), 128–132.

334. HARTMAN, S. *Sur les bases statistiques.* Studia Math., **10** (1948), 120–139.

335. ———*Sur une méthode d'estimation des moyennes de Weyl pour les fonctions périodiques et presque-périodiques.* Studia Math., **12** (1951), 1–24.

336. ———*Über die Verteilung der Fastperioden von fastperiodischen Funktionen auf Gruppen.* Studia Math., **15** (1955), 56–61.

337. ———*Funkcje prawie okresowe (szkic teorii).* Roczn. Polsk. Towarz. Mat. ser. 1, **1** (1955), 323–343.

338. ———*Über Niveaulinien fastperiodischer Funktionen.* Studia Math., **20** (1961), 313–325.

339. ———*On interpolation by almost-periodic functions.* Colloq. Math., **8** (1961), 99–101.

340. ———*Les intégrales de fonctions presque-périodiques et les sections de séries de Fourier.* Studia Math., **22** (1962/63), 147–160.

341. ———*R-fastperiodische Funktionen.* Studia Math. (Ser. Specjalna), **1** (1963), 39–40.

342. ———*Verallgemeinerte harmonische Analysis.* Ann. Polon. Math., **16** (1965), 341–452.

343. HARTMAN, S., RYLL-NARDZEWSKI, C. *Théorèmes abstraits de Kronecker et les fonctions presque-périodiques.* Studia Math., **13** (1953), 296–310.

344. ———*Über die Spaltung von Fourrierreihen fastperiodischer Funktionen.* Studia Math., **19** (1960), 287–295.

345. ———*Almost-periodic extensions of functions.* Bull. Acad. Polon. Sci. Sér. Sci., Math. Astr. Phys., **11** (1963), 427–429.

346. ———*Almost-periodic extensions of functions.* Colloq. Math., **12** (1964), 23–29.

347. HAVILAND, E. K. *On statistical methods in the theory of almost-periodic functions.* Proc. Nat. Acad. Sci. U.S., **19** (1933), 549–555.

348. HAYASHI, C., SHIBAYAMA, H., UEDA, Y. *Quasi-periodic oscillations in a self-oscillatory system with external force. Analytic methods in the theory of non-linear vibrations.* Proc. Intern. Symp. non-linear Vibrations, vol. I, 1961, 495–509.

349. HELMBERG, GILBERT. *Zerlegungen des Mittelwertes fastperiodischer Funktionen.* I. J. f. reine u. angew. Math., **207** (1961), 31–52.

350. ———*Zerlegungen des Mittlewertes fastperiodischer Funktionen.* II. J. f. reine u. angew. Math., **208** (1961), 1–21.

351. ———*Ein Zusammenhang zwischen Fourier-Reihen und Wertverteilungen fastperiodischer Funktionen.* Math. Z., **81** (1963), 300–307.

352. ——*Almost-periodic functions and Dini's theorem.* Nederl. Akad. Wetensch. Proc., Ser. A, **67**. Indag. Math., **26** (1964), 173–177.

353. HEWITT, E. *Linear functions on almost periodic functions.* Trans. Am. Math. Soc., **74** (1953), 303–322.

354. ——*A new proof of Plancherel's theorem for locally compact Abelian groups.* Acta Sci. Math. (Szeged), **24** (1963), 219–227.

355. HEWITT, E., ROSS, K. A. *Abstract harmonic analysis.* Springer-Verlag, Berlin, 1963.

356. HIRSCHFELD, R. *Sur l'analyse harmonique dans les groupes localement compactes.* C.R. Acad. Sci. Paris, **246** (1958), 1138–1140.

357. ISEKI, K. *Vector-space valued functions on semigroups.* I, II, III. Proc. Japan. Acad., **31** (1955), 16–19; 152–155; 699–710.

358. IVANOV, V. N. *On linear differential operators in the space of almost-periodic functions.* (Russian). Trudy Saratovsk. Inst. Meh. Selsk., fasc., **38** (1965), 141–149.

359. IWASAWA, K. *Einige Sätze über freie Gruppen.* Proc. Imp. Acad. Tokyo, **19** (1943), 272–274.

360. JACOBS, K. *Ergodentheorie und fastperiodische Funktionen auf Halbgruppen.* Math. Z., **64** (1956), 298–338.

361. ——*Fastperiodizitätseigenschaften allgemeiner Halbgruppen in Banach-Räumen.* Math. Z., **67** (1957), 83–92.

362. JAKUBOVIČ, V. A. *The method of matrix inequalities in the theory of stability of nonlinear control systems.* I. (Russian). Avtomat. i Telemeh., **25** (1964), 1017–1029.

363. ——*Periodic and almost-periodic limit regimes of automatic control systems with some discontinuous non-linearities.* (Russian). Doklady Akad. Nauk SSSR, **171** (1966), 533–537.

364. JESSEN, BØRGE. *Ueber die Nullstellen einer analytischen fastperiodischen Funktion. Eine Verallgemeinerung der Jensenschen Formel.* Math. Ann., **108** (1933), 485–516.

365. ——*Remark on the theorems of R. Petersen and S. Takahashi.* Mat. Tidsskrift (1935), 85–86.

366. ——*Ueber die Sekularkonstanten einer fastperiodischen Funktion.* Math. Ann., **111** (1935), 355–363.

367. ——*Om Sekularkonstanten for en naestenperiodisk Funktion.* Mat. Tidsskrift, B (1937), 45–48.

368. ——*Mouvement moyen et distribution des valeurs des fonctions presque-périodiques.* Comptes Rendus du dixième Congrès des Math. Scandinaves, København, 1946, 301–312.

369. ——*On the proofs of the fundamental theorem on almost periodic functions.* Mat.-fysiske Medd., **25**, 8 (1949), 1–12.

370. ——*Mean motion and almost periodic functions.* Proc. Second Canadian Math. Congress, Vancouver, 1949, 76–92.

371. ——*Some aspects of the theory of almost-periodic functions.* Proc. Intern. Congr. Math., Amsterdam, 1954, vol. I, 305–314.

372. JESSEN, B., TORNEHAVE, H. *Mean motions and zeros of almost-periodic functions.* Acta Math., **77** (1945), 137–279.

373. KAC, M. *Almost-periodicity and the representation of integers as sums of squares.* Am. J. Math., **62** (1940), 122–126.

374. ———*Convergence and divergence of non-harmonic gap series.* Duke Math. J., **8** (1941), 544–545.

375. KAC, M., VAN KAMPEN, E. R., WINTNER, A. *On the distribution of the values of real almost-periodic functions.* Am. J. Math., **61** (1939), 985–991.

376. ———*Ramanujan sums and almost-periodic functions.* Am. J. Math., **62** (1940), 107–114.

377. KAHANE, JEAN-PIERRE. *Quelques généralisations de la notion de périodicité.* Textos de mat. No. 6, conf. 3, 12 p. Instituto de Física e Matemática, Universidade do Recife, 1960.

378. ———*Sur les fonctions presque-périodiques généralisées dont le spectre est vide.* Studia Math., **21** (1961/62), 231–236.

379. KALUZA, THEODOR (jr.). *Untersuchung fastperiodischer Funktionen mittels äquidistanter Zahlmengen.* J. f. reine u. angew. Math., **181** (1939), 153–176.

380. ———*Zur Existenz stetiger grenzperiodischer Funktionen mit formal vorgegebener Fourierreihe,* Archiv der Math., **5** (1954), 344–346.

381. KAMPEN, E. R. VAN. *Almost periodic functions and compact groups.* Ann. Math., **37** (1936), 78–91.

382. ———*On almost periodic functions of constant absolute value.* J. London Math. Soc., **12** (1937), 3–16.

383. ———*On the asymptotic distribution of a uniformly almost-periodic function.* Am. J. Math., **61** (1939), 729–732.

384. ———*On uniformly almost-periodic multiplicative and additive functions.* Am. J. Math., **62** (1940), 627–634.

385. KAMPEN, E. R. VAN, WINTNER, A. *On the almost-periodic behavior of multiplicative number theoretical functions.* Am. J. Math., **62** (1940), 613–626.

386. KAWADA, Y. *Über den Mittelwert der messbaren fastperiodischen Funktionen auf einer Gruppe.* Proc. Imp. Acad. Tokyo, **19** (1943), 264–266.

387. ———*Two remarks on H. Weyl's theorems.* Kōdai Math. Sem. Rep., **3** (1949), 3–6.

388. KAWATA, T. *A gap theorem for the Fourier series of an almost periodic function.* Tôhoku Math. J., **43** (1937), 274–276.

389. ———*A remark on the non-vanishing of almost-periodic functions.* Proc. Imp. Acad. Tokyo, **16** (1940), 157–160.

390. KAWATA, T., TAKAHASHI, S. *On the convergency of an almost-periodic Fourier series.* Tôhoku Math. J., **45** (1938), 145–153.

391. KAWATA, T., UDAGAWA, M. *Some gap theorems.* Kōdai Math. Sem. Rep., **5–6** (1949), 19–22.

392. KEINER, H. *Verallgemeinerte fastperiodische Funktionen auf Halbgruppen.* Arch. Math., **8** (1957), 129–134.

393. KERSHNER, R., WINTNER, A. *On the asymptotic distribution of almost-periodic functions with linearly independent frequencies.* Am. J. Math., **58** (1936), 91–94.

394. KESTELMAN, H. *Measurable almost-periodic functions.* Mathematika, **3** (1956), 140–143.

395. KONDÔ, M. *La structure d'un flot topologique.* I. Proc. Japan. Acad., **25** (1949), 1–10.

396. KOPEĆ, J. *On vector-valued almost-periodic functions.* Ann. Soc. Polon. Math., **25** (1952), 100–105.

397. KOVANKO, A. S. *Sur une généralisation des fonctions presque-périodiques.* C.R. Acad. Sci. Paris, **186** (1928), 354–355.

398. ——*Sur quelques généralisations des fonctions presque-périodiques.* C.R. Acad. Sci. Paris, **186** (1928), 729–730.

399. ——*Sur l'approximation des fonctions presque-périodiques généralisées.* Rec. Math., **36** (1929), 142–145.

400. ——*Sur l'approximation des fonctions presque-périodiques généralisées.* C.R. Acad. Sci. Paris, **188** (1929), 142–145.

401. ——*Sur une classe de fonctions presque-périodiques qui engendre les classes de fonctions presque-périodiques de W. Stepanoff, H. Weyl et A. Besikovitch.* C.R. Acad. Sci. Paris, **189** (1929), 393–396.

402. ——*Sur les classes de fonctions presque-périodiques généralisées.* Annali Mat. pura ed appl., **9** (1931), 1–21.

403. ——*Sur la structure des fonctions presque-périodiques généralisées.* C. R. Acad. Sci. Paris, **198** (1934), 792–794.

404. ——*Sur quelques espaces de fonctions presque-périodiques généralisées.* Bulletin (Izvestija) de l'Acad. Sci. de la R.S.S.U., **1** (1935), 75–96.

405. ——*Sur la structure des fonctions presque-périodiques généralisées.* Rec. Math., **42** (1935), 3–18.

406. ——*On compactness of systems of generalized almost-periodic functions in the sense of V. V. Stepanov.* (Russian). Doklady Akad. Nauk SSSR, **26** (1940), 219–221.

407. ——*Sur les systèmes compacts de fonctions presque-périodiques généralisées de W. Stepanoff.* Rec. Math., **9 (51)** (1941), 389–401.

408. ——*On compactness of systems of generalized almost-periodic functions in the sense of A. S. Besicovitch.* (Russian). Doklady Akad. Nauk SSSR, **32** (1941), 118–119.

409. ——*Sur la compacité des systèmes de fonctions presque-périodiques généralisées de A. Besicovitch.* Comptes Rendus (Doklady) Acad. Sci. U.R.S.S., **43** (1944), 49–50.

410. ——*Sur la compacité des systèmes de fonctions presque-périodiques généralisées de H. Weyl.* Comptes Rendus (Doklady) Acad. Sci. U.R.S.S., **43** (1944), 275–276.

411. ——*On compactness of systems of generalized almost-periodic functions in the sense of Besicovitch.* (Russian). Mat. Sb., **16 (58)** (1945), 366–382.

412. ——*On the problem of completeness of the Weyl's space (W_p) and the applicability of Fisher-Riesz's theorem.* (Russian). Uspehi Mat. Nauk, 3, 5 (1948), 161–162.

413. ——*On the problem of convergence of function sequences in the sense of Weyl's metric D_{W_ω}.* (Russian). Ukr. Mat. Ž., **3** (1951), 465–476.

414. ——*On the problem of expansion of almost-periodic functions in a finite sum of almost-periodic functions.* (Russian). Učen. Zap. Univ. Lwow, **5** (1953), 47–48.

415. ——*On compactness of systems of generalized almost-periodic functions in the sense of Weyl.* (Russian). Ukr. Mat. Ž., **5** (1953), 185–195.

416. ——*On compactness of systems of Levitan's almost-periodic functions.* (Russian). Učen. Zap. Univ. 29, Lwow, **6** (1954), 45–49.

417. ——*On a new property and a new definition of generalized almost-periodic functions in the sense of A. S. Besicovitch.* (Russian). Doklady Polyt. Inst. Lwow, **I**, fasc. 1 (1955), 3–4.

418. ———*Application of the Riesz-Fischer theorem to the almost-periodic functions of Weyl.* (Russian). Dopovidi ta povid. Lwow Univ., No. 5 (1955), 93.

419. ———*A note on a property of B_p-uniformly integrable functions.* (Russian). Dopovidi ta povid. Lwow. Univ. No. 6, part 2 (1955), 74–78.

420. ———*On the problem of applicability of the Fischer-Riesz's theorem to almost-periodic functions in the sense of Weyl.* (Russian). Učen. Zap. Univ. 38 Lwow, **7** (1956), 42–47.

421. ———*On a new property and a new definition of generalized almost-periodic functions in the sense of A. S. Besicovitch.* (Russian). Ukr. Mat. Ž., **8** (1956), Polon. 273–288.

422. ———*Sur une propriété périodique des fonctions $\tilde{B}$-presque-périodiques.* Ann. Math., **8** (1960), 271–275.

423. ———*On the expansibility of S_p and W_p almost-periodic functions into finite sums of the same.* (Russian). Ukrain. Mat. Ž., **13** (1961), 226–231.

424. KOVANKO, A. S., LISEVIČ, L. M. *Some properties of the indefinite integral and derivative of an S_p almost-periodic function.* (Ukr.). Dopovidi Akad. Nauk Ukr. R.S.R. (1963), 705–706.

425. ———*Almost-periodic solutions of certain differential equations with almost-periodic right sides.* (Ukr.). Visnik Lwow. Univ. Ser. Mat. (1965), fasc. 2, 3–8.

426. KRASNOSELSKIĬ, M. A., PEROV, A.I. *A principle concerning the existence of bounded, periodic and almost-periodic solutions for systems of ordinary differential equations.* (Russian). Doklady Akad. Nauk SSSR, **123** (1958), 235–238.

427. KREIN, M. G. *On the theory of almost-periodic functions on a topological group.* (Russian). Doklady Akad. Nauk SSSR, **30** (1941), 5–8.

428. ———*On positive functionals on almost-periodic functions.* (Russian). Doklady Akad. Nauk SSSR, **30** (1941), 912.

429. ———*Lectures on stability theory of solutions of differential equations in Banach space.* (Russian). Kiev, 1964, 186 p.

430. KREIN, M. G., LEVIN, B. JA. *On entire almost-periodic functions of exponential type.* (Russian). Doklady Akad. Nauk SSSR, **64** (1949), 285–287.

431. KREIN, M. G., LEVITAN, B. M. *On some minimum problems in the Stepanov's space of almost periodic functions.* (Russian). Zap. Inst. Mat. Mech. Kharkovsk. Gos. Univ., **17** (1940), 111–124.

432. KULTZE, ROLF. *Fastperiodische Kompaktifikation von Halbgruppen.* Math. Ann., **139** (1959), 44–50.

433. KUPCOV, N. P. *On absolute and uniform convergence of Fourier series of almost-periodic functions.* (Russian). Mat. Sb., **40 (82)**, (1956) 157–178.

434. LADYŽENSKAJA, O. A. *The mixed problem for the hyperbolic equation.* (Russian). Moscow, 1953.

435. LALAGUË, P. *Classes de fonctions indéfiniment dérivables presque-périodiques de spectre donné.* C.R. Acad. Sci. Paris, **236** (1953), 2473–2475.

436. ———*Sur certaines classes de fonctions indéfiniment derivables.* Ann. Scientif. École Norm. Sup. Paris, **72** (1959), 237–298.

437. LANGENHOP, C. E., SEIFERT, G. *Almost-periodic solutions of second order non-linear differential equations with almost periodic forcing.* Proc. Am. Math. Soc., **10** (1959), 425–432.

438. DE LEEUW, K. *Linear spaces with compact group of operators.* Illinois J. Math., **2** (1958), 367–377.

439. LEVIN, B. JA. *On the secular constant of a holomorphic almost-periodic function.* (Russian). Doklady Akad. Nauk SSSR, **33** (1941), 182–184.

440. ———*On the almost-periodic classes* (B_p) *and* (W_p). (Russian). Trudy Odesskogo Univ., **3** (1941), 135–139.

441. ———*A new construction of the theory of almost-periodic functions in the sense of Levitan.* (Russian). Doklady Akad. Nauk SSSR, **62** (1948), 585–588.

442. ———*On almost-periodic functions in the sense of Levitan.* (Russian). Ukr. Mat. Ž., **1** (1949), 49–101.

443. ———*On functions which are defined by their values on an interval.* (Russian). Doklady Akad. Nauk SSSR, **70** (1951), 757–760.

444. ———*On quasi-analytic classes of almost-periodic functions.* (Russian). Doklady Akad. Nauk SSSR, **70** (1950), 949–952.

445. LEVIN, B. JA., LEVITAN, B. M. *On Fourier series of generalized almost-periodic functions.* (Russian). Doklady Akad. Nauk SSSR, **22** (1939), 543–546.

446. ———*Supplement to the paper "On Fourier series of generalized almost-periodic functions."* (Russian). Zap. Inst. Mat. Mech. Kharkovsk. Gos. Univ., **17** (1940), 109–110.

447. LEVITAN, B. M. *A new generalization of almost-periodic functions.* I. (Russian). Doklady Akad. Nauk SSSR, **17** (1937), 283–284.

448. ———*On linear differential equations with almost-periodic coefficients.* (Russian). Doklady Akad. Nauk SSSR, **17** (1937), 285–286.

449. ———*On Fourier series of a class of almost-periodic functions.* (Ukrainian). Zap. Inst. Mat. Mech. Kharkovsk. Gos. Univ., **14** (1937), 105–116.

450. ———*A new generalization of almost-periodic functions.* (Russian). Zap. Inst. Mat. Mech. Kharkovsk. Gos. Univ., **15** (1938), 3–32.

451. ———*Über eine Verallgemeinerung der stetigen fastperiodischen Funktionen von H. Bohr.* Ann. Math., **40** (1939), 805–815.

452. ———*Die Verallgemeinerung der Operation der Verschiebung im Zusammenhang mit fastperiodischen Funktionen.* Rec. Math., **7** (1940), 449–478.

453. ———*On functions with pure point spectrum.* (Russian). Zap. Inst. Mat. Mech. Kharkovsk. Gos. Univ., **16** (1940), 89–101.

454. ———*A new generalization of almost-periodic functions in the sense of H. Bohr.* (Russian). Zap. Inst. Mat. Mech. Kharkovsk. Gos. Univ., **17** (1940), 125–126.

455. ———*A generalization of the operation of translation, in connexion with almost-periodic functions.* (Russian). Doklady Akad. Nauk SSSR, **26** (1940), 639–642.

456. ———*A generalization of the operation of translation and infinite hypercomplex systems.* Rec. Math., **16** (1945), 259–280.

457. ———*A generalization of the operation of translation and infinite hypercomplex systems.* Rec. Math., **17** (1945), 9–44.

458. ———*A generalization of the operation of translation and infinite hypercomplex systems.* Rec. Math., **17** (1945), 163–192.

459. ———*On the approximation of N-almost-periodic functions by trigonometric polynomials.* (Russian). Doklady Akad. Nauk SSSR, **56** (1947), 907–909.

460. ———*On generalized positive definite and generalized almost-periodic sequences* (Russian). Doklady Akad. Nauk SSSR, **58** (1947), 977–980.

461. ———*Generalized positive definite and generalized almost-periodic functions.* (Russian). Doklady Akad. Nauk SSSR, **58** (1947), 1593–1596.

462. ———*Some problems of the theory of almost-periodic functions*. I. (Russian). Usp. Mat. Nauk, **2**, 5 (1947), 132–192.

463. ———*Some problems of the theory of almost-periodic functions*. II. (Russian). Usp. Mat. Nauk, **2**, 6 (1947), 174–214.

464. ———*Generalized almost-periodic functions*. (Russian). Mat. Sb., **24** (1949), 321–346.

465. ———*Almost-periodic functions*. (Russian). Moscow, 1953.

466. ———*Operators of generalized translation and some of their applications*. (Russian). Moscow, 1962, 323 p.

467. ———*On the integration of almost-periodic functions with values in a Banach space*. (Russian). Ivz. Akad. Nauk SSSR, Ser. Mat., **30** (1966), 1101–1110.

468. LEVITAN, B. M., STEPANOV, V. V. *On almost-periodic functions, essentially belonging to the class W*. (Russian). Doklady Akad. Nauk SSSR, **22** (1948), 229–232.

469. LEE, KWOK-PING. *Sur les séries de Fourier et les classes quasi-analytiques des fonctions presque-périodiques*. Quarterly J. Sci., Wu Han Univ., **9** (1948), 1–16.

470. ———*Sur l'approximation des fonctions analytiques presque-périodiques*. Quarterly J. Sci., Wu Han Univ., **9** (1948), 17–31.

471. LINÉS, ESCARDÓ E. *Aplicaciones de la teoria de redes regulares al estudio de las funciones casi-periódicas*. Consejo Superior de Investigaciones Científicas, Madrid, 1943, 1–79.

472. LINFOOT, E. H. *Generalization of two theorems of H. Bohr*. J. London Math. Soc., **3** (1928), 177–182.

473. ———*A remark on Bohr-Fourier series*. J. London Math. Soc., **4** (1929), 121–123.

474. LISEVIČ, L. N. *Almost periodic-solutions of a hyperbolic system of linear differential equations with almost-periodic coefficients*. (Ukrainian). Dopovidi Akad. Nauk Ukr. RSR (1956), 220–222.

475. ———*Extension of Favard's theorems to the case of a linear system of differential equations with analytic almost-periodic coefficients*. (Ukrainian). Dovopidi Akad. Nauk Ukrain. RSR (1960), 148–149.

476. LJUBARSKIJ, G. JA. *On integration in the mean of almost-periodic functions on a topological group*. (Russian). Usp. Mat. Nauk, **3**, 3 (1948), 195–201.

477. LOOMIS, L. H. *Abstract Harmonic Analysis*. New York, 1953.

478. ———*The spectral characterization of a class of almost-periodic functions*. Ann. Math. (2), **72** (1950), 362–369.

479. LOONSTRA, F. *Sur les mouvements presque-périodiques*. Indagationes Math., **8** (1946), 447–454.

480. LOVE, E. R. *More-than-uniform almost-periodicity*. J. London Math. Soc., **26** (1951), 14–25.

481. LYKOVA, O. B., MITROPOLSYK, Y. A. *Sur l'existence de solutions quasi-périodiques d'un système canonique troublé*. Colloques Int. CNRS. Les vibrations forcées dans les systèmes non-linéaires (1965), 415–422.

482. MAAK, W. *Eine neue Definition der fastperiodischen Funktionen*. Abh. Math. Sem. Hans. Universität, **11** (1936), 240–244.

483. ———*Abstrakte fastperiodische Funktionen*. Abh. Math. Sem. Hans. Universität. **11** (1936), 367–380.

484. ———*Über den Begriff der fastperiodischen Funktionen*. Mat. Tidsskrift B (1938), 7–12.

485. ———*Moduln fastperiodischer Funktionen*. Abh. Math. Sem. Univ. Hamburg, **16** (1949), 56–71.

486. ———*Almost-periodic invariant vector sets in a metric vector space.* Proc. Nat. Acad. Sci. U.S., **36** (1950), 208–210.

487. ———*Summierung der Fourierreihen gleichartig fastperiodischer Funktionen auf Gruppen.* Math. Z., **52** (1950), 770–778.

488. ———*Fastperiodische invariante Vektormoduln in einem metrischen Vektorraum.* Math. Ann., **122** (1950), 157–166.

489. ———*Fastperiodische Funktionen.* Springer-Verlag, Berlin, 1950.

490. ———*Integralmittelwerte von Funktionen auf Gruppen und Halbgruppen.* Journal f. reine u. angew. Math., **190** (1952), 34–48.

491. ———*Fastperiodische Funktionen auf Halbgruppen.* Acta Math., **87** (1952), 33–58.

492. ———*Periodizitätseigenschaften unitärer Gruppen in Hilberträumen.* Math. Scand., **2** (1954), 334–344.

493. ———*Fastperiodische Funktionen auf der Modulgruppe.* Math. Scand., **3** (1955), 44–48.

494. ———*Zur Theorie der Modulgruppe.* Abh. Math. Sem. Univ. Hamburg, **22** (1958), 267–275.

495. ———*Fastautomorphe Funktionen.* Sitzungsberichte Bayerische Akad. Wiss., Math.-Naturwiss. Klasse, 1959, 289–319.

496. MALKIN, I. G. *On almost-periodic oscillations of non-linear non-autonomous systems.* (Russian). Prikl. Mat. Mech., **18** (1954), 681–704.

497. ———*Some problems of the theory of non-linear oscillations.* (Russian). Moscow, 1956.

498. MARČENKO, V. A. *Application of the method of Fejér-Bochner to generalized Fourier series.* (Russian). Doklady Akad. Nauk SSSR, **53** (1946), 7–10.

499. ———*Transform-operators.* (Russian). Doklady Akad. Nauk SSSR, **74** (1950), 185–187.

500. ———*Generalized almost-periodic functions.* (Russian). Doklady Akad. Nauk SSSR, **74** (1950), 893–895.

501. ———*Summation methods for generalized Fourier series.* (Russian). Zap. Mat. Obšč. Kharkov, **20** (1950), 3–32.

502. ———*On the functions which are normal with respect to a symmetric operation of translation.* (Russian). Zap. Mat. Obšč. Kharkov, **20** (1950), 33–42.

503. MARKOFF, A. *Stabilität im Liapounoffschen Sinne und Fastperiodizität.* Math. Z., **36** (1933), 708–738.

504. MARCUS, L., MOORE, R. A. *Oscillations and disconjugacy for linear differential equations with almost periodic coefficients.* Acta. Math., **96** (1956), 99–123.

505. MASSERA, J. L. *Un criterio de existencia de soluciones casiperiódicas de ciertos sistemas de ecuaciones diferenciales casi-periódicas.* Publ. Inst. Mat. y Estad. Fac. Ing. y Agrimensura, Montevideo, **3** (1958), 89–102.

506. MASSERA, J. L., SCHÄFFER, J. J. *Linear Differential equations and functional analysis.* I. Ann. Math., **67** (1958), 517–572.

507. ———*Linear differential equations and function spaces.* Academic Press, New York, 1966.

508. MATSUSHITA, SHIN-ICHI. *Fonctions presque-périodiques du type spécial.* I, II, III, IV, V, VI. Proc. Japan. Acad., **31** (1955), 70–75, 157–160, 214–219, 278–283, 334–339, 436–440.

509. MAURIN, K. *On Parseval equation for almost periodic vectors.* Studia Math., **13** (1953), 83–86.

510. ———*Metody przestrzeni Hilberta.* Warszawa, 1959.

511. MAZUR, S., ORLICZ, W. *Sur quelques propriétés des fonctions périodiques et presque-périodiques.* Studia Math., **9** (1940), 1–16.

512. MEISTERS, G. H. *Quotients of almost periodic functions.* Bull. Acad. Polon. Sci., Sér. sci. math. astron. phys., **13** (1965), 621–624.

513. MILLER, RICHARD K. *On almost-periodic differential equations.* Bull. Am. Math. Soc., **70** (1964), 792–795.

514. ———*Almost periodic differential equations as dynamical systems with applications to the existence of almost-periodic solutions.* J. Differential Equations, **1** (1965), 337–345.

515. MILLIONŠČIKOV, V. M. *Recurrent and almost-periodic limit trajectories of non-autonomous systems of differential equations.* (Russian). Doklady Akad. Nauk SSSR, **161** (1965), 43–44.

516. MINORSKY, N. *Non-linear oscillations.* Van Nostrand, Princeton, N.J., 1962.

517. MONTEL, P. *Lecţiuni despre funcţiunile aproape-periodice.* Cluj, 1937.

518. MUCKENHOUPT, C. F. *Almost-periodic functions and vibrating systems.* J. Math. Phys., **8** (1928–1929), 163–199.

520. MUSIELAK, J. *On absolute convergence of Fourier series of some almost-periodic functions.* (Polish). Zeszyty Nauk. Uniw. Mickiewicza, Math.-Chem., **1** (1957), 9–17.

521. ———*On absolute convergence of Fourier series of some almost-periodic functions.* Ann. Polon. Math., **6**, No. 2 (1959), 145–156.

522. NASYROV, R. M. *On stability of almost-periodic motions in certain critical cases.* (Russian). Trudy Univ. Družby Narodov im. P. Lumumby, **5** (1964), 30–44.

523. NEMYTSKIĬ, V. V., STEPANOV, V. V. *Qualitative theory of differential equations.* Princeton Univ. Press, 1960, VIII + 523 p.

524. VON NEUMANN, J. *Almost-periodic functions in a group.* I. Trans. Am. Math. Soc., **36** (1934), 445–492.

525. NICOLESCU, M. *Funzioni poliarmoniche in un semipiano.* Rend. Sem. Mat. Univ. Roma, **1** (1936), 271–276.

526. ———*Funcţii poliarmonice aproape-periodice.* Bul. şt. Acad. R.P.R., **5** (1953), 273–283.

527. ———*Analiză matematică.* Vol. II. Ed. tehnică, Bucureşti, 1958.

528. NIKAIDÔ HUKUKANE. *A proof of the invariant mean-value theorem on almost-periodic functions.* Proc. Am. Math. Soc., **6** (1955), 361–363.

529. NORGIL, R. *Undersøgelser of naestenperiodiske Funktioner.* Mat. Tidsskrift (1931), 73–91.

530. OGASAWARA, T. *Almost periodic-functions in groups.* Coll. Papers Fac. Sci., Osaka Univ., **11** (1942), 115–123.

531. OMAROV, E. O. *Almost-periodic solutions of linear partial differential equations with constant coefficients with a free term.* (Russian). Izv. Akad. Nauk Kazah. SSR, Ser. Mat. Mech. No. 8 (12) (1959), 28–36.

532. OPIAL, Z. *Sur les solutions presque périodiques des équations différentielles du premier et du second ordre.* Ann. Polonici Math., **7** (1959), 51–61.

533. ———*Sur la stabilité des solutions périodiques et presque-périodiques de l'équation différentielle* $x'' + F(x') + g(x) = p(t)$. Bull. Acad. Polon. Sci., Sér. Sci. Math. Astr. Phys., **7** (1959), 495–500.

534. ———*Sur les solutions périodiques et presque-périodiques de l'équation différentielle* $x'' + kf(x)x' + g(x) = kp(t)$. Ann. Polon. Math., **7** (1960), 309–319.

535. ———*La presque-périodicité et les trajectoires sur le tore.* C.R. Acad. Sci. Paris, **250** (1960), 3565–3566.

536. ———*Sur les solutions presque-périodiques d'une classe d'équations différentielles.* Ann. Polon. Math., **9** (1960/61), 157–181.

537. ———*Sur une équation différentielle presque-périodique sans solution presque-périodique.* Bull. Acad. Polon. Sci., Sér. Sci. Math. Astr. Phys., **9** (1961), 673–676.

538. PECK, J. E. L. *Almost-periodic functions.* Proc. Am. Math. Soc., **3** (1952), 107–110.

539. PEDERSEN, E. *Über einige besondere Klassen von fastperiodischen Funktionen.* Mat.-fysiske Medd., **8** (1928), 1–22.

540. PEROV, A. I. *Periodic, almost-periodic, and bounded solutions of the differential equation dx/dt = f(t, x).* (Russian). Doklady Akad. Nauk SSSR, **132** (1960), 531–534.

541. PETER, F., WEYL, H. *Die Vollständigkeit der primitiven Darstellungen einer geschlossenen kontinuierlichen Gruppe.* Math. Ann., **97** (1927), 737–755.

542. PETERSEN, R. *En analytisk Funktion med specielle naestenperiodiske Egenskaber.* Mat. Tidsskrift (1933), 33–44.

543. ———*Untersuchungen über eine analytische Funktion mit speziellen fastperiodischen Eigenschaften.* Mat.-fysiske Medd., **12**, 14 (1934), 1–30.

544. ———*Om den formelle Differentiation of Fourierudviklingen for en naestenperiodisk Funktion.* Mat. Tidsskrift (1935), 37–41.

545. ———*Om Laguerre-Polynomier og naestenperiodiske Funktioner.* Mat. Tidsskrift (1945), 145–150.

546. ———*Laplace-transformation of almost-periodic functions.* Den 11-te Skandinaviske Matematikerkongress, Trondheim, 1949, 158–165.

547. PHILLIPS, R. S. *On linear transformations.* Trans. Am. Math. Soc., **48** (1940), 516–541.

548. PITT, H. R. *On the Fourier coefficients of almost-periodic functions.* J. London Math. Soc., **14** (1939), 143–150.

549. PONTRJAGIN, L. *Les fonctions presque-périodiques et l'analysis situs.* C.R. Acad. Sci. Paris, **196** (1933), 1201–1203.

550. PROUSE, GIOVANNI. *Soluzioni quasi-periodiche delle equazioni lineari di tipo parabolico.* Rend. Ist. Lomb. (A), **96** (1962), 847–860.

551. ———*Soluzioni quasi-periodiche di un'equazione iperbolica a coefficienti variabili.* Ric. di Mat., **XI** (1962), 123–138.

552. ———*Sulle equazioni differenziali astratte quasi-periodiche secondo Stepanoff.* Ric. di Mat., **XI** (1962), 254–270.

553. ———*Soluzioni quasi-periodiche dell'equazione differenziale di Navier-Stokes in due dimensioni.* Rend. Sem. Mat. Padova, **33** (1963), 186–212.

554. ———*Soluzioni quasi-periodiche dell'equazione non omogenea della membrana vibrante, con termine dissipativo quadratico.* Rend. Acad. Naz. Lincei, **37** (1964), 246–252.

555. ———*Soluzioni quasi-periodiche dell'equazione non omogenea delle onde con termine dissipativo non lineare. I, II, III, IV.* Rend. Accad. Naz. Lincei, **38** (1965), 804–807; **39** (1965), 11–18, 155–160, 240–144.

556. PUTNAM, C. R. *Stability and almost-periodicity in dynamical systems.* Proc. Am. Math Soc., **5** (1954), 352–356.

557. ———*Unilateral stability and almost periodicity.* J. Math. Mech., **9** (1960), 915–917.

558. RELLICH, F. *Ueber die von Neumannschen fastperiodischen Funktionen auf einer Gruppe.* Math. Ann., **111** (1935), 560–567.

559. RICCI, MARIA LAVINIA. *Sulle equazioni differenziali con termine noto quasi-periodico secondo Stepanoff.* Rend. Ist. Lombardo Sci. Lettere, **96** (1962), 861–882.

560. RICCI, MARIA LAVINIA, RIZZONELLI, PIERANITA. *Alcuni complementi alla teoria delle funzioni quasi-periodiche astratte.* Ist. Lombardo Accad. Sci. Lett. Rend. A, **95** (1961), 525–534.

561. ———*Sulle funzioni l^1-quasi-periodiche.* Ist. Lombardo Accad. Sci. Lett. Rend., **95** (1961), 941–946.

562. RICCI, M. L., VAGHI, C. *Soluzioni quasi-periodiche di un'equazione funzionale.* Ric. Mat., **13** (1964), 242–260.

563. RIESZ, M. *Zum Eindeutigkeitssatz der fastperiodischen Funktionen.* Medd. Lunds Univ. Mat. Sem., **1** (1934), 1–9.

564. ———*Eine Bemerkung über den Eindeutigkeitssatz der Theorie der fastperiodischen Funktionen.* Mat. Tidsskrift, (1934), 11–13.

565. RJABOV, JU. A. *On a method of finding a bound for the region of existence of periodic and almost-periodic solutions of quasi-linear differential equations with a small parameter and a non-linear characteristic of non-linearity.* (Russian). Izv. Vysš. Učebn. Zaved. Matematika nr. 2 (33) (1963), 101–107.

566. RODRIQUEZ, VIDAL R. *Contribución al Estudio de las Succesiones casi-periódicas y sus Generalizaciones.* Univ. de Barcelona (1948), 1–73.

567. ROETTINGER, IDA. *Note on use of almost-periodic functions in the solution of certain boundary value problems.* J. Math. Phys., **27** (1948), 232–239.

568. RYLL-NARDZEWSKI, C. *Concerning almost-periodic extensions of functions.* Colloq. Math., **12** (1964), 235–237.

569. SABBIONI, C. *Sopra un esempio di funzione quasi-periódica.* Boll. Un. Mat. Italiana, **8** (1953), 301–303.

570. SANSONE, G. *Equazioni differenziali nel campo reale.* I, II. (Seconda edizione). Bologna, 1948, 1949.

571. SANDOR, S. T. *Asupra ecuaţiilor diferenţiale liniare de ordin superior cu coeficieţni aproape-periodici.* Bul. ştiinţ. Acad. R.P.R., **7** (1955), 329–346.

572. ———*Ecuaţiile diferenţiale liniare neomogene cu coeficienţi aproape-periodici şi ecuaţiile cvasiliniare cu parametru mic.* Bul. Ştiinţ. Acad. R.P.R., **7** (1955), 623–698.

573. SCHÄFFER, JUAN JORGE. *Analytische Parameterabhängigkeit der fastperiodischen Lösungen.* Rend. Circ. Mat. Palermo, **5** (1956), 204–236.

574. ———*Sobre ecuaciones diferenciales nolineales casi-periódicas.* Publ. Inst. Mat. Estad. Fac. Ing. Agr., Montevideo, **3** (1957), 17–52.

575. ———*Linear differential equations and functional analysis. IX. Almost-periodic equations.* Math. Ann., **150** (1963), 111–118.

576. SCHMIDT, R. *Die trigonometrische Approximation für eine Klasse von verallgemeinerten fastperiodischen Funktionen.* Math. Ann., **100** (1928), 333–356.

577. SCHWARTZ, L. *Sur les fonctions moyenne-périodiques.* C.R. Acad. Sci. Paris, **223**, (1946), 68–70.

578. ———*Théorie générale des fonctions moyenne-périodiques.* Ann. Math., **48** (1947), 857–929.

579. SEIFERT, GEORGE. *Almost-periodic solutions for systems of differential equations near points of nonlinear first approximations.* Proc. Am. Math. Soc., **11** (1960), 429–435.

580. ———*Stability conditions for separation and almost-periodicity of solutions of differential equations.* Contr. Differential Equations, **1** (1963), 483–487.

581. ———*Uniform stability of almost-periodic solutions of almost-periodic systems of differential equations.* Contr. Differential Equations, **II** (1963), 269–276.

582. ———*A condition for almost periodicity with some applications to functional-differential equations.* J. Differential Equations, **1** (1965), 393–408.

583. ———*Stability conditions for the existence of almost-periodic solutions of almost-periodic systems.* J. Math. Anal. Appl., **10** (1965), 409–418.

584. ———*Almost-periodic solutions for almost-periodic systems of ordinary differential equations.* J. Differential Equations, **2** (1966), 305–319.

585. SEMADENI, Z. *Generalizations of Bohr's theorem on Fourier series with independent characters.* Studia Math., **23** (1964), 159–179.

586. SEYNSCHE, INGEBORG. *Zur Theorie der fastperiodischen Zahlfolgen.* Rend. Circ. Mat. Palermo, **55** (1931), 395–421.

587. SHIGA, K. *Bounded representations on a topological vector space and weak almost-periodicity.* Japan. J. Math., **25** (1955), 21–35.

588. SIBUYA, YASUTAKA. *Sur les solutions bornées d'un système des équations différentielles ordinaires non-linéaires à coefficients périodiques.* J. Fac. Sci. Univ. Tokyo, **7** (1956), 333–341.

589. ———*Sur un système d'équations différentielles ordinaires à coefficients presque-périodiques et contenant des paramètres.* J. Fac. Sci. Univ. Tokyo, **7** (1957), 407–417.

590. ———*On bounded solutions of ordinary differential equations with almost-periodic coefficients.* Bol. Soc. Mat. Mexicana (2), **5** (1960), 290–293.

591. ———*Almost-periodic solutions of a system of ordinary differential equations with periodic coefficients.* Natur. Sci. Rep. Ochanomizu Univ., **13** (1962), no. 2, 21–30.

592. ŠIMANOV, S. N. *Almost-periodic solutions of nonhomogeneous linear differential equations with lag.* (Russian). Izv. Vysš. Učebn. Zaved. Matematika (1958), no. 4 (5), 270–274.

593. ———*Almost-periodic solutions in non-linear systems with retardation* (Russian). Doklady Akad. Nauk SSSR, **125** (1959), 1203–1206.

594. SMOLICKIĬ, KH. L. *On almost-periodicity of generalized solutions of wave equation.* (Russian). Doklady Akad. Nauk SSSR, **60** (1948), 353–356.

595. SOBOLEV, S. L. *Sur la presque-périodicité des solutions de l'équation des ondes,* I, II, III. Comptes Rendus (Doklady) de l'Acad. Sci. de l'U.R.S.S., **48** (1945), 542–545; 618–620; **49** (1945), 12–15.

596. STEPANOFF, W. W. *Sur quelques généralisations des fonctions presque-périodiques.* C.R. Acad. Sci. Paris, **181** (1925), 90–94.

597. ———*Über einige Verallgemeinerungen der fastperiodischen Funktionen.* Math. Ann., **90** (1925), 473–492.

598. ———*On the Fischer-Riesz theory for some classes of almost-periodic functions.* (Russian). Usp. Mat. Nauk, **3**, 6 (1948), 186–187.

599. ———*On a metric in the space of S_2-almost-periodic-functions.* (Russian). Doklady Akad. Nauk SSSR, **64** (1949), 171–174.

600. ———*On a class of almost-periodic functions.* (Russian). Doklady Akad. Nauk SSSR, **64** (1949), 297–300.

601. STEPANOFF, W. W., TYCHONOFF, A. *Sur les espaces des fonctions presque-périodiques.* C.R. Acad. Sci. Paris, **196** (1933), 1199–1201.

602. ———*Über die Räume der fastperiodischen Funktionen*. Rec. Math., **41** (1934), 166–178.

603. STRUBLE, R. A. *Almost-periodic functions on locally compact groups*. Proc. Nat. Sci. U.S., **39** (1953), 122–126.

604. STRZELECKI, E. *On a problem of interpolation by periodic and almost-periodic functions*. Colloq. Math., **11** (1963), 91–99.

605. SUNYER I BALAGUER, F. *Une généralisation des fonctions presque-périodiques*. C.R. Acad. Sci. Paris, **228** (1949), 732–734.

606. ———*Une généralisation des fonctions presque-périodiques: fonctions presque-elliptiques*. C.R. Acad. Sci. Paris, **228** (1949), 797–799.

607. SZ.-NAGY, BÉLA. *Über gewisse Extremalfragen bei transformierten trigonometrischen Entwicklungen. I. Periodischer Fall*. Ber. Verh. Sachs. Akad. Leipzig, **90** (1938), 103–134.

608. ———*Über gewisse Extremalfragen bei transformierten trigonometrischen Entwicklungen. II. Nichtperiodischer Fall*. Ber. Verh. Sachs. Akad. Leipzig, **91** (1939), 3–24.

609. ———*Über die Ungleichung von H. Bohr*. Math. Nachr., **9** (1953), 255–259.

610. SZ.-NAGY, B., RIESZ, F. *Leçons d'Analyse fonctionnelle*. Budapest, 1965.

611. SZÁSZ, O. *The jump of almost-periodic functions and of Fourier integrals*. Duke Math. J., **7** (1940), 360–366.

612. SZEGÖ, G. *Zur Theorie der fastperiodischen Funktionèn*. Math. Ann., **96** (1926), 378–382.

613. ŠTOKALO, I. Z. *Stability and instability criteria for the solutions of linear differential equations with quasi-periodic coefficients*. (Russian). Mat. Sb., **19** (1946), 263–286.

614. ———*On the theory of linear differential equations with quasi-periodic coefficients*. (Russian). Sb. Trudov Inst. Mat. Akad. Nauk USSR, **8** (1946).

615. ———*Theory of symbolic representation of solutions of linear differential equations with quasi-periodic coefficients*. (Russian). Sb. Trudov Inst. Mat. Akad. Nauk USSR, **11** (1948), 43–59.

616. TAKAHASHI, S. *On the property of the Fourier series of an almost periodic function*. Proc. Imp. Acad. Tokyo, **11** (1935), 90–92.

617. ———*On the formal integration of the Fourier series of an almost-periodic function*. Mat. Tidsskrift (1935), 82–85.

618. ———*On an almost-periodic Fourier series*. Tôhoku Math. J., **44** (1937), 12–17.

619. ———*An almost-periodic function in the mean*. Proc. Imp. Acad. Tokyo, **13** (1937), 129–133.

620. ———*On the convergency of the Fourier series of an almost-periodic function*. Proc. Physico-Math. Soc. Japan, **20** (1939), 611–617.

621. ———*An application of the Fourier transform to almost-periodic functions*. Proc. Imp. Acad. Tokyo, **14** (1939), 87–89.

622. ———*Some new properties of the Bohr almost-periodic Fourier series*. Japan. J. Math., **16** (1939), 99–133.

623. THEILER, G. *O demonstraţie directă a teoremei lui Kronecker din teoria funcţiilor aproape-periodice şi cîteva consecinţe ale acesteia*. Lucrările Inst. petrol, gaze şi geologie, Bucureşti, **4** (1958), 279–291.

624. TORNEHAVE, HANS. *A theorem on the mean motions of almost-periodic functions*. Mat.-fysiske Medd., **25**, 20 (1950), 1–18.

625. ———*Recent investigations on almost-periodic movements.* Tolfte Scandinaviska Matematiker-kongressen, Lund, 1953, 302–309.
626. ———*On almost-periodic movements.* Mat.-fysiske Medd., **28**, 13 (1954), 1–42.
627. ———*On the Fourier series of Stepanov almost-periodic functions.* Math. Scand., **2** (1954), 237–242.
628. ———*On entire functions almost-periodic in two directions.* Math. Scand., **6** (1958), 160–174.
629. TORTRAT, ALBERT. *Sur les fonctions presque-périodiques des classes B_p de Besicovitch* C.R. Acad. Sci. Paris, **252** (1961), 1723–1725.
630. ———*Sur les fonctions presque-périodiques des classes B_p de Besicovitch.* C.R. Acad. Sci. Paris, **254** (1962), 2709–2711.
631. TREJO, C. A. *Sobre funciones con media local casi periódica.* Ann. Soc. Cient. Argentina, **132** (1941), 137–138.
632. TULEGENOV, B. *On the existence of quasi-periodic solutions of a non-linear first-order differential equation.* (Russian). Izv. Akad. Nauk Kazah. SSSR. Ser. fiz.-mat. nauk no. 2 (1964), 72–76.
633. TURCU, AUREL. *Soluţii aproape-periodice ale ecuaţiei lui Duffing în caz de nerezonanţă.* Studia Univ. Babeş-Bolyai, Ser. Math.-Phys., **9**, No. 2 (1964), 61–76.
634. ———*Almost-periodic solutions of the equation of Duffing in the case of resonance.* Studia Univ. Babeş-Bolyai, Ser. Math.-Phys., **10** (1965), no. 1, 83–94.

635. UPTON, C. I. F. *Riesz almost-periodicity.* J. London Math. Soc., **31** (1956), 407–426.
636. URSELL, H. D. *Normality and almost-periodic functions.* J. London Math. Soc., **4** (1929), 123–127.
637. ———*Normality of almost-periodic functions.* II. J. London Math. Soc., **5** (1930), 47–50
638. ———*Parseval's theorem for almost-periodic functions.* Proc. London Math. Soc., **32** (1931), 402–440.
639. ———*On the convergence almost everywhere of Rademacher's series and of the Bochner-Fejér sums of a function almost-periodic in the sense of Stepanoff.* Proc. London Math. Soc., **33** (1932), 457–466.
640. UTTLEY, ROBERT G. *Almost-periodicity and trigonometric series.* Trans. Am. Math. Soc., **99** (1961), 414–424.

641. VAGHI, CARLA. *Sulla regolarizzazione delle soluzioni quasi-periodiche dell'equazione non omogenea delle onde.* Ist. Lombardo Accad. Sci. Lett. Rend. A, **96** (1962), 267–285.
642. ———*Soluzioni C-quasi-periodiche dell'equazione non omogenea delle onde.* Ric. Mat., **12** (1963), 195–215.
643. ———*Su un'equazione iperbolica con coefficienti periodici e termine noto quasi-periodico.* Rend. Ist. Lomb. (A), **100** (1966), 155–180.
644. DE LA VALLÉE-POUSSIN, C. J. *Sur les fonctions presque-périodiques de H. Bohr.* Ann. Soc. Sci. Bruxelles, **47** (1927), 141–158.
645. ———*Sur les fonctions presque-périodiques de H. Bohr. Note complémentaire et explicative.* Ann. Soc. Sci. Bruxelles, **48** (1928), 56–57.
646. VASCONI, A. *Sull'integrazione delle funzioni quasi-periodiche secondo Stepanoff negli spazi di Clarkson.* Rend. Ist. Lombardo Sci. Lettere, **95** (1961), 1024–1029.
647. ———*Sull'equazione delle onde con termine noto quasi-periodico secondo Stepanoff.* Ist. Lombardo Accad. Sci. Lett. Rend. A, **96** (1962), 903–914.

648. WALTHER, A. *Über lineare Differenzengleichungen mit konstanten Koeffizienten und fastperiodischer rechter Seite.* Nachr. Ges. Wiss. Göttingen. Math.-Phys. Klasse (1927), 196–216.

649. ———*Fastperiodische Folgen und Potenzreihen mit fastperiodischen Koeffizienten.* Abh. Math. Sem. Hamburg Universität, **VI** (1928), 217–234.

650. ———*Fastperiodische Folgen und ihre Fouriersche Analyse.* Atti del Congresso Internazionale dei Matematici, Bologna, **2** (1928), 289–298.

651. ———*Algebraische Funktionen von fastperiodischen Funktionen.* Monatshefte f. Math., **40** (1933), 444–457.

652. WECKEN, FRANZ. *Abstrakte Integrale und fastperiodische Funktionen.* Math. Z., **45** (1949), 377–404.

653. WEIL, A. *Sur les fonctions presque-périodiques de v. Neumann.* C.R. Acad. Sci. Paris, **200** (1935), 38–40.

654. ———*L'intégration dans les groupes topologiques et ses applications.* Herman, Paris, 1965.

655. WEINSTEIN, ALESSANDRO. *Sulle soluzioni quasi-periodiche di una classe di equazioni ellittiche.* Rend. Accad. Naz. Lincei (8) **32** (1962), 863–866.

656. WESTON, J. D. *Almost-periodic functions.* Mathematika, **2** (1955), 128–131.

657. WEXLER, DINU. *Solutions périodiques et presque-périodiques des systèmes d'équations différentielles aux impulsions.* C.R. Acad. Sci. Paris, **259** (1964), 287–289.

658. ———*Solutions presque-périodiques des systèmes d'équations différentielles à perturbations distributions.* C.R. Acad. Sci. Paris, **262** (1966), A, 436–439.

659. ———*Solutions périodiques et presque-périodiques des systèmes d'équations différentielles linéaires en distributions.* Journal Diff. Equations, **2** (1966), 12–32.

660. WEYL, H. *Beweis des Fundamentalsatzes in der Theorie der fastperiodischen Funktionen.* Sitzungsb. der Preuss. Akad. Wiss. (1926), 211–214.

661. ———*Integralgleichungen und fastperiodische Funktionen.* Math. Ann., **97** (1927), 338–356.

662. ———*Harmonics on homogeneous manifolds.* Annals of Math., **35** (1934), 486–499.

663. ———*Mean motion.* Am. J. Math., **60** (1938), 889–896.

664. ———*Mean motion. II.* Am. J. Math., **61** (1939), 143–148.

665. ———*Almost-periodic invariant vector sets in a metric vector space.* Am. J. Math., **71** (1949), 178–205.

666. WIENER, N. *Verallgemeinerte trigonometrische Entwicklungen.* Nachr. Ges. Wiss. Göttingen, Math.-Phys. Klasse (1925), 151–158.

667. ———*The spectrum of an arbitrary function.* Proc. London Math. Soc., **27** (1926), 483–496.

668. ———*On the closure of certain assemblages of trigonometrical functions.* Proc. Nat. Acad. Sci. U.S., **13** (1927), 27–29.

669. ———*A new definition of almost-periodic functions.* Ann. Math., **28** (1927), 365–367.

670. ———*Generalized harmonic analysis.* Acta Math., **55** (1930), 117–258.

671. ———*The Fourier Integral and certain of its Applications.* Dover Publications, New York, 1959.

672. WIENER, N., WINTNER, A. *On the ergodic dynamics of almost periodic systems.* Am. J. Math., **63** (1941), 794–824.

673. WINTNER, A. *Über einige Anwendungen der Theorie der fastperiodischen Funktionen auf das Levi-Civitasche Problem der mittleren Bewegung.* Annali di Mat. pura ed appl., **10** (1932), 277–282.

674. ———*On the asymptotic repartition of the values of real almost periodic functions.* Am. J. Math., **54** (1932), 334–345.

675. ———*Über die Dichte fastperiodischer Zahlenfolgen.* Studia Math., **4** (1933), 1–3.

676. ———*On the asymptotic differential distribution of almost-periodic and related functions.* Am. J. Math., **56** (1934), 401–406.

677. ———*Über die asymptotische Verteilung von fastperiodischen Funktionen mit linear unabhängigen Exponenten.* Prace Matematyczno-fizyczne, **43** (1936), 55–62.

678. ———*Almost-periodic functions and Hill's theory of lunar perigee.* Am. J. Math., **59** (1937), 795–802.

679. ———*On the almost-periodic behavior of the lunar node.* Am. J. Math., **61** (1939), 49–60.

680. ———*On an ergodic analysis of the remainder term of mean motions.* Proc. Nat. Acad. Sci. U.S., **26** (1940), 126–129.

681. ———*Prime divisors and almost-periodicity.* J. Math. Phys., **21** (1942), 52–56.

682. ———*Number-theoretical almost-periodicities.* Am. J. Math., **67** (1945), 173–193.

683. WOLF, F. *Contribution à la théorie des séries trigonométriques généralisées et des séries à fonctions orthogonales.* Publ. Fac. Sci. Univ. Masaryk, **130** (1931), 1–34.

684. ———*Approximation by trigonometrical polynomials and almost-periodicity.* Proc. London Math. Soc., **44** (1938), 100–114.

685. YOSIDA, K. *Asymptotic almost-periodicities and ergodic theorems.* Proc. Imp. Acad. Tokyo, **15** (1939), 255–259.

686. ———*Functional analysis.* Springer-Verlag, Berlin, 1965.

687. YOSHIZAWA, TARO. *Extreme stability and almost-periodic solutions of functional-differential equations.* Arch. Rat. Mech. Anal., **17** (1964), 149–170.

688. ———*Asymptotic stability of solutions of an almost-periodic system of functional-differential equations.* Rend. Circ. Math. Palermo (2), **13** (1964), 209–221.

689. ———*Stability theory by Liapunov's second method.* Chap. VII. Math. Soc. Japan, Tokyo, 1966.

690. ZAIDMAN, SAMUEL. *Sur la presque-périodicité des solutions de l'équation des ondes non homogène.* J. Math. Mech., **8** (1959), 369–382.

691. ———*Solutions presque-périodiques des équations hyperboliques.* C.R. Acad. Sci. Paris, **250** (1960), 2112–2114.

692. ———*Solutions presque-périodiques dans le problème de Cauchy, pour l'équation non-homogène des ondes.* I, II. Rend. Accad. Naz. Lincei (8), **30** (1961), 677–681, 823–827.

693. ———*Soluzioni limitate e quasi-periodiche dell'equazione del calore non-omogenea.* I, II. Rend. Accad. Naz. Lincei (8), **31** (1961), 362–368; (8) **32** (1962), 30–37.

694. ———*Solutions presque-périodiques des équations hyperboliques.* Ann. Sci. École Norm. Sup. Paris (3), **79** (1962), 151–198.

695. ———*Teoremi di quasi-periodicità per alcune equazioni differenziali operazionali.* Rend. Sem. Mat. Fis. Milano, **33** (1963), 220–235.

696. ———*Quasi-periodicità per l'equazione di Poisson.* Rend. Accad. Naz. Lincei (8), **34** (1963), 241–245.

697. ———*Un teorema di esistenza per un problema non bene posto.* Rend. Accad. Naz. Lincei (8), **35** (1963), 17–22.

698. ———*Quasi-periodicità per una equazione operazionale del primo ordine.* Rend. Accad. Naz. Lincei, **35** (1963), 152–157.

699. ——Soluzioni quasi-periodiche per alcune equazioni differenziali in spazi Hilbertiani. Ric. mat., **13** (1964), 118–134.

700. ——Almost-periodicity for some differential equations in Hilbert spaces. Rend. Accad. Naz. Lincei (8), **37** (1964), 253–257.

701. ——Soluzioni limitate o quasi-periodiche dell'equazione di Poisson. Ann. di Mat. pura ed appl., **64** (1965), 365–405.

702. ZUBOV, V. I. On almost periodic solutions of systems of differential equations. (Russian). Vestnik Leningrad. Univ., **15** (1960), No. 1, 104–106.

703. ——Periodic and almost-periodic forced oscillations arising from the action of an external force. (Russian). Izv. Vysš. Učebn. Zaved. Mat. No. 3 (19), (1960), 93–102.

704. ——Oscillations in non-linear and control systems. (Russian), Leningrad, 1962.

Further References

705. AMERIO, L., PROUSE, G. *Almost Periodic Functions and Functional Equations.* Van Nostrand, New York, 1971.

706. BASIT, BOLIS. *Note on a theorem of Levitan for the integral of almost periodic functions.* Rend. Ist. Mat. Univ. Trieste, **V** (1973), 9–14.

707. ——*On the indefinite integrals of abstract functions.* Analele Şt. Univ. Iaşi, **XXIX** (1984), Fasc. 3, 50–54.

708. BERGLUND, J. F., JUNGHENN, H. D., MILNES, P. *Compact Right Topological Semigroups and Generalizations of Almost Periodicity.* Springer-Verlag, 1978.

709. BURCKEL, R. B. *Weakly Almost Periodic Functions on Semigroups.* Gordon and Breach, New York, London, Paris, 1970.

710. CENUSA, G. *Properties of the mean of a random function almost periodic in probability.* Bulletin Mathematique (Bucharest), 24 (1980), Fasc. 4.

711. CENUSA, G., SACUIU, I. *Some properties of random functions almost periodic in probability.* Rev Roumaine Math. Pures Appl., **XXV** (1980), 1217–1325.

712. CLAASEN, T. A. C. M., MECKLENBRÄUKER, W. F. G. *On stationary linear time-varying systems.* IEEE Trans. Circuits and Systems, **29** (1982), 169–184.

713. COOKE, R. L. *Almost-periodic functions.* Amer. Math. Monthly, 1981, 515–526.

714. CORDUNEANU, C. *Periodic and almost periodic solutions of some convolution equations.* Trudy (Proc.) Mezhd. Konf. Nel. Kolebanjam, Kiev, Vol. I, 1970, 311–320.

715. ——*Bounded and almost periodic solutions of certain nonlinear elliptic equations.* Tohoku Math. Journal, **32** (1980), 265–278.

716. ——*Almost periodic discrete processes.* Libertas Mathematica, **II** (1982), 159–169.

717. ——*Integrodifferential equations with almost periodic solutions.* Volterra and Functional Differential Equations. M. Dekker, New York, 1982, 233–244.

718. ——*Almost periodic solutions to nonlinear elliptic and parabolic equations.* Nonlinear Analysis—TMA, **7** (1983), 357–363.

719. ———*Some almost periodicity criteria for ordinary differential equations.* Libertas Mathematica, **III** (1983), 21–43.

720. ———*Periodic and almost periodic oscillations in nonlinear systems.* Control Problems for Systems Described by Partial Differential Equations and Applications, Springer-Verlag, 1987, 196–203.

721. CORDUNEANU, C., GOLDSTEIN, J. A. *Almost periodic solutions to nonlinear abstract equations.* Differential Equations, North Holland, 1984, 115–121.

722. DAFERMOS, C. M. *Almost periodic processes and almost periodic solutions of evolution equations.* Dynamic Systems (Proc. Univ. Florida Int. Symp.), Academic Press, New York, 1977.

723. DAVIS, A. M. *Almost periodic extensions of band-limited functions and its application to nonuniform sampling.* IEEE Trans. Circuits and Systems, **33** (1986), 933–938.

724. DEMIDOVICH, B. P. *Lectures on Mathematical Theory of Stability* (Russian). Nauka, Moscow, 1967.

725. DUNKEL, C. F., RAMIREZ, D. E. *Topics in Harmonic Analysis.* Appleton-Century-Crofts, New York, 1971.

726. FATTORINI, H. D. *Second Order Linear Differential Equations in Banach Spaces.* North Holland, Amsterdam, 1985.

727. FINK, A. M. *Almost Periodic Differential Equations.* Springer-Verlag, Berlin, 1974

728. ———*Extensions of almost automorphic sequences.* J. Math. Analysis Appl., **27** (1969), 519–523.

729. ———*Almost periodic functions invented for special purposes.* SIAM Review, **14** (1972), 572–581.

730. ———*A Bibliography of Almost Periodic Functions and Applications.* This list contains approximately 1,000 titles of papers in the field (dated Feb. 20, 1987).

731. HALANAY, A., WEXLER, D. *The Qualitative Theory of Pulse Systems* (Romanian). Ed. Acad., Bucharest, 1968. A Russian translation is available (1971).

732. HANEBALY, E. *Solutions presque périodiques d'équations différentielles monotones.* C. R. Acad. Sci. Paris, **297** (1983), Serie I, 263–265.

733. HANSEL, G., TROALLIC, J. P. *Suites uniformement distribuées et fonctions faiblement presque-périodiques.* Bull. Soc. Math. France, **108** (1980), 207–212.

734. HARASAHAL, V. H. *Almost Periodic Solutions of Ordinary Differential Equations* (Russian). Nauka, Kazah. SSR, Alma-Ata, 1970.

735. HARAUX, A. *Nonlinear Evolution Equations; Global Behavior of Solutions.* Springer-Verlag, New York, 1981.

736. ———*A simple almost-periodicity criterion and applications.* J. Differential Equations, **66** (1987), 51–61.

737. ———*Asymptotic behavior for two-dimensional quasi-autonomous almost-periodic evolution equations.* J. Differential Equations, **66** (1987), 62–70.

738. HARRIS, C. J., MILES, J. F. *Stability of Linear Systems.* Academic Press, New York, 1980.

739. ISOKAWA, Y. *An identification problem in almost and asymptotically almost periodically correlated processes.* J. Appl. Probability, **19** (1982), 456–462.

740. JOHNSON, RUSSEL. *Almost-periodic functions with unbounded integral.* Pacific J. Math., **14** (1980), 347–362.

741. ———*Bounded solutions of scalar almost periodic linear equations.* Illinois J. Math., **25** (1981), 632–643.

742. JOHNSON, R., MOSER, J. *The rotation number for almost periodic potentials.* Commun. Math. Physics, **84** (1982), 403–438.
743. KATZNELSON, Y. *An Introduction to Harmonic Analysis.* Dover, New York, 1976.
744. KRASNOSELSKII, M. A., BURD, V. S., KOLESOV, YU. S. *Nonlinear Almost Periodic Oscillations.* John Wiley, New York, 1973.
745. LEVITAN, B. M., ZHIKOV, V. V. *Almost Periodic Functions and Differential Equations.* Cambridge Univ. Press, Cambridge, England, 1982.
746. MAUCLAIRE, J. L. *Sur la théorie des suites presque-periodiques.* Proc. Japan Acad. Sci., Ser. A, **61** (1985), 153–155 and 190–192.
747. MEISTERS, G. *Bibliography of Almost Periodic Functions,* 1923–1969. This list contains more than 400 entries.
748. MILLER, R. K. *Almost periodic behavior of solutions of a nonlinear Volterra system.* Quart. Appl. Math., **28** (1971), 553–570.
749. MILNES, PAUL. *On vector-valued weakly almost periodic functions.* J. London Math. Soc., (2), **22** (1980), 467–472.
750. ———*The Bohr almost periodic functions do not form a linear space.* Math. Zts., **188** (1984), 1–2.
751. ———*On Bohr almost periodicity.* Math. Proc. Cambridge Phil. Soc., **99** (1986), 489–493.
752. ONICESCU, O., CENUSĂ, G, SĂCUIU, I. *Random Functions Almost Periodic in Probability* (Romanian). Ed. Academiei, Bucharest, 1983.
753. ONICESCU, O., ISTRĂTESCU, V. Approximation theorems for random functions. Rend. Mat., (VI), **8** (1975), 65–81.
754. PANKOV, A. A. *Bounded and Almost Periodic Solutions of Nonlinear Differential Equations* (Russian). Naukova Dumka, Kiev, 1985.
755. PETROVANU, D. *Periodic and almost periodic solutions of parabolic equations and E. Rothe's method.* Bull. Ins. Polit. Iaşi, (N.S.), **23** (1977), nos. 1–2, 17–22.
756. POPPE, H., STARK, A. *A compactness criterion for the space of almost periodic functions,* II. Math. Nach., **108** (1982), 153–157.
757. PRECUPANU, A. M. *Fonctions F-presque-périodiques et F-asymptotiquement presque-pérodiques.* Analele Sti. Univ. Iaşi, **XIX** (1973), 321–339.
758. ———*Some properties of the almost periodic functions in probability.* Ibidem, **XXV** (1979), 93–97.
759. ———*Sur les suites asymptotiquement presque-périodiques en probabilité.* Ibidem, **XXVI** (1980), 19–22.
760. ———*On the almost periodic function in probability.* Rend. Mat. Appl. (VII), **2** (1982), 613–626.
761. SEIFERT, G. *Almost periodic solutions for a certain class of almost periodic systems.* Proc. Amer. Math. Soc., **84** (1982), 47–51.
762. SELL, G. R. *Almost periodic solutions of linear partial differential equations.* J. Math. An. Appl., **42** (1973), 302–312.
763. SIBUYA, Y. *Almost periodic solutions of Poisson's equation.* Proc. Amer. Math. Soc., **28** (1971), 195–198.
764. SIMIRAD, C. *Sur les solutions périodiques et presque-périodiques des systèmes d'équations différentielles du premier ordre.* Analele Sti. Univ. Iaşi, (Mathematica), **XXIV** (1978), 299–305.
765. STAFFANS, O. *On almost periodicity of solutions of an integro-differential equation.* J. Integral Equations, **8** (1985), 249–260.

766. TRAPLE, JANUSZ. *Weak almost periodic solutions of differential equations.* J. Diff. Equations, **45** (1982), 199–206.
767. UPTON, C. J. F. *On classes of continuous almost periodic functions.* Proc. Lond Math. Soc., **XXXV** (1977), 159–179.
768. ———*Some new classes of almost periodic functions.* Mathematics Research Report, University of Melbourne, No. 8, 1980.
769. WU, T.-S. *Left almost periodicity does not imply right almost periodicity.* Bull. Amer. Math. Soc., **72** (1966), 314–316.
770. YAMAGUCHI, M., NISHIHARA, K. *Almost periodic and asymptotically almost periodic solutions of some wave equations with quasiperiodic coefficients.* Proc. Fac. Sci. Tokai Univ., **XVII** (1982), 45–59.
771. YOSHIZAWA, T. *Stability Theory and Existence of Periodic Solutions and Almost Periodic Solutions.* Springer-Verlag, 1975.
772. ZAIDMAN, S. *Solutions presque-periodiques des equations differentielles abstraites.* L'Enseign. Math., **XXIV** (1978), 87–110.
773. ———*Abstract Differential Equations.* Pitman Publ. Ltd., London, 1979.
774. ———*Notes on abstract almost-periodicity.* Riv. Mat. Univ. Parma, (4), **5** (1979), 837–845.
775. ———*Almost Periodic Functions in Abstract Spaces.* Pitman Publ. Ltd., London, 1985.
778. ———*Periodic solutions of abstract differential equations with periodic right-hand side.* Rend. Ist. Lombardo, Sci. Matematiche e Appl., A, **118** (1987), 95–98.
779. ———*Integration of Poisson stable functions of two variables.* Libertas Mathematica, **VII** (1987), 59–63.

INDEX